Embedded Systems

Series editors

Nikil D. Dutt, Irvine, CA, USA
Grant Martin, Santa Clara, CA, USA
Peter Marwedel, Dortmund, Germany

This Series addresses current and future challenges pertaining to embedded hardware, software, specifications and techniques. Titles in the Series cover a focused set of embedded topics relating to traditional computing devices as well as high-tech appliances used in newer, personal devices, and related topics. The material will vary by topic but in general most volumes will include fundamental material (when appropriate), methods, designs and techniques.

More information about this series at http://www.springer.com/series/8563

Peter Marwedel

Embedded System Design

Embedded Systems Foundations
of Cyber-Physical Systems,
and the Internet of Things

Third Edition

 Springer

Peter Marwedel
TU Dortmund
Dortmund
Germany

ISSN 2193-0155 ISSN 2193-0163 (electronic)
Embedded Systems
ISBN 978-3-319-56043-4 ISBN 978-3-319-56045-8 (eBook)
DOI 10.1007/978-3-319-56045-8

Library of Congress Control Number: 2017938320

Printed on acid-free paper

This Springer imprint is published by Springer Nature
The registered company is Springer International Publishing AG
The registered company address is: Gewerbestrasse 11, 6330 Cham, Switzerland

This book is dedicated to my family members Veronika, Malte, Gesine, and Ronja.

Preface

What Does This Book Present?

"Information technology (IT) is on the verge of another revolution. ... networked systems of embedded computers ... have the potential to change radically the way people interact with their environment by linking together a range of devices and sensors that will allow information to be collected, shared, and processed in unprecedented ways. ... The use ... throughout society could well dwarf previous milestones in the information revolution."

This citation from a report of the National Research Council in the USA [392] describes very nicely the dramatic impact of information technology in embedded systems. Such systems can be understood as information processing embedded into an enclosing product [355], such as a car or an airplane. This revolution has already had a major impact and is still continuing. For example, the availability of mobile devices has had an impact on societies. Due to the increasing integration of computing and physical objects, the term cyber-physical systems (CPS) has been introduced. Such systems *"are engineered systems that are built from and depend upon the synergy of computational and physical components"* [394]. Objects or "things" play a key rule in the definition of the popular term "Internet of Things" (IoT). IoT *"... describes ... a variety of devices ... able to interact and cooperate with each other to reach common goals"* [179]. Terms such as pervasive and ubiquitous computing, ambient intelligence, and "Industry 4.0" are also referring to the dramatic impact of changes caused by information technology. More changes like the use of self-driving cars and more remotely controlled airborne devices are on the horizon.

This importance of embedded/cyber-physical systems and IoT is so far not well reflected in many of the current curricula. Designing the mentioned systems requires knowledge and skills from areas beyond traditional programming and algorithm design. Obtaining an overview of such knowledge is very difficult, due to the wide range of relevant areas. This book aims at facilitating the acquisition of knowledge from the relevant areas. It provides material for a first course on such systems and includes an overview of key concepts for the integration of information technology with physical objects. It covers hardware as well as software aspects. This is in-line with the ARTIST guidelines for curricula of embedded systems: *"The development of embedded systems cannot ignore the underlying hardware characteristics. Timing, memory usage, power consumption, and physical failures are important"* [86].

This book has been designed as a textbook. However, this book provides more references than typical textbooks do and also helps to structure the area. Hence, this book should also be useful for faculty members and engineers. For students, the inclusion of a rich set of references facilitates access to relevant sources of information.

This book focuses on the fundamental bases of software and hardware. Specific products and tools are mentioned only if they have outstanding characteristics. Again, this is in-line with the ARTIST guidelines: *"It seems that fundamental bases are really difficult to acquire during continuous training if they haven't been initially learned, and we must focus on them"* [86]. As a consequence, this book goes beyond teaching embedded system design by programming microcontrollers. This book presents the **fundamentals of embedded systems design which are needed for the design of CPS and IoT systems**. With this approach, we would like to make sure that the material taught will not be outdated too soon. The concepts covered in this book should be relevant for a number of years to come.

The proposed positioning of the current textbook in computer science and computer engineering curricula is explained in a paper [356]. We want to relate the most important topics in this area to each other. This way, we avoid a problem mentioned in the ARTIST guidelines: *"The lack of maturity of the domain results in a large variety of industrial practices, often due to cultural habits. ... curricula ... concentrate on one technique and do not present a sufficiently wide perspective. ... As a result, industry has difficulty finding adequately trained engineers, fully aware of design choices"* [86].

This book should also help to bridge the gap between practical experiences with programming microcontrollers and more theoretical issues. Furthermore, it should help to motivate students and teachers to look at more details. While this book covers a number of topics in detail, others are covered only briefly. These brief sections have been included in order to put a number of related issues into perspective. Furthermore, this approach allows lecturers to have appropriate links in this book for adding complementary material of their choice. Due to the rich set of references, this book can also be used as a comprehensive tutorial, providing pointers for additional reading. Such references can also stimulate taking benefit of this book during laboratories, projects, and independent studies as well as a starting point for research.

The scope of this book includes specification techniques, hardware components, system software, application mapping, evaluation and validation, as well as exemplary optimizations and test methods. This book covers embedded systems and their interface to the physical environment from a wide perspective, but cannot cover every related area. Legal and socioeconomic aspects, human interfaces, data analysis, application-specific aspects, and a detailed presentation of physics and communication are beyond the scope of this book. The coverage of the Internet of Things is limited to areas linked to embedded systems.

Who Should Read the Book?

This book is intended for the following audience:

- Computer science (CS), computer engineering (CE), and electrical engineering (EE) students as well as students in other information and communication technology (ICT)-related areas who would like to specialize in embedded/cyber-physical systems or IoT. This book should be appropriate for third-year students who do have a basic knowledge of computer hardware and software. This means that this book primarily targets senior undergraduate students[1]. However, it can also be used at the graduate level if embedded system design is not part of the undergraduate program or if the discussion of some topics is postponed. This book is intended to pave the way for **more advanced topics** that should be **covered in follow-up courses**. This book assumes a basic knowledge of computer science. EE students may have to read some additional material in order to fully understand the topics of this book. This should be compensated by the fact that some material covered in this book may already be known to EE students.
- Engineers who have so far worked on systems hardware and who have to move more toward software of embedded systems. This book should provide enough background to understand the relevant technical publications.
- Ph.D. students who would like to get a quick, broad overview of key concepts in embedded system technology before focusing on a specific research area.
- Professors designing a new curriculum for the mentioned areas.

How Is This Book Different from Earlier Editions?

The first edition of this book was published in 2003. The field of embedded systems is moving fast, and many new results have become available since then. Also, there are areas for which the emphasis has shifted. In some cases, a more detailed treatment of the topic became desirable. New developments have been taken up when the first German edition of this book was published in 2007. Therefore, it became necessary to publish a second English edition in late 2010/early 2011. Since

---

[1]This is consistent with the curriculum described by T. Abdelzaher in a recent report on CPS education [393].

then, some technological changes occurred. There was a clear shift from single core systems toward multi-core systems. Cyber-physical systems (CPS) and the Internet of Things (IoT) gained more attention. Power consumption and thermal issues have become more important, and many designs now have tight power and thermal constraints. Furthermore, safety and security have gained importance as well. Overall, it became necessary to publish the current third edition.

The changes just described had an impact on several chapters. We are now including and linking those aspects of embedded systems that provide foundations for the design of CPS and IoT systems. The preface and the introduction have been rewritten to reflect these changes. (a) The chapter on specifications and modeling now includes partial differential equations and transaction-level modeling (TLM). The use of this book in flipped-classroom-based teaching led to the consideration of more details, in particular of specification techniques. (b) The chapter on embedded system hardware now includes multi-cores, a rewritten section on memories, more information on the conversion between the analog and the digital domain, including pulse-width modulation (PWM). There are updated descriptions of field pro-grammable gate arrays (FPGAs) and a brief section on security issues in embedded systems. (c) The chapter on system software now contains a section on Linux in embedded systems and more information on resource access protocols. (d) In the context of system evaluation, there are new subsections on quality metrics, on safety/security, on energy models, and on thermal issues. (e) The chapter on mapping to execution platforms has been restructured: A standard classification of scheduling problems has been introduced, and multi-core scheduling algorithms have been added. There is now a clearer distinction between jobs and tasks. The description of hardware/software codesign has been dropped. (f) The chapter on optimizations has been updated. Some information on compilation for specialized processors has been dropped.

All chapters have been carefully reviewed and updated if required. There are also an improved integration of (color) graphics, more assignments (problems), as well as a clearer distinction between definitions, theorems, proofs, code, and examples. We adopted the Springer book layout style.

For this edition, it is typically not feasible to cover the complete book in a single course for undergraduates and lecturers can select a subset which fits the local needs and preferences.

Dortmund, Germany   Peter Marwedel
February 2017

Acknowledgements

My Ph.D. students, in particular Lars Wehmeyer, did an excellent job in proof-reading the second edition of this book. The same applies to Michael Engel for the third edition. Also, the students attending my courses provided valuable help. Corrections were contributed by David Hec, Thomas Wiederkehr, Thorsten Wilmer, and Henning Garus. In addition, the following colleagues and students gave comments or hints which were incorporated into this book: R. Dömer, N. Dutt (UC Irvine), A.B. Kahng (UC San Diego), T. Mitra (Nat. Univ. Singapore), W. Kluge, R. von Hanxleden (U. Kiel), P. Buchholz, J.J. Chen, M. Engel, H. Krumm, K. Morik, O. Spinczyk (TU Dortmund), W. Müller, F. Rammig (U. Paderborn), W. Rosenstiel (U. Tübingen), L. Thiele (ETH Zürich), and R. Wilhelm (Saarland University). Material from the following persons was used to prepare this book: G.C. Buttazzo, D. Gajski, R. Gupta, J.P. Hayes, H. Kopetz, R. Leupers, R. Niemann, W. Rosenstiel, H. Takada, L. Thiele, and R. Wilhelm. M. Engel provided invaluable help during various iterations of our course, including the generation of videos which are now available on YouTube. Ph.D. students of my group contributed to the assignments included in this book. Of course, the author is responsible for all errors and mistakes contained in the final manuscript.

I do acknowledge the support of the European Commission through projects MORE, Artist2, ArtistDesign, Hipeac(2), PREDATOR, MNEMEE, and MADNESS. Furthermore, support by Deutsche Forschungsgemeinschaft (DFG) with grants for the Collaborative Research Center SFB 876 and the FEHLER (grant Ma 943/10) project is also acknowledged. These projects provided an excellent context for writing the third edition of this book. Synopsys Inc. provided access to their Virtualizer virtual platform.

This book has been produced using the LATEX typesetting system from the TeXnicCenter user interface. Some functions have been plotted with GNU Octave. I would like to thank the authors of this software for their contribution to this work.

Acknowledgments also go to all those who have patiently accepted the author's additional workload during the writing of this book and his resulting reduced availability for professional as well as personal partners.

Names used in this book without any reference to copyrights or trademarks may still be legally protected.

Please enjoy reading the book!

Contents

About the Author

Peter Marwedel was born in Hamburg, Germany. He received a Dr. rer. nat. degree in Physics in 1974 and a Dr. habil. degree in Computer Science in 1987, both from the University of Kiel (Germany). From 1974 to 1989, he was a faculty member of the Institute for Computer Science and Applied Mathematics at the same university. He has been a professor at TU Dortmund, Germany, since 1989. He held a chair for embedded systems at the Computer Science Department from 1989 until 2014. He is chairing ICD e.V., a local company specializing in technology transfer. Dr. Marwedel was a visiting professor of the University of Paderborn in 1985/1986 and of the University of California at Irvine in 1995. He served as Dean of the Computer Science Department from 1992 to 1995. Dr. Marwedel has been active in making the DATE conference and the WESE workshop successful and in initiating the SCOPES series of workshops. He started to work on high-level synthesis in 1975 (in the context of the MIMOLA project) and focused on the synthesis of very long instruction word (VLIW) machines. Later, he added efficient compilation for embedded processors to his scope, with an emphasis on retargetability, the memory architecture, and optimization for the worst-case execution time. His scope also includes synthesis of self-test programs for processors, reliable computing, multimedia-based teaching, and cyber-physical systems. He served as a cluster leader for ArtistDesign, a European Network of Excellence on Embedded and Real-Time Systems. He was the vice-chair of the collaborative research center SFB

876 on resource-constrained machine learning[2] until 2015. He is an IEEE and a DATE Fellow. Dr. Marwedel won the teaching award of his university, the ACM SIGDA Distinguished Service Award, the EDAA Lifetime Achievement Award, and the ESWEEK Lifetime Achievement Award. Dr. Marwedel is a member of IEEE, ACM and the Gesellschaft für Informatik (GI).

He is married and has two daughters and a son. His hobbies include hiking, photography, bike riding, and model railways.

E-mail: peter.marwedel@tu-dortmund.de
Web sites: https://www.marwedel.eu/peter,
http://ls12-www.cs.tu-dortmund.de/∼marwedel

---

[2]See http://www.sfb876.tu-dortmund.de

Frequently Used Mathematical Symbols

Due to covering many areas in this book, there is a high risk of using the same symbol for different purposes. Therefore, symbols have been selected such that the risk of confusion is low. This table is supposed to help maintaining a consistent notation.

a	Weight
a	Allocation
A	Availability ($\rightarrow$ reliability)
A	Area
A	Ampere
$b_{..}$	Communication bandwidth
B	Communication bandwidth
c_R	Characteristic vector for Petri net
c_p	Specific thermal capacitance
c_v	Volumetric heat capacity
C_i	Execution time
C	Capacitance
C	Set of Petri net conditions
C_{th}	Thermal capacity
C	Celsius
d_i	Absolute deadline
D_i	Relative deadline
$e(t)$	Input signal
e	Euler's number (2.71828..)
E	Energy
E	Graph edge
f	Frequency
$f()$	General function
f	Probability density
f_i	Finishing time of task/job i

F	Probability distribution
F	Flow relation of Petri net
g	Gravity
g	Gain of operational amplifier
G	Graph
h	Height
i	Index, task/job number
I	Current
j	Index, dependent task/job
J	Set of jobs
J	Joule
J_j	Job j
J	Jitter
k	Index, processor number
k	Boltzmann constant ($\sim 1.3807 * 10^{-23}$ J/K)
K	Kelvin
l	Processor number
l_i	Laxity of task/job i
L	Processor type
L	Length of conductor
L_i	Lateness of task τ_i
L_{max}	Maximum lateness
m	Number of processors
m	Mass
m	Meter
m	Milli-prefix
M	Marking of Petri net
MS_{max}	Makespan
n	Index
n	Number of tasks/jobs
N	Net
$\mathbb{N}$	Natural numbers
$O(\ )$	Landauer's notation
p_i	Priority of task τ_i
p_i	Place i of Petri net
P	Power
$P(S)$	Semaphore operation
Q	Resolution
Q	Charge
r_i	Release time of task/job i
R	Reliability
R_{th}	Thermal resistance
$\mathbb{R}$	Real numbers
s	Time index

s	Restitution
s_j	Starting time of task/job j
s	Second
S	State
S	Semaphore
$\mathcal{S}$	Schedule
S_j	Size of memory j
t	Time
t_i	Transition i of Petri net
T	Period
T	Timer (for SDL)
T_i	Period of task τ_i
T	Temperature
u_i	Utilization of task τ_i
$U_{..}$	Utilization
U_{max}	Maximum utilization
v	Velocity
$w(t)$	Signal
V	Graph nodes
V	Voltage
V	Volt
V_t	Threshold voltage
$V(S)$	Semaphore operation
$\mathcal{V}$	Volume
$w(t)$	Signal
$W(p,t)$	Weight in Petri net
W	Watt
x	Input variable
$x(t)$	Signal
$X_{..}$	Decision variable
$Y_{..}$	Decision variable
$z(t)$	Signal
Z	High impedance
$\mathbb{Z}$	Integer numbers
$\alpha^{..}$	Arrival curve in real-time calculus
α	Switching activity
α	1st component in Pinedo's triplet
$\beta^{..}$	Service function in real-time calculus
β	2nd component in Pinedo's triplet
β	Reciprocal of max. utilization
$\gamma^{..}$	Work load in real-time calculus
γ	3rd component in Pinedo's triplet
Δ	Time interval
κ	Thermal conductivity

λ	Failure rate
π	Number pi (3.1415926..)
π	Set of processors
π_i	Processor i
ρ	Mass density
τ_i	Task τ_i
τ	Set of tasks
ξ	Threshold for *RM-US* scheduling

Chapter 1
Introduction

This chapter presents terms used in the context of embedded systems together with their history as well as opportunities, challenges and common characteristic of embedded and cyber-physical systems. Furthermore, educational aspects, design flows and the structure of this book are introduced.

1.1 History of Terms

Until the late 1980s, information processing was associated with large mainframe computers and huge tape drives. Later, miniaturization allowed information processing with personal computers (PCs). Office applications were dominating, but some computers were also controlling the physical environment, typically in the form of some feedback loop.

Later, Mark Weiser created the term "**ubiquitous computing**" [548]. This term reflects Weiser's prediction to have computing (and information) *anytime, anywhere*. Weiser also predicted that computers are going to be integrated into products such that they will become invisible. Hence, he created the term "**invisible computer.**" With a similar vision, the predicted penetration of our day-to-day life with computing devices led to the terms "**pervasive computing**" and "**ambient intelligence**." These three terms focus on only slightly different aspects of future information technology. Ubiquitous computing focuses more on the long-term goal of providing information anytime, anywhere, whereas pervasive computing focuses more on practical aspects and the exploitation of already available technology. For ambient intelligence, there is some emphasis on communication technology in future homes and smart buildings.

Miniaturization also enabled the integration of information processing and the physical environment using computers. This type of information processing has been called an "**embedded system**":

© Springer International Publishing AG 2018
P. Marwedel, *Embedded System Design*, Embedded Systems,
DOI 10.1007/978-3-319-56045-8_1

Definition 1.1 (Marwedel [355]): *"Embedded systems are information processing systems embedded into enclosing products".*

Examples include embedded systems in cars, trains, planes, and telecommunication or fabrication equipment. Such systems come with a large number of common characteristics, including real-time constraints, and dependability as well as efficiency requirements. For such systems, the link to physics and physical systems is rather important. This link is emphasized in the following citation [316]:

"Embedded software is software integrated with physical processes. The technical problem is managing time and concurrency in computational systems."

This citation could be used as a definition of the term "embedded software" and could be extended into a definition of "embedded systems" by just replacing "software" by "system".

However, the strong link to physics has recently been stressed even more by the introduction of the term **"cyber-physical systems"** (CPS for short). CPS can be defined as follows:

Definition 1.2 (Lee [317]): *"Cyber-Physical Systems (CPS) are integrations of computation and physical processes".*

The new term emphasizes the link to physical quantities such as time, energy, and space. Emphasizing this link makes sense, since it is frequently ignored in a world of applications running on servers and on PCs. For CPS, we may be expecting models to include models of the physical environment as well. In this sense, we may think of CPS to comprise embedded systems (the information processing part) and the physical environment, or **CPS = ES + physics**. This is also reflected in Fig. 1.1.

Fig. 1.1 Relationship between embedded systems and CPS according to Lee's definition

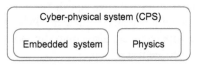

In their call for proposals, the National Science Foundation in the US mentions also communication [394]:

*"Emerging CPS will be coordinated, distributed, and **connected**, and must be robust and responsive."*

This is also done in the acatech report on CPS [11]: CPS ... *"represent **networked**, software-intensive embedded systems in a control loop, provide networked and distributed services."*

Interconnection and collaboration are also explicitly mentioned in a call for proposals by the European Commission [149]: *"Cyber-Physical Systems (CPS) refer to next generation embedded ICT systems that are **interconnected** and collaborating including through the **Internet of Things**, and providing citizens and businesses with a wide range of innovative applications and services."*

The importance of communication was visualized by the European Commission earlier, as shown in Fig. 1.2.

Fig. 1.2 Importance of communication (© European Commission)

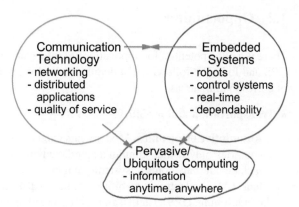

From these citations, it is clear that the authors do not only associate the integration of the cyber- and the physical world with the term CPS. Rather, there is also a strong communication aspect. Actually, the term CPS is not always used consistently. Some authors emphasize the integration with the physical environment, and others emphasize communication.

Communication is more explicit in the term "**Internet of Things**" (IoT), which can be defined as follows:

Definition 1.3 ([179]): The term Internet of Things "*... describes the pervasive presence of a variety of devices – such as sensors, actuators, and mobile phones – which, through unique addressing schemes, are able to interact and cooperate with each other to reach common goals.*"

This term is linking sensors (such that sensed information is available on the Internet) and actuators (such that things can be controlled from the Internet). Sensor examples include smoke detectors, heaters, and refrigerators. The Internet of Things is expected to allow the communication between trillions of devices in the world. This vision affects a large amount of businesses.

The exploitation of IoT technology for production has been called "Industry 4.0" [70]. Industry 4.0 targets a more flexible production for which the entire life cycle from the design phase onward is supported by the IoT.

To some extent, it is a matter of preferences whether the linking of physical objects to the cyber-world is called CPS or IoT. Taken together, CPS and IoT include most of the future applications of IT.

The design of these future applications requires knowing fundamental design techniques for embedded systems. This book focuses on such fundamental techniques and foundations of embedded system design. Please remember that these are used in IoT and CPS designs though this is not repeatedly stated in each and every context. This book discusses interfaces of embedded systems to CPS and IoT systems. However, application-specific aspects of CPS and IoT are usually not covered.

1.2   Opportunities

There is a huge potential for applications of information processing in the context of CPS and IoT. The following list demonstrates this potential and the large variation of corresponding areas:

- **Transportation and mobility**:

  - **Automotive electronics**: Modern cars can be sold in technologically advanced countries only if they contain a significant amount of electronics [396]. These include air bag control systems, engine control systems, anti-braking systems (ABS), electronic stability programs (ESP) and other safety features, air-conditioning, GPS systems, anti-theft protection, driver assistance systems, and many more. In the future, self-driving cars may be a reality. These systems can help to reduce accidents and the impact on the environment.
  - **Avionics**: A significant amount of the total value of airplanes is due to the information processing equipment, including flight control systems, anti-collision systems, pilot information systems, autopilots, and others. Dependability is of utmost importance. Embedded systems can decrease emissions (such as carbon dioxide) from airplanes. Autonomous flying is also becoming reality, at least for drones.
  - **Railways**: For railways, the situation is similar to the one discussed for cars and airplanes. Again, safety features contribute significantly to the total value of trains, and dependability is extremely important. Advanced signaling aims at safe operation of trains at high speed and short intervals between trains. The European Train Control System (ETCS) [422] is one step in this direction. Autonomous rail-based transportation is already used in restricted contexts like shuttle trains at airports.
  - **Ships, ocean technology, and maritime systems**: In a similar way, modern ships use large amounts of IT-equipment, e.g., for navigation, for safety, for optimizing the operation in general, and for bookkeeping (see, e.g., http://www.smtcsingapore.com/ and https://dupress.deloitte.com/dup-us-en/focus/internet-of-things/iot-in-shipping-industry.html).

- **Factory automation**: Fabrication equipment is a very traditional area in which embedded systems have been employed for decades. In order to optimize production technologies further, CPS/IoT technology can be used. CPS/IoT technology is the key toward a more flexible production, being the target of "Industry 4.0" [70].
  Factory automation is enabled by logistics. There are several ways in which CPS and IoT systems can be applied to **logistics** [285]. For example, radio frequency identification (RFID) technology provides easy identification of each and every object, worldwide. Mobile communication allows unprecedented interaction.

- **Health sector**: The importance of healthcare products is increasing, in particular in aging societies. Opportunities start with new sensors, detecting diseases faster and more reliably. Available information can be stored in patient information systems.

New **data analysis** techniques can be used to detect increased risks and improve chances for healing. Therapies can be supported with personalized medication. New devices can be designed to help patients, e.g., handicapped patients. Also, surgery can be supported with new devices. Embedded system technologies also allow for a significantly improved result monitoring, giving doctors much better means for checking whether or not a certain treatment has a positive impact. This monitoring also applies to remotely located patients. Lists of projects in this area can be found, for example, at http://cps-vo.org/group/medical-cps and at http://www.nano-tera.ch/program/health.html.

- **Smart buildings**: Information processing can be used to increase the comfort level in buildings, can reduce the energy consumption within buildings, and can improve safety and security. Sub-systems which traditionally were unrelated must be connected for this purpose. There is a trend toward integrating air-conditioning, lighting, access control, accounting, safety features, and distribution of information into a single system. Tolerance levels of air-conditioning sub-systems can be increased for empty rooms, and the lighting can be automatically reduced. Air condition noise can be reduced to a level required for the actual operating conditions. Intelligent usage of blinds can also optimize lighting and air-conditioning. Available rooms can be displayed at appropriate places, simplifying ad hoc meetings and cleaning. Lists of non-empty rooms can be displayed at the entrance of the building in emergency situations (provided the required power is still available). This way, energy can be saved on cooling, heating, and lighting. Also, safety can be improved. Initially, such systems might mostly be present in high-tech office buildings, but the trend toward energy-efficient buildings also affects the design of private homes. One of the goals is to design so-called **zero-energy-buildings** (buildings which produce as much energy as they consume) [404]. Such a design would be one contribution toward a reduction of the global carbon dioxide footprint and global warming.
- **Smart grid**: In the future, the production of energy is supposed to be much more decentralized than in the past. Providing stability in such a scenario is difficult. IT technology is required in order to achieve a sufficiently stable system. Information on the smart grid can be found, for example, at https://www.smartgrid.gov/the_smart_grid and at http://www.smartgrids.eu/.
- **Scientific experiments**: Many contemporary experiments in sciences, in particular in physics, require the observation of experiment outcomes with IT devices. The combination of physical experiments and IT devices can be seen as a special case of CPS.
- **Public safety**: The interest in various kinds of safety is also increasing. Embedded and cyber-physical systems and the Internet of Things can be used to improve safety in many ways. This includes identification/authentication of people, for example with fingerprint sensors or face recognition systems.
- **Structural health monitoring**: Natural and artificial structures such as mountains, volcanoes, bridges, and dams (see, e.g., Fig. 1.3) are potentially threatening lives. We can use embedded system technology to enable advance warnings in case of increased dangers like avalanches or collapsing bridges.

Fig. 1.3 Example of a dam to be monitored (Möhnesee dam), © P. Marwedel

- **Disaster recovery**: In the case of major disasters such as earthquakes or flooding, it is essential to save lives and provide relief to survivors. Flexible communication infrastructures are essential for this.
- **Robotics**:Robotics is also a traditional area in which embedded/cyber-physical systems have been used. Mechanical aspects are very important for robots. Robots, modeled after animals or human beings, have been designed. Figure 1.4 shows such a robot.

Fig. 1.4 Robot "Johnnie"
(courtesy H. Ulbrich, F.
Pfeiffer, Lehrstuhl für
Angewandte Mechanik, TU
München), © TU München

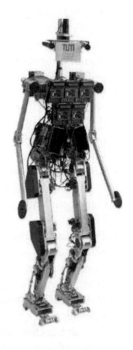

- **Agriculture and breeding**: There are many applications in these areas. For example, the *"regulations for traceability of agricultural animals and their movements require the use of technologies like IoT, making possible the real-time detection of animals, for example, during outbreaks of (a) contagious disease"* [493].
- **Military applications**: Information processing has been used in military equipment for many years. In fact, some of the very first computers analyzed military radar signals.
- **Telecommunication**: Mobile phones have been one of the fastest growing markets in the recent years. For mobile phones, radio frequency (RF) design, digital signal processing, and low-power design are key aspects. Telecommunication is a key aspect of IoT. Other forms of telecommunication are also important.
- **Consumer electronics**: Video and audio equipment is a very important sector of the electronics industry. The information processing integrated into such equipment is steadily growing. New services and better quality are implemented using advanced digital signal processing techniques. Many TV sets (in particular, high-definition TV sets), multimedia phones, and game consoles comprise powerful high-performance processors and memory systems. They represent special cases of embedded systems.

Even more applications are listed in a report on opportunities and challenges of the IoT [493]. The large set of examples demonstrates the huge variety of embedded systems with applications in CPS and IoT systems. In a way, many of the future applications of IT technology can be linked to such systems.

The long list of application areas of embedded systems is resulting in a corresponding economic importance of such systems. The acatech report [11] mentions that at the time of writing the report, 98% of all microprocessors are used in these systems. In a way, embedded system design is an enabler for many products and has an impact on the combined market volume size of all the areas mentioned. However, it is difficult to quantify the size of the CPS/IoT market since the total market volume of all these areas is significantly larger than the market volume of their IT components. Referring to the value of semiconductors in the CPS/IoT market would also be misleading, since that value is only a fraction of the overall value.

The economic importance of CPS and the IoT is reflected in calls for proposals by funding organizations, such as the NSF [394] and the European Commission [149].

1.3 Challenges

Unfortunately, the design of embedded systems and their integration in CPS and IoT systems comes with a large number of difficult design issues. Commonly found issues include the following:

- Cyber-physical and IoT systems must be **dependable**.

Definition 1.4: A system is **dependable** if it provides its intended service with a high probability and does not cause any harm.

A key reason for the need of being dependable is that these systems are directly connected to the physical environment and have an immediate impact on that environment. The issue needs to be considered during the entire design process. Dependability encompasses the following aspects of a system:

1. **Safety**:

Definition 1.5: *"Safety is the state of being 'safe' (from French sauf), the condition of being protected from harm or other non-desirable outcomes"* [557].

Typically, we consider harm or threats from **within** the system we are using during its **normal operation**. For example, no software malfunction should be a threat to lives.

2. **Security**:

Definition 1.6: A system is **secure** if it is protected against harm caused by **attacks** originating from **outside** the system.

Connecting components in IoT systems enables attacks from the outside. Cyber-crime and cyber-warfare are special cases of such attacks, with significant amounts of potential damages. Connecting more and more components enables even more attacks and leads to more damages. This has become a serious issue in the design and proliferation of IoT systems.

The only really secure solution is to disconnect components, which contradicts the original motivation of using connected systems. Related research is therefore expected to be one of the fastest growing areas in IT-related research.

According to Ravi et al. [289], the following typical elements of security requirements exist:

– A **user identification process** validates users before allowing them to use the system.
– **Secure network access** provides a network connection or service access only if the device is authorized.
– **Secure communications** include a number of communication-related features.
– **Secure storage** requires confidentiality and integrity of data.
– **Content security** enforces usage restrictions.

3. **Confidentiality**:

Definition 1.7: A system provides **confidentiality** if information can only be accessed by the intended recipients.

Confidentiality is typically implemented using techniques which are found in secure systems, i.e., encryption.

4. **Reliability**: This term refers to malfunctions of systems resulting from components not working according to their specification at design time. Lack of reliability can be caused by breaking components. Reliability is the probability that

a system will not fail[1]. For an evaluation of reliability, we are not considering malicious attacks from the outside but only effects occurring within the system itself during normal, intended operation.

5. **Repairability**: Repairability (also spelled reparability) is the probability that a failing system can be repaired within a certain time.
6. **Availability**: Availability is the probability that the system is available. Reliability and repairability must be high and security hazards absent in order to achieve a high availability.

Designers may be tempted to focus just on the functionality of systems initially, assuming that dependability can be added once the design is working. Typically, this approach does not work, since certain design decisions will not allow achieving the required dependability in the aftermath. For example, if the physical partitioning is done in the wrong way, redundancy may be impossible. Therefore, *"making the system dependable must not be an after-thought,"* it must be considered from the very beginning [292]. Good compromises achieving an acceptable level of safety, security, confidentiality, and reliability have to be found [284].

Even perfectly designed systems can fail if the assumptions about the workload and possible errors turn out to be wrong [292]. For example, a system might fail if it is operated outside the initially assumed temperature range.

• If we look closely at the interface between the physical and the cyber-world, we observe a **mismatch between physical and cyber-models**. The following list shows examples:

– Many cyber-physical systems must meet **real-time constraints**. Not completing computations within a given time frame can result in a serious loss of the quality provided by the system (e.g., if the audio or video quality is affected) or may cause harm to the user (e.g., if cars, trains, or planes do not operate in the predicted way). Some time, constraints are called hard time constraints:

Definition 1.8 (**Kopetz [292]**): *"A time-constraint is called **hard** if not meeting that constraint could result in a catastrophe".*

All other time constraints are called **soft time constraints**.

Many of today's information processing systems are using techniques for speeding-up information processing *on the average*. For example, caches improve the average performance of a system. In other cases, reliable communication is achieved by repeating certain transmissions. These cases include Ethernet protocols: They typically rely on resending messages whenever the original messages have been lost. On average, such repetitions result in a (hopefully only) small loss of performance, even though for a certain message the communication delay can be orders of magnitude larger than the normal delay. In the context of real-time systems, arguments about the average performance or delay cannot be accepted. *"A guaranteed system response has to be explained without statistical arguments"* [292]. Many modeling techniques in computer science do not model real time. Frequently, time is modeled

---

[1] A formal definition of this term is provided in Definition 5.35 on p. 270 of this book.

without any physical units attached to it, which means that no distinction is made between picoseconds and centuries. The resulting problems are very clearly formulated in a statement made by E. Lee: *"The lack of timing in the core abstraction (of computer science) is a flaw, from the perspective of embedded software"* [318].

- Many embedded systems are **hybrid systems** in the sense that they include analog and digital parts. Analog parts use continuous signal values in continuous time, whereas digital parts use discrete signal values in discrete time. Many physical quantities are represented by a pair, consisting of a real number and a unit. The set of real numbers is uncountable. In the cyber-world, the set of representable values for each number is finite. Hence, **almost all physical quantities can only be approximated in digital computers**. During simulations of physical systems on digital computers, we are typically assuming that this approximation gives us meaningful results. In a paper, Taha considered consequences of the non-availability of real numbers in the cyber-world [499].
- Physical systems can exhibit the so-called **Zeno effect**. The Zeno effect can be introduced with the help of the bouncing ball example. Suppose that we are dropping a bouncing ball onto the floor from a particular height. After releasing the ball, it will start to fall, being accelerated by the gravitation of the earth. When it hits the floor, it will bounce, i.e., it will start to move in the opposite direction. However, we assume that bouncing will have some damping effect and that the initial speed of the ball after the bouncing will be reduced by a factor of $s < 1$, compared to the speed right before the bouncing. s is called the restitution. Due to this, the ball will not reach its initial height. Furthermore, the time to reach the floor a second time will be shorter than for the initial case. This process will be repeated, with smaller and smaller intervals between the bounces. However, according to the ideal model of partially elastic bounces, this process will go on and on. Figure 1.5 visualizes the trajectory (the height as a function of time) of the bouncing ball.

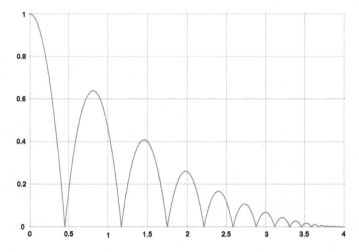

Fig. 1.5 Trajectory of a bouncing ball (© Openmodelica)

Now, let Δ be an arbitrary time interval, anywhere in the time domain. Would there be an upper bound on the number of bounces in this time interval? No, there would not be an upper bound, since bouncing is repeated in shorter and shorter intervals.

This is a special case of the Zeno effect. A system is said to exhibit a Zeno effect, when it is possible to have an unlimited number of events in an interval of finite length [386]. Mathematically speaking, this is feasible since infinite series may be converging to a finite value. In this case, the infinite series of times at which bouncing occurs is converging to a finite instance in time. In the ideal model, bouncing happens again and again, but intervals between bounces get shorter and shorter. See the discussion on p. 43 for more details.

On digital computers, the unlimited number of events can only be approximated.
– Many CPS comprise control loops, like the one shown in Fig. 1.6.

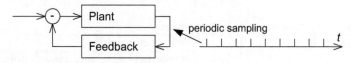

Fig. 1.6 Control loop

Control theory was initially based on analog feedback systems. For digital, discrete time feedback, periodic sampling of signals has been the default assumption for decades and it worked reasonably well. However, periodic sampling is possibly not the best approach. We could save resources if we would extend sampling intervals during times of relatively constant signals. This is the idea of **adaptive sampling**. Adaptive sampling is an area of active research [202].
– Traditional sequential programming languages are not the best way to describe concurrent, timed systems.
– Traditionally, the process of verifying whether or not some product is a correct implementation of the specification is generating a Boolean result: either the product is correct or not. However, two physically existing products will never be exactly identical. Hence, we can only check with some level of imprecision whether a product is a correct implementation of the design. This introduces fuzziness, and Boolean verification is replaced by fuzzy verification [178, 424].
– Edward Lee pointed out that the combination of a deterministic physical model and a deterministic cyber-model will possibly be a non-deterministic model [314]. Non-deterministic sampling can be one reason for this.

Overall, we observe a mismatch between the physical and the cyber-world. Effectively, we are still looking for appropriate models for CPS, but cannot expect to completely eliminate the mismatch.

• Embedded systems must be **resource aware**. The following resources can be considered:

1. **Energy**: *Electronic* information processing technology (IT) uses electrical energy for information processing. The amount of electrical energy used is frequently called "consumed energy." Strictly speaking, this term is not correct, since the total amount of energy is invariant. Actually, we are **converting** electrical energy into some other form of energy, typically thermal energy. For embedded systems, the availability of **electrical** power and energy (as the integral of power over time) is a deciding factor. This was already observed in a Dutch road mapping effort: *"Power is considered as the most important constraint in embedded systems"* [144].

Why should we care about the amount of electrical energy converted, i.e., why should there be energy-awareness? There are many reasons for this. Most reasons are applicable to most types of systems, but there are exceptions, as shown in Table 1.1.

Table 1.1 Relevance of reasons to be energy-aware

System type example	Relevant during use?		
	Plugged factory	Charged laptop	Unplugged sensor network
Global warming	Yes	Hardly	No
Cost of energy	Yes	Hardly	Typically not
Increasing performance	Yes	Yes	Yes
Problems with cooling, avoiding hot spots	Yes	Yes	Yes
Avoiding high currents, metal migration	Yes	Yes	Yes
Energy a very scarce resource	No	Hardly	Yes

Global warming is of course a very important reason for trying to be energy-aware. However, typically, very limited amounts of energy are available to unplugged systems, and hence, their contribution to global warming is small.

The cost of energy is relevant whenever the amount of energy needed is expensive. For plugged systems, this could happen due to large amounts of consumed energy. For unplugged systems, these amounts are typically small, but there could be cases for which it is very expensive to provide even a small amount.

Increased computing performance usually requires additional energy and, hence, has an impact on the resulting energy consumption.

Thermal effects are becoming more important and have to be considered as well. The reliability of circuits decreases with increasing temperatures. Hence, increased energy consumptions are typically decreasing the reliability.

In some cases (like remote sensor nodes), energy is a really scarce resource.

It is interesting to look at those cases where certain reasons to save energy can be considered irrelevant: Unplugged systems, due to the limited capacity of batteries, consume very small amounts of energy and are therefore not relevant for global warming. For systems connected to the power grid, energy is not a really scarce resource. Systems which are only temporarily connected to the power grid are somewhere between their plugged and unplugged counterparts.

The importance of power and energy efficiency was initially recognized for embedded systems. The focus on these objectives was later taken up for general-purpose computing as well and led to initiatives such as the **green computing initiative** [10].

2. **Run-time**: Embedded systems should exploit the available hardware architecture as much as possible. Inefficient use of execution time (e.g., wasted processor cycles) should be avoided. This implies an optimization of execution times across all levels, from algorithms down to hardware implementations.

3. **Code size**: For some embedded systems, code typically has to be stored on the system itself. There may be tight constraints on the storage capacity of the system. This is especially true for **systems on a chip** (SoCs), systems for which all the information processing circuits are included on a single chip. If the instruction memory is to be integrated onto this chip, it should be used very efficiently. For example, there may be medical devices implanted into the human body. Due to size and communication constraints of such devices, code has to be very compact.

   However, the importance of this design goal might change, when dynamically loading code becomes acceptable or when larger memory densities (measured in bits per volume unit) become available. Flash-based memories and new memory technologies will potentially have a large impact.

4. **Weight**: All portable systems must be lightweight. A low weight is frequently an important argument for buying a particular system.

5. **Cost**: For high-volume embedded systems in mass markets, especially in consumer electronics, competitiveness on the market is an extremely crucial issue, and efficient use of hardware components and the software development budget are required. A minimum amount of resources should be used for implementing the required functionality. We should be able to meet requirements using the least amount of hardware resources and energy. In order to reduce the energy consumption, clock frequencies and supply voltages should be as low as possible. Also, only the necessary hardware components should be present, and overprovisioning should be avoided. Components which do not improve the worst-case execution time (such as many caches or memory management units) can sometimes be omitted.

Due to resource awareness targets, software designs cannot be done independently of the underlying hardware. Therefore, software and hardware must be taken into account during the design steps. This, however, is difficult, since such integrated approaches are typically not taught at educational institutes. The cooperation between electrical engineering and computer science has not yet reached the required level.

A mapping of specifications to custom hardware would provide the best energy efficiency. However, hardware implementations are very expensive and require long design times. Therefore, hardware designs do not provide the flexibility to change designs as needed. We need to find a good compromise between efficiency and flexibility.

- CPS and IoT systems are frequently collecting huge amounts of data. These large amounts of data have to be stored, and they have to be analyzed. Hence, there is a strong link between the problems of **big data** and CPS/IoT. This is exactly the topic of our collaborative research center SFB 876[2]. SFB 876 focuses on machine learning under resource constraints.
- **Impact beyond technical issues**: Due to the major impact on society, legal, economic, social, human, and environmental impacts must be considered as well:

  - The integration of many components, possibly by different providers, raises serious issues concerning liability. These issues are being discussed, for example, for self-driving cars. Also, ownership issues must be solved. It is unacceptable to have one of the involved companies own all rights.
  - Social issues include the impact of new IT devices on society. Currently, this impact is frequently only detected long after the technology became available.
  - Human issues comprise user-friendly man–machine interfaces.
  - Contributions to global warming and the production of waste should be at an acceptable level. The same applies to the consumption of resources.

- Real systems are concurrent. Managing **concurrency** is therefore another major challenge.
- Cyber-physical and IoT systems are typically consisting of heterogeneous hardware and software components from various providers and have to operate in a changing environment. The resulting **heterogeneity** poses challenges for the correct cooperation of components. It is not sufficient to consider only software or only hardware design. Design complexity requires adopting a hierarchical approach. Furthermore, real embedded systems consist of many components, and we are interested in **compositional design**. This means we would like to study the impact of combining components. For example, we would like to know whether we could add a GPS system to the sources of information in a car without overloading the communication bus.
- Embedded system design involves knowledge from many areas.
  It is difficult to find staff members with a sufficient amount of knowledge in all relevant areas. Even organizing the knowledge transfer between relevant areas is already challenging. Designing a curriculum for a program in embedded system design is even more challenging, due to the tight ceilings for the total workload for students. Overall, **tearing down walls between disciplines** and departments or at least lowering them would be required.

A list of challenges is also included in a report on IoT by Sundmaeker et al. [493].

---

[2]See http://www.sfb876.tu-dortmund.de.

1.4 Common Characteristics

In addition to the challenges listed above, there are more common characteristics of embedded, cyber-physical, and IoT systems, independently of the application area.

- CPS and IoT systems use **sensors** and **actuators** to connect the embedded system to the physical world. For IoT, these components are connected to the Internet.

Definition 1.9: Actuators are devices converting numbers into physical effects.

- Typically, embedded systems are **reactive systems**, which are defined as follows:

Definition 1.10 (Bergé [542]): *"A reactive system is one that is in continual interaction with its environment and executes at a pace determined by that environment."*

Reactive systems are modeled as being in a certain state, waiting for an input. For each input, they perform some computation and generate an output and a new state. Hence, automata are good models of such systems. Mathematical functions, describing the problems solved by most algorithms, would be an inappropriate model.

- Embedded systems are **underrepresented in teaching and in public discussions**. Real embedded systems are complex. Hence, comprehensive equipment is required for realistically teaching embedded systems design. However, teaching CPS design can be appealing, due to the visible impact on the physical behavior.
- These systems are frequently **dedicated toward a certain application**. For example, processors running control software in a car or a train will typically always run that software, and there will be no attempt to run a game or spreadsheet program on the same processor. There are mainly two reasons for this:
  1. Running additional programs would make those systems less dependable.
  2. Running additional programs is only feasible if resources such as memory are unused. No unused resources should be present in an efficient system.

  However, the situation is slowly changing. For example, the AUTOSAR initiative [26] demonstrates more dynamism in the automotive industry.
- Most embedded systems do not use keyboards, mice, and large computer monitors for their user interface. Instead, there is a **dedicated user interface** consisting of push buttons, steering wheels, pedals, etc. Because of this, the user hardly recognizes that information processing is involved. This is consistent with the introduction of the term **disappearing computer**.

Table 1.2 highlights some distinguishing features between the design of PC-like or data center server-like systems and embedded systems.

Compatibility with traditional instruction sets employed for PCs is less important for embedded systems, since it is typically possible to compile software applications for architectures at hand. Sequential programming languages do not match well with the need to describe concurrent real-time systems, and other ways of modeling applications may be preferred. Several objectives must be considered during the

Table 1.2 Distinction between PC-like and embedded system design

	Embedded	PC-/Server-like
Architectures	Frequently heterogeneous very compact	Mostly homogeneous not compact ($\times 86$ etc.)
$\times 86$ compatibility	Less relevant	Very relevant
Architecture fixed?	Sometimes not	Yes
Model of computation (MoCs)	C+multiple models (data flow, discrete events, …)	Mostly von Neumann (C, C++, Java)
Optimization objectives	Multiple (energy, size, …)	Average performance dominates
Safety-critical?	Possibly	Usually not
Real-time relevant	Frequently	Hardly
Applications	Guarantees for several concurrent apps. needed	Best effort approaches to run application
Apps. known at design time	Yes, for real-time systems	Only some (e.g., WORD)

design of embedded/cyber-physical systems. In addition to the average performance, the worst-case execution time, energy consumption, weight, reliability, operating temperatures, etc., may have to be optimized. Meeting real-time constraints is very important for CPS, but hardly so for PC-like systems. Time constraints can be verified at design time only if all the applications are known at this time. Also, it must be known, which applications should run concurrently. For example, designers must ensure that a GPS-application, a phone call, and data transfers can be executed at the same time without losing voice samples. For PC-like systems, knowledge about concurrently running software is almost never available and best effort approaches are typically used.

Why does it make sense to consider all types of embedded systems in one book? It makes sense because information processing in embedded systems has many common characteristics, despite being physically so different.

Actually, not every embedded system will have all the above characteristics. We can define the term "embedded system" also in the following way:

Definition 1.11: Information processing systems meeting **most of the characteristics** listed above are called **embedded systems**.

This definition includes some fuzziness. However, it seems to be neither necessary nor possible to remove this fuzziness.

1.5   Curriculum Integration of Embedded Systems

Unfortunately, embedded systems are hardly covered even in the latest edition of the Computer Science Curriculum, as published by ACM and the IEEE Computer Society [9]. However, the growing number of applications results in the need for more education in this area. This education should help to overcome the limitations of currently available design technologies for embedded systems. For example, there is still a need for better specification languages, models, tools generating implementations from specifications, timing verifiers, system software, real-time operating systems, low-power design techniques, and design techniques for dependable systems. This book should help teaching the essential issues and should be a stepping stone for starting more research in the area. **Additional information related to the book can be obtained from the following Web page**:

http://ls12-www.cs.tu-dortmund.de/~marwedel/es-book

This page includes links to slides, videos, simulation tools, error corrections, and other related material. Videos are directly accessible from

https://www.youtube.com/user/cyphysystems

Users of this material who discover errors or who would like to make comments on how to improve the material should send an e-mail to:

peter.marwedel@tu-dortmund.de

Due to the availability of this book and of videos, it is feasible and recommended to try out flipped classroom teaching [358]. With this style of teaching, students are requested to watch the videos (or read the book) at home. The presence of the students in the classroom is then used to interactively solve problems. This helps to strengthen problem-solving competences, team work, and social skills. In this way, the availability of the Internet is exploited to improve teaching methods for students actually present at their university. Assignments could use the information in this or in complementary books (e.g., [82, 168, 565]).

With flipped classroom teaching, existing laboratory session slots can be completely dedicated to gaining some practical experience with CPS. Toward this end, a course using this textbook should be complemented by an exciting laboratory, using, for example, small robots, such as Lego Mindstorms™or microcontrollers (e.g., Raspberry Pie, Arduino, or Odroid). For microcontroller boards which are available on the market, educational material is typically available. Another option is to let students gain some practical experience with finite state machine tools.

Prerequisites

The book assumes a basic understanding in several areas:

- electrical networks at the high-school level (e.g., Kirchhoff's laws),
- operational amplifiers (optional),
- computer organization, for example, at the level of the introductory book by J.L. Hennessy and D.A. Patterson [204],
- fundamental digital circuits such as gates and registers,
- computer programming (including foundations of software engineering),
- fundamentals of operating systems,
- fundamentals of computer networks (important for IoT!),
- finite state machines,

- some first experience with programming microcontrollers,
- fundamental mathematical concepts (tuples, integrals, and linear)
- algorithms (graph algorithms and optimization algorithms such as branch and bound), and
- the concept of NP-completeness.

Knowledge in statistics and Fourier series would be useful.

These prerequisites can be grouped into courses as shown in the top row in Fig. 1.7.

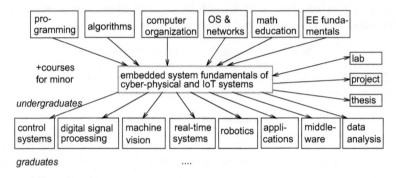

Fig. 1.7 Positioning of the topics of this book

Recommended Additional Teaching

The book should be complemented by follow-up courses providing a more specialized knowledge in some of the following areas (see the bottom row in Fig. 1.7)[3]:

- control systems,
- digital signal processing,
- machine vision,
- real-time systems, real-time operating systems, and scheduling,
- robotics,
- application areas such as telecommunications, automotive, medical equipment, and smart homes,
- middleware,
- specification languages and models for embedded systems,
- sensors and actuators,
- security and dependability of computer systems,
- data analysis techniques for CPS and IoT systems,
- low-power design techniques,
- physical aspects of CPS,
- computer-aided design tools for application-specific hardware,
- formal verification of hardware systems,

---

[3]The partitioning between undergraduate courses and graduate courses may differ between universities.

- testing of hardware and software systems,
- performance evaluation of computer systems,
- ubiquitous computing,
- advanced communication techniques for IoT,
- the Internet of Things (IoT),
- impact of embedded, CPS, and IoT systems, and
- legal aspects of embedded CPS and IoT systems.

1.6   Design Flows

The design of embedded/cyber-physical systems and the Internet of Things is a rather complex task, which has to be broken down into a number of sub-tasks to be tractable. These sub-tasks must be performed one after the other, and some of them must be repeated.

The design information flow starts with ideas in people's heads. These ideas should incorporate knowledge about the application area. They must be captured in a design specification. In addition, standard hardware and system software components are typically available and should be reused whenever possible (see Fig. 1.8).

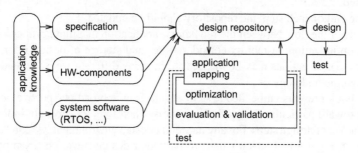

Fig. 1.8 Simplified design information flow

In Fig. 1.8 (as well as in other similar diagrams in this book), we are using **boxes with rounded corners for stored information** and **rectangles for transformations on data**. In particular, information is stored in the **design repository**. The repository allows keeping track of design models. In most cases, the repository should provide version management or "revision control," such as CVS [88], SVN [108], or "git" (see https://www.git-scm.com). A good design repository should also come with a design management interface which would also keep track of the applicability of design tools and sequences, all integrated into a comfortable graphical user interface (GUI). The design repository and the GUI can be extended into an **integrated development environment** (IDE), also called **design framework** (see, e.g.,[329]). An integrated development environment keeps track of dependencies between tools and design information.

Using the repository, design decisions can be taken in an iterative fashion. At each step, design model information must be retrieved. This information is then considered.

During design iterations, **applications are mapped** to execution platforms and new (partial) design information is generated. The generation comprises the mapping of operations to concurrent tasks, the mapping of operations to either hardware or software (called hardware/software partitioning), compilation, and scheduling.

Designs should be **evaluated** with respect to various objectives including performance, dependability, energy consumption, and manufacturability. At the current state of the art, usually none of the design steps can be guaranteed to be correct. Therefore, it is also necessary to **validate** the design. Validation consists of checking intermediate or final design descriptions against other descriptions. Thus, each design decision should be evaluated and validated.

Due to the importance of the efficiency of embedded systems, **optimizations** are important. There is a large number of possible optimizations, including high-level transformations (such as advanced loop transformations) and energy-oriented optimizations.

Design iterations could also include **test** generation and an evaluation of the testability. Testing needs to be included in the design iterations if testability issues are already considered during the design steps. In Fig. 1.8, test generation has been included as optional step of design iterations (see the dashed box). If test generation is not included in the iterations, it must be performed after the design has been completed.

At the end of each step, the repository should be updated.

Details of the flow between the repository, application mapping, evaluation, validation, optimization, testability considerations, and storage of design information may vary. These actions may be interleaved in many different ways, depending on the design methodology used.

This book presents embedded system design from a broad perspective, and it is not tied toward particular design flows or tools. Therefore, we have not indicated a particular list of design steps. For any particular design environment, we can "unroll" the loop in Fig. 1.8 and attach names to particular design steps. For example, this leads to the particular case of the SpecC [167] design flow shown in Fig. 1.9.

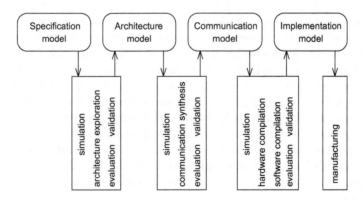

Fig. 1.9 Design flow for SpecC tools (simplified)

In this case, a particular set of design steps, such as architecture exploration, communication synthesis, and software and hardware compilation, are included. The precise meaning of these terms is not relevant in this book. In the case of Fig. 1.9, validation and evaluation are explicitly shown for each of the steps, but are wrapped into one larger box.

A second instance of an unfolded Fig. 1.8 is shown in Fig. 1.10. It is the V-model of design flows [527], which has to be adhered to for many German IT projects, especially in the public sector, but also beyond.

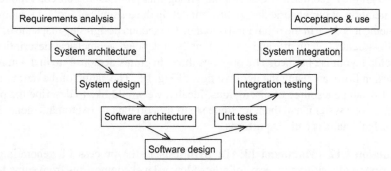

Fig. 1.10 Design flow for the V-model

Figure 1.10 very clearly shows the different steps that must be performed. The steps correspond to certain phases during the software development process (the precise meaning is again not relevant in the context of this book). Note that taking design decisions, evaluating and validating designs are lumped into a single box in this diagram. Application knowledge, system software, and system hardware are not explicitly shown. The V-model also includes a model of the integration and testing phase (right "wing") of the diagram. This corresponds to an inclusion of testing in the loop of Fig. 1.8. The shown model corresponds to the V-model version "97". The more recent V-model XT allows a more general set of design steps. This change matches very well our interpretation of design flows in Fig. 1.8. Other iterative approaches include the **waterfall model** and the **spiral model**. More information about software engineering for embedded systems can be found in a book by J. Cooling [109].

Our generic design flow model is also consistent with flow models used in hardware design. For example, Gajski's Y-chart [165] (see Fig. 1.11) is a very popular model.

Fig. 1.11 Gajski's Y-chart and design path (red, bold)

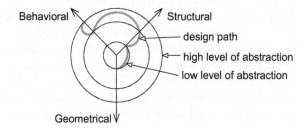

Gajski considers design information in three dimensions: behavior, structure, and layout. The first dimension just reflects the behavior. A high-level model would describe the overall behavior, while finer-grained models would describe the behavior of components. Models at the second dimension include structural information, such as information about hardware components. High-level descriptions in this dimension could correspond to processors and low-level descriptions to transistors. The third dimension represents geometrical layout information of chips. Design paths will typically start with a coarse-grained behavioral description and finish with a fine-grained geometrical description. Along this path, each step corresponds to one iteration of our generic design flow model. In the example of Fig. 1.11, an initial refinement is done in the behavioral domain. The second design step maps the behavior to structural elements, and so on. Finally, a detailed geometrical description of the chip layout is obtained. The previous three diagrams demonstrate that a number of design flows are using the iterative flow of Fig. 1.8. The nature of the iterations in Fig. 1.8 can be a source of discussions. Ideally, we would like to describe the properties of our system, and then, let some smart tool do the rest. Automatic generation of design details is called **synthesis**.

Definition 1.12 (Marwedel [354]): Synthesis is the process of generating the description of a system in terms of related lower-level components from some high-level description of the expected behavior.

Automatic synthesis is assumed to perform this process automatically. Automatic synthesis, if successful, avoids many manual design steps. The goal of using automatic synthesis for the design of systems has been considered in the "describe-and-synthesize" paradigm by Gajski [166]. This paradigm is in contrast to the more traditional "specify-explore-refine" approach, also known as "design-and-simulate" approach. The second term stresses the fact that manual design typically has to be combined with simulation, for example, for catching design errors. In the traditional approach, simulation is more important than in automatic synthesis.

1.7 Structure of This Book

Consistent with the design information flow shown above, this book is structured as follows: Chap. 2 provides an overview of specification techniques, languages, and models. Key hardware components of embedded systems are presented in Chap. 3. Chap. 4 deals with system software components, particularly embedded operating systems. Chap. 5 contains the essentials of embedded system design evaluation and verification. Mapping applications to execution platforms is one of the key steps in the design process of embedded systems. Standard techniques (including scheduling) for achieving such a mapping are listed in Chap. 6. Due to the need for generating efficient designs, many optimization techniques are needed. From among the abundant set of available optimization techniques, several groups are mentioned in Chap. 7.

Chap. 8 contains a brief introduction to testing mixed hardware/software systems. The Appendix comprises prerequisites for understanding the book, and it can be skipped by students familiar with the topics covered there.

It may be necessary to design special purpose hardware or to optimize processor architectures for a given application. However, hardware design is not covered in this book. Coussy and Morawiec [112] provide an overview of high-level hardware synthesis techniques.

The content of this book is different from the content of most other books on embedded systems design. Traditionally, the focus of many books on embedded systems is on explaining the use of microcontrollers, including their memory, I/O, and interrupt structure. There are many such books [34, 35, 38, 39, 169–171, 201, 267, 303, 403].

We believe that due to the increasing complexity of embedded and cyber-physical systems, this focus has to be extended to include at least different specification paradigms, fundamentals of hardware building blocks, the mapping of applications to execution platforms, as well as evaluation, validation, and optimization techniques. In the current book, we will be covering all these areas. The goal is to provide students with an introduction to embedded systems, enabling students to put the different areas into perspective.

For further details, we recommend a number of sources (some of which have also been used in preparing this book):

- There is a large number of sources of information on specification languages. These include earlier books by Young [581], Burns and Wellings [81], Bergé [542], and de Micheli [119]. There is a huge amount of information on languages such as SystemC [390], SpecC [167], and Java [16, 73, 126, 258, 549].
- A book written by Edward Lee et al. also includes physical aspects of cyber-physical systems [320].
- Approaches for designing and using real-time operating systems (RTOSes) are presented in a book by Kopetz [292].
- Real-time scheduling is covered comprehensively in the books by Buttazzo [82], by Krishna and Shin [297], and by Baruah et al. [40].
- Other sources of information about embedded systems include books by Laplante [308], Vahid [529], and the ARTIST road map [65], the "Embedded Systems Handbook" [586], and books by Gajski et al. [168], and Popovici et al. [432].
- Approaches for embedded system education are covered in the Workshops on Embedded Systems Education (WESE); see [183] for results from the workshop held in 2015.
- The Web site of the European network of excellence on embedded and real-time systems [23] provides numerous links for the area.
- The Web page of a special interest group of ACM [8] focuses on embedded systems.
- Symposia dedicated toward embedded/cyber-physical systems include the Embedded Systems Week (see http://www.esweek.org) and the Cyber-Physical Systems Week (see http://www.cpsweek.org).

- Robotics is an area that is closely linked to embedded and cyber-physical systems. We recommend the book by Siciliano et al. [462] for information on robotics.
- There are specialized books and articles on the Internet of Things [179, 185, 186].

1.8   Problems

We suggest solving the following problems either at home or during a flipped classroom session:

1.1   Please list possible definitions of the term "embedded system"!

1.2   How would you define the term "cyber-physical system (CPS)"? Do you see any difference between the terms "embedded systems" and "cyber-physical systems"?

1.3   What is the "Internet of Things" (IoT)?

1.4   What is the goal of "Industry 4.0"?

1.5   In which way does this book cover CPS and IoT design?

1.6   In which application areas do you see opportunities for CPS and IoT systems? Where do you expect major changes caused by information technology?

1.7   Use the sources available to you to demonstrate the importance of embedded systems!

1.8   Which challenges must be overcome in order to fully take advantage of the opportunities?

1.9   What is a hard timing constraint? What is a soft timing constraint?

1.10   What is the "Zeno effect"?

1.11   What is adaptive sampling?

1.12   Which objectives must be considered during the design of embedded systems?

1.13   Why are we interested in energy-aware computing?

1.14   What are the main differences between PC-based applications and embedded/CPS applications?

1.15   What is a reactive system?

1.16   On which Web sites do you find companion material for this book?

1.17 Compare the curriculum of your educational program with the description of the curriculum in this introduction. Which prerequisites are missing in your program? Which advanced courses are available?

1.18 What is flipped classroom teaching?

1.19 How could we model design flows?

1.20 What is the "V-model"?

1.21 How could we define the term "synthesis"?

Chapter 2
Specifications and Modeling

How can we describe the system which we would like to design and how can we represent intermediate design information? Models and description techniques allowing us to capture the initial specification as well as intermediate design information will be presented in this chapter.

2.1 Requirements

Consistent with the simplified design flow (see Fig. 1.8), we will first of all describe requirements and approaches for specifying embedded systems. Specifications for embedded systems provide **models** of the system under design (SUD). Models can be defined as follows:

Definition 2.1 (Jantsch [256]): *"A model is a simplification of another entity, which can be a physical thing or another model. The model contains exactly those characteristics and properties of the modeled entity that are relevant for a given task. A model is minimal with respect to a task if it does not contain any other characteristics than those relevant for the task."*

Models are described in languages. Languages should be capable of representing the following features[1]:

- **Hierarchy**: Human beings are generally not capable of comprehending systems containing many objects (states, components) having complex relations with each other. The description of all real-life systems needs more objects than human beings can understand. Hierarchy (in combination with **abstraction**) is a key mechanism

---

[1]Information from the books of Burns et al. [81], Bergé et al. [542], and Gajski et al. [166] is used in this list.

© Springer International Publishing AG 2018
P. Marwedel, *Embedded System Design*, Embedded Systems,
DOI 10.1007/978-3-319-56045-8_2

helping to solve this dilemma. Hierarchies can be introduced such that humans need to handle only a small number of objects at any time.

There are two kinds of hierarchies:

- **Behavioral hierarchies**: Behavioral hierarchies are hierarchies containing objects necessary to describe the system behavior. States, events, and output signals are examples of such objects.
- **Structural hierarchies**: Structural hierarchies describe how systems are composed of physical components.

  For example, embedded systems can be comprised of components such as processors, memories, actuators, and sensors. Processors, in turn, include registers, multiplexers, and adders. Multiplexers are composed of gates.

- **Component-based design** [464]: It must be "easy" to derive the behavior of a system from the behavior of its components. If two components are connected, the resulting new behavior should be predictable. For example, suppose that we add another component (say, some GPS unit) to a car. The impact of the additional component on the overall behavior of the system (including buses) should be predictable.
- **Concurrency**: Real-life systems are distributed, concurrent systems composed of components. It is therefore necessary being able to specify concurrency conveniently. Unfortunately, human beings are not very good at understanding concurrent systems and many problems with real systems are actually a result of an incomplete understanding of possible behaviors of concurrent systems.
- **Synchronization and communication**: Components must be able to communicate and to synchronize. Without communication, components could not cooperate and we would use each of them in isolation. It must also be possible to agree on the use of resources. For example, it is necessary to express mutual exclusion.
- **Timing behavior**: Many embedded systems are real-time systems. Therefore, explicit timing requirements are one of the characteristics of embedded systems. The need for explicit modeling of time is even more obvious from the term "cyber-physical system." Time is one of the key dimensions of physics. Hence, timing requirements **must** be captured in the specification of embedded/cyber-physical systems.

  However, standard theories in computer science model time only in a very abstract way. The O-notation is one of the examples. This notation just reflects growth rates of functions. It is frequently used to model run-times of algorithms, but it fails to describe real execution times. In physics, quantities have units, but the O-notation does not even have units. So, it would not distinguish between femtoseconds and centuries. A similar remark applies to termination properties of algorithms. Standard theories are concerned with proving that a certain algorithm *eventually* terminates. For real-time systems, we need to show that certain computations are completed in a given amount of time, but the algorithm as a whole should possibly run until power is turned off.

According to Burns and Wellings [81], modeling time must be possible in the following four contexts:

- Techniques for measuring **elapsed time**:
  For many applications, it is necessary to check how much time has elapsed since some computation was performed. Access to a timer would provide a mechanism for this.
- Means for **delaying of processes** for a specified time:
  Typically, real-time languages provide some delay construct. Unfortunately, typical implementations of embedded systems in software do not guarantee precise delays. Let us assume that task τ should be delayed by some amount Δ. Usually, this delay is implemented by changing task τ's state in the operating system from "ready" or "run" to "suspended." At the end of this time interval, τ's state is changed from "suspended" to "ready." This does not mean that the task actually executes. If some higher priority task is executing or if preemption is not used, the delayed task will be delayed longer than Δ.
- Possibility to specify **timeouts**:
  There are many situations in which we must wait for a certain event to occur. However, this event may actually not occur within a given time interval and we would like to be notified about this. For example, we might be waiting for a response from some network connection. We would like to be notified if this response is not received within some amount of time, say Δ. This is the purpose of **timeouts**. Real-time languages usually also provide some timeout construct. Implementations of timeouts frequently come with the same problems which we mentioned for delays.
- Methods for specifying **deadlines** and **schedules**:
  For many applications, it is necessary to complete certain computations in a limited amount of time. For example, if the sensors of some car signal an accident, air bags must be ignited within about ten milliseconds. In this context, we must guarantee that the software will decide whether or not to ignite the air bags in that given amount of time. The air bags could harm passengers if they go off too late. Unfortunately, most languages do not allow to specify timing constraints. If they can be specified at all, they must be specified in separate control files, pop-up menus, etc. But the situation is still bad even if we are able to specify these constraints: Many modern hardware platforms do not have a very predictable timing behavior. Caches, stalled pipelines, speculative execution, task preemption, interrupts, etc. may have an impact on the execution time which is very difficult to predict. Accordingly, **timing analysis** (verifying the timing constraints) is a very hard design task.

- **State-oriented behavior**: It was already mentioned in Chap. 1 on p. 15 that automata provide a good mechanism for modeling reactive systems. Therefore, the state-oriented behavior provided by automata should be easy to describe. However, classical automata models are insufficient, since they cannot model timing and since hierarchy is not supported.

- **Event handling**: Due to the reactive nature of embedded systems, mechanisms for describing events must exist. Such events may be external events (caused by the environment) or internal events (caused by components of the system under design).
- **Exception-oriented behavior**: In many practical systems, exceptions do occur. In order to design dependable systems, it must be possible to describe actions to handle exceptions easily. It is not acceptable that exceptions must be indicated for each and every state (such as in the case of classical state diagrams).

Example 2.1: In Fig. 2.1, input *k* might correspond to an exception.

Fig. 2.1 State diagram with exception *k*

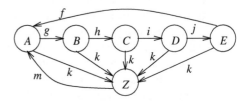

Specifying this exception at each state makes the diagram very complex. The situation would get worse for larger state diagrams with many transitions. On p. 47, we will show how all the transitions can be replaced by a single one (see Fig. 2.12).   ∇

- **Presence of programming elements**: Popular programming languages have proven to be a convenient means of expressing computations that must be performed. Hence, programming language elements should be available in the specification technique used. Classical state diagrams do not meet this requirement.
- **Executability**: Specifications are not automatically consistent with the ideas in people's heads. Executing the specification is a means of plausibility checking. Specifications using programming languages have a clear advantage in this context.
- **Support for the design of large systems**: There is a trend toward large and complex embedded software programs. Software technology has found mechanisms for designing such large systems. For example, object orientation is one such mechanism. It should be available in the specification methodology.
- **Domain-specific support**: It would of course be nice if the same specification technique could be applied to all the different types of embedded systems, since this would minimize the effort for developing specification techniques and tool support. However, due to the wide range of application domains including those listed in Sect. 1.2, there is little hope that one language can be used to efficiently represent specifications in all such domains. For example, control-dominated, data-dominated, centralized and distributed application domains can all benefit from language features dedicated toward those domains.
- **Readability**: Of course, specifications must be readable by human beings. Otherwise, it would not be feasible to validate whether or not the specification meets

the real intent of the persons specifying the system under design. All design documents should also be machine-readable into order to process them in a computer. Therefore, specifications should be captured in languages which are readable by humans and by computers.

Initially, such specifications could use a natural language such as English or Japanese. Even this natural language description should be captured in a design document, so that the final implementation can be checked against the original document. However, natural languages are not sufficient for later design phases, since natural languages lack key requirements for specification techniques: It is necessary to check specifications for completeness and absence of contradictions. Furthermore, it should be possible to derive implementations from the specification in a systematic way. Natural languages do not meet these requirements.

- **Portability and flexibility**: Specifications should be independent of specific hardware platforms so that they can be easily used for a variety of target platforms. Ideally, changing the hardware platform should have no impact on the specification. In practice, small changes may have to be tolerated.
- **Termination**: It should be feasible to identify terminating processes from the specification. This means that we would like to use specifications for which the halting problem (the problem of figuring out whether or not a certain algorithm will terminate; see, for example, [469]) is decidable.
- **Support for non-standard I/O devices**: Many embedded systems use I/O devices other than those typically found in a PC. It should be possible to describe inputs and outputs for those devices conveniently.
- **Non-functional properties**: Actual systems under design must exhibit a number of non-functional properties, such as fault tolerance, size, extendability, expected lifetime, power consumption, weight, disposability, user friendliness, and electro-magnetic compatibility (EMC). There is no hope that all these properties can be defined in a formal way.
- **Support for the design of dependable systems**: Specification techniques should provide support for designing dependable systems. For example, specification languages should have unambiguous semantics, facilitate formal verification, and be capable of describing security and safety requirements.
- **No obstacles to the generation of efficient implementations**: Since embedded systems must be efficient, no obstacles prohibiting the generation of efficient realizations should be present in the specification.
- **Appropriate model of computation (MoC)**: The von-Neumann model of sequential execution combined with some communication techniques is a commonly used MoC. However, this model has a number of serious problems, in particular for embedded system applications. Problems include the following:

  - Facilities for describing timing are lacking.
  - Von-Neumann computing is implicitly based on accesses to globally shared memory (such as in Java). It has to guarantee mutually exclusive access to shared resources. Otherwise, multi-threaded applications allowing preemptions

at any time can lead to very unexpected program behaviors[2]. Using primitives for ensuring mutually exclusive access can, however, very easily lead to deadlocks. Possible deadlocks may be difficult to detect and may remain undetected for many years.

Example 2.2: Lee [316] provided a very alarming example in this direction. Lee studied implementations of a simple observer pattern in Java. For this pattern, changes of values must be propagated from some producer to a set of subscribed observers. This is a very frequent pattern in embedded systems, but is difficult to implement correctly in a multi-threaded von-Neumann environment with preemptions. Lee's code is a possible implementation of the observer pattern in Java for a multi-threaded environment:

```
public synchronized void addListener(listener) {...}
public synchronized void setValue(newvalue)
{
  myvalue=newvalue;
  for (int i=0; i<mylisteners.length; i++) {
    myListeners[i].valueChanged(newvalue)    }
}
```

Method addListener subscribes new observers, and method setValue propagates new values to subscribed observers. In general, in a multi-threaded environment, threads can be preempted any time, resulting in an arbitrarily interleaved execution of these threads. Adding observers while setValue is already active could result in complications, i.e., we would not know if the new value had reached the new listener. Moreover, the set of observers constitutes a global data structure of this class. Therefore, these methods are synchronized in order to avoid changing the set of observers while values are already partially propagated. This way, only one of the two methods can be active at a given time. This mutual exclusion is necessary to prevent unwanted interleavings of the execution of methods in a multi-threaded environment. Why is this code problematic? It is problematic since valueChanged could attempt to get exclusive access to some resource (say, R). If that resource is allocated to some other method (say, A), then this access is delayed until A releases R. If A calls (possibly indirectly) addListener or setValue before releasing R, then these methods will be in a deadlock: setValue waits for R, releasing R requires A to proceed, and A cannot proceed before its call of setValue or addListener is serviced. Hence, we will have a deadlock.

This example demonstrates the existence of deadlocks resulting from using multiple threads which can be arbitrarily preempted and therefore require mutual exclusion for their access to critical resources. Lee showed [316] that many of the proposed "solutions" of the problem are problematic themselves. So, even this very simple pattern is difficult to implement correctly in a multi-threaded von-Neumann environment. This example shows that concurrency is really difficult to understand for humans and there may be the risk of oversights, even after very rigorous code inspections.   ∇

---

[2]Examples are typically provided in courses on operating systems.

Lee came to the drastic conclusion that "*nontrivial software written with threads, semaphores, and mutexes is incomprehensible to humans*" and that "*threads as a concurrency model are a poor match for embedded systems. ... they work well only ... where best-effort scheduling policies are sufficient*" [318].

The underlying reasons for deadlocks have been studied in detail in the context of operating systems (see, for example, [483]). From this context, it is well-known that four conditions must hold at run-time to get into a deadlock: mutual exclusion, no preemption of resources, holding resources while waiting for more, and a cyclic dependency between threads. Obviously, all four conditions are met in the above example. The theory of operating systems provides no general way out of this problem. Rare deadlocks may be acceptable for a PC, but they are clearly unacceptable for a safety-critical system.

We would like to specify our SUD such that we do not have to care about possible deadlocks. Therefore, it makes sense to study non-von-Neumann MoCs avoiding this problem. We will study such MoCs from the next section onward. It will be shown that the observer pattern can be easily implemented in other MoCs.

From the list of requirements, it is already obvious that there will not be any single formal language meeting all these requirements. Therefore, in practice, we must live with compromises and possibly also with a mixture of languages (each of which would be appropriate for describing a certain type of problems). The choice of the language used for an actual design will depend on the application domain and the environment in which the design has to be performed. In the following, we will present a survey of languages that can be used for actual designs. These languages will demonstrate the essential features of the corresponding MoC.

2.2 Models of Computation

Models of computation (MoCs) describe the mechanism assumed for performing computations. In the general case, we must consider systems comprising components. It is now common practice to strictly distinguish between the computations performed in the components and communication. This distinction paves the way for **reusing components** in different contexts and enables *plug-and-play* for system components. Accordingly, we define models of computation as follows [255–257, 315]:

Definition 2.2: Models of computation (MoCs) define

- **Components** and the organization of computations in such components: Procedures, processes, functions, finite state machines are possible components.
- **Communication protocols**: These protocols describe methods for communication between components. Asynchronous message passing and rendez-vous based communication are examples of communication protocols.

Relations between components can be captured in graphs. In such graphs, we will refer to the computations also as processes or tasks. Accordingly, relations between these will be captured by **task graphs** and **process networks**. Nodes in the graph represent components performing computations. Computations map input data streams to output data streams. Computations are sometimes implemented in high-level programming languages. Typical computations contain (possibly non-terminating) iterations. In each cycle of the iteration, they consume data from their inputs, process the data received, and generate data on their output streams. Edges represent relations between components. We will now introduce these graphs at a more detailed level.

The most obvious relation between computations is their causal dependence: Many computations can only be executed after other computations have terminated. This dependence is typically captured in **dependence graphs**. Fig. 2.2 shows a dependence graph for a set of computations.

Fig. 2.2 Dependence graph

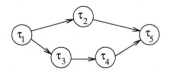

Definition 2.3: A dependence graph is a directed graph $G = (\tau, E)$, where τ is the set of **vertices** or **nodes** and E is the set of **edges**. $E \subseteq \tau \times \tau$ imposes a relation on τ. If $(\tau_1, \tau_2) \in E$ with $\tau_1, \tau_2 \in \tau$, then τ_1 is called an **immediate predecessor** of τ_2 and τ_2 is called an **immediate successor** of τ_1. Suppose E^* is the transitive closure of E. If $(\tau_1, \tau_2) \in E^*$, then τ_1 is called a **predecessor** of τ_2 and τ_2 is called a **successor** of τ_1.

Such dependence graphs form a special case of task graphs. Task graphs may contain more information than modeled in Fig. 2.2. For example, task graphs may include the following extensions of dependence graphs:

1. **Timing information**: Tasks may have arrival times, deadlines, periods, and execution times. In order to show them graphically, it may be useful to include this information in the graphs. However, we will indicate such information separately from the graphs in this book.
2. **Distinction between different types of relations** between computations: Precedence relations just model constraints for possible execution sequences. At a more detailed level, it may be useful to distinguish between constraints for scheduling and communication between computations. Communication can also be described by edges, but additional information may be available for each of the edges, such as the time of the communication and the amount of information exchanged. Precedence edges may be kept as a separate type of edges, since there could be situations in which computations must execute sequentially even though they do not exchange information.
   In Fig. 2.2, input and output (I/O) are not explicitly described. Implicitly, it is assumed that computations without any predecessor in the graph might be receiving input at some time. Also, they might generate output for the successor and

that this output could be available only after the computation has terminated. It is often useful to describe input and output more explicitly. In order to do this, another kind of relation is required. Using the same symbols as Thoen [515], we use partially filled circles for denoting input and output. In Fig. 2.3, partially filled circles identify I/O edges.

Fig. 2.3 Graph including I/O nodes and edges

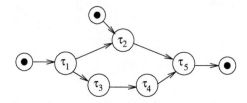

3. **Exclusive access to resources**: Computations may be requesting exclusive access to some resource, for example, to some input/output device or some communication area in memory. Information about necessary exclusive access should be taken into account during scheduling. Exploiting this information might, for example, be used to avoid the priority inversion problem (see p. 206). Information concerning exclusive access to resources can be included in the graphs.
4. **Periodic schedules**: Many computations, especially in digital signal processing, are periodic. This means that we must distinguish more carefully between a task and its execution (the latter is frequently called a **job** [332]). Graphs for such schedules are infinite. Figure 2.4 shows a graph including jobs J_{n-1} to J_{n+1} of a periodic task.

Fig. 2.4 Graph including jobs

5. **Hierarchical graph nodes**: The complexity of the computations denoted by graph nodes may be quite different. On the one hand, specified computations may be quite involved and contain thousands of lines of program code. On the other hand, programs can be split into small pieces of code so that in the extreme case, each of the nodes corresponds only to a single operation. The graph node complexity is also called their **granularity**. Which granularity should be used? There is no universal answer to this. For some purposes, the granularity should be as large as possible. For example, if we consider each of the nodes as one process to be scheduled by a real-time operating system (RTOS), it may be wise to work with large nodes in order to minimize context switches between different processes. For other purposes, it may be better to work with nodes modeling just a single operation. For example, nodes must be mapped to hardware or to software. If a certain operation (such as the frequently used discrete cosine transform, or DCT) can be mapped to special purpose hardware, then it should not be buried in a complex node that contains many other operations. It should rather be modeled as its own node. In order to avoid frequent changes of the granularity, hierarchical graph nodes are very useful. For example, at a high hierarchical level, the nodes

may denote complex tasks, at a lower level basic blocks[3], and at an even lower level individual arithmetic operations. Figure 2.5 shows a hierarchical version of the dependence graph in Fig. 2.2, using a rectangle to denote a hierarchical node.

Fig. 2.5 Hierarchical task graph

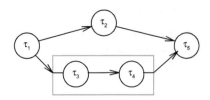

As indicated above, MoCs can be classified according to the models of communication (reflected by edges in the task graphs) and the model of computations within the components (reflected by the nodes in the task graphs). In the following, we will explain prominent examples of such models:

• **Models of communication**:
  We distinguish between two communication paradigms: **shared memory** and **message passing**. Other communication paradigms exist (e.g., entangled states in quantum mechanics [64]), but are not considered in this book.

  – **Shared memory**:
    For shared memory, communication is performed by accesses to the same memory from all components. Access to shared memory should be protected, unless access is totally restricted to reads. If writes are involved, exclusive access to the memory must be guaranteed while components are accessing shared memories. Segments of program code, for which exclusive access must be guaranteed, are called **critical sections**. Several mechanisms for guaranteeing exclusive access to resources have been proposed. These include semaphores, conditional critical regions, monitors, and spin locks (see books on operating systems like Stallings [483]). Shared memory-based communication can be very fast, but is difficult to implement in multiprocessor systems if no common memory is physically available.

  – **Message passing**: In this case, messages are sent and received. Message passing can be implemented easily even if no common memory is available. However, message passing is generally slower than shared memory-based communication. We distinguish between three kinds of message passing:

    · **Asynchronous message passing**, also called **non-blocking communication**:
      In asynchronous message passing, components communicate by sending messages through channels which can buffer the messages. The sender does not need to wait for the recipient to be ready to receive the message. In real life, this corresponds to sending a letter or an e-mail. A potential problem is the fact that messages must be stored and that message buffers can overflow. There

---

[3]Basic blocks are code blocks of maximum length not including any branch except possibly at their end and not being branched into.

are variations of this scheme, including communicating finite state machines (see p. 58) and data flow models (see p. 64).

- **Synchronous message passing** or **blocking communication, *rendez-vous* based communication**: In synchronous message passing, available components communicate in atomic, instantaneous actions called *rendez-vous*.
The component reaching the point of communication first has to wait until the partner has also reached its point of communication. In real life, this corresponds to physical meetings or phone calls. There is no risk of overflows, but performance may suffer. Examples of languages following this model of computation include CSP (see p. 106) and Ada (see p. 107).

- **Extended *rendez-vous*, remote invocation**: In this case, the sender is allowed to continue only after an acknowledgment has been received from the recipient. The recipient does not have to send this acknowledgment immediately after receiving the message, but can do some preliminary checking before actually sending the acknowledgment.

- **Organization of computations within the components**:

  - **Differential equations**: Differential equations are capable of modeling analog circuits and physical systems. Hence, they can find applications in cyber-physical system modeling.
  - **Finite state machines (FSMs)**: This model is based on the notion of a finite set of states, inputs, outputs, and transitions between states. Several of these machines may need to communicate, forming so-called **communicating finite state machines (CFSMs)**.
  - **Data flow**: In the data flow model, the availability of data triggers the possible execution of operations.
  - **Discrete event model**: In this model, there are events carrying a totally ordered time stamp, indicating the time at which the event occurs. Discrete event simulators typically contain a global event queue sorted by time. Entries from this queue are processed according to this order. The disadvantage is that this model relies on a global notion of event queues, making it difficult to map the semantic model onto parallel implementations. Examples include VHDL (see p. 94), SystemC (see p. 93), and Verilog (see p. 104).
  - **Von-Neumann model**: This model is based on the sequential execution of sequences of primitive computations.

- **Combined models**: Actual languages are typically combining a certain model of communication with an organization of computations within components. For example, StateCharts (see p. 47) combines finite state machines with shared memories. SDL (see p. 58) combines finite state machines with asynchronous message passing. Ada (see p. 107) and CSP (see p. 106) combine von-Neumann execution with synchronous message passing. Table 2.1 gives an overview of combined models most of which we will consider in this chapter. This table also includes examples of languages for many of the MoCs.

Let us look at MoCs with a defined model for computations within components. For differential equations, Modelica [382], commercial languages such as

Table 2.1 Overview of MoCs and languages considered

Communication/organization of components	Shared memory	Message passing	
		Synchronous	Asynchronous
Undefined components	Plain text or graphics, use cases		
			(Message) sequence charts
Differential equations	Modelica, Simulink®, VHDL-AMS		
Communicating finite state machines (CFSMs)	StateCharts		SDL
Data flow	Scoreboarding, Tomasulo algorithm		Kahn networks, SDF
Petri nets	C/E nets, P/T nets, ...		
Discrete event (DE) model[a]	VHDL, Verilog, and SystemC	(Only experimental systems) Distributed DE in Ptolemy	
Von-Neumann model	C, C++, Java	C, C++, Java, ... with libraries	
		CSP, Ada	

[a]The classification of VHDL, Verilog, and SystemC is based on the implementation of these languages in simulators. Message passing can be modeled in these languages "on top" of the simulation kernel.

Simulink® [510], and the extension VHDL-AMS [236] of the hardware description language VHDL are examples of languages.

Scoreboarding and the Tomasulo algorithm are data flow-driven techniques for dynamically scheduling instructions in computer architectures. They are described in books in computer architecture (see, for example, Hennessy and Patterson [205]) and not presented in this book.

Some MoCs have advantages in certain application areas, while others have advantages in others. Choosing the "best" MoC for a certain application may be difficult. Being able to mix MoCs (such as in the Ptolemy framework [435]) can be a way out of this dilemma. Also, models may be translated from one MoC into another one. Non-von-Neumann models are frequently translated into von-Neumann models. The distinction between the different models is blurring if the translation between them is easy.

Designs starting from non-von-Neumann models are frequently called **model-based designs** [400]. The key idea of model-based design is to have some abstract model of the system under design (SUD). Properties of the SUD can then be studied at the level of this model, without having to care about software code. Software code is generated only after the behavior of the model has been studied in detail, and this software is generated automatically [453]. The term "model-based design" is usually associated with models of control systems, comprising traditional control system elements such as integrators and differentiators. However, this view may be too restricted, since we could also start with abstract models of consumer systems.

In the following, we will present different MoCs, using existing languages as examples for demonstrating their features. A related (but shorter) survey is provided by Edwards [141]. For a more comprehensive presentation, see [180].

2.3 Early Design Phases

The very first ideas about systems are frequently captured in a very informal way, possibly on paper. Frequently, only descriptions of the SUD in a natural language such as English or Japanese exist in the early phases of design projects. They are typically using a very informal style. These descriptions should be captured in some machine-readable document. They should be encoded in the format of some word processor and stored by a tool managing design documents. A good tool would allow link between the requirements, a dependence analysis as well version management. DOORS® [221] exemplifies such a tool.

2.3.1 Use Cases

For many applications, it is beneficial to envision potential usages of the SUD. Such usages are captured in **use cases**. Use cases describe possible applications of the SUD. Different notations for use cases could be used.

Support for a systematic approach to early specification phases is the goal of the so-called UML™standardization effort [161, 199, 413]. UML stands for "Unified Modeling Language." UML was designed by leading software technology experts and is supported by commercial tools. UML primarily aims at the support of the software design process. UML provides a standardized form for use cases.

For use cases, there is neither a precisely specified model of the computations nor is there a precisely specified model of the communication. It is frequently argued that this is done intentionally in order to avoid caring about too many details during the early design phases.

Example 2.3: For example, Fig. 2.6 shows some use cases for an answering machine[4]. There are five use cases for the owner of the answering machine and one for potential callers. We have to make sure that all six use cases can be implemented correctly.

Fig. 2.6 Use case example

∇

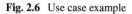

[4]We assume that UML is covered in-depth in a software engineering course included in the curriculum. Therefore, UML is only briefly discussed in this book.

Use cases identify different classes of users as well as the applications to be supported by the SUD. In this way, it is possible to capture expectations at a very high level.

2.3.2 (Message) Sequence Charts and Time/Distance Diagrams

At a more detailed level, we might want to explicitly indicate the sequences of messages which must be exchanged between components in order to implement some use of the SUD. **Sequence charts** (SCs)—earlier called **message sequence charts** (MSCs)—provide a mechanism for this. Sequence charts use one dimension (usually the vertical dimension) of a two-dimensional chart to denote sequences and the second dimension to reflect the different communication components. SCs describe partial orders between message transmissions, and they display a possible behavior of a SUD. SCs are also standardized in UML. UML 2.0 has extended SCs with elements allowing a more detailed description than UML 1.0.

Example 2.4: Figure 2.7 shows one of the use cases of the answering machine as an example. Dashed lines are so-called lifelines. Messages are assumed to be ordered

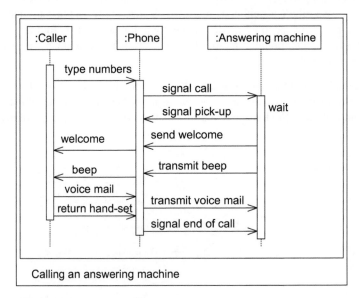

Fig. 2.7 Answering machine in UML™

according to their sequence along the lifeline. We assume that, in this example, all information is sent in the form of messages. Arrows used in this diagram denote asynchronous messages. This means several messages can be sent by a sender without waiting for the receipt to be confirmed. Boxes on top of lifelines represent active control at the corresponding component. In the example, the answering machine is waiting for the user to pick up the phone within a certain amount of time. If he or she fails to do so, the machine signals a pickup itself and sends a welcome message to the caller. The caller is then supposed to leave a voice mail message. Alternative sequences (e.g., an early termination of the call by the caller or the callee picking up the phone) are not shown.   ∇

Complex control-dependent actions cannot be described by SCs. Other MoCs must be used for this. Frequently, certain preconditions must be met for a SC to apply. Such preconditions, a distinction between sequences which might happen and those which must happen, as well as other extensions are available in the so-called Live Sequence Charts [114].

Time/distance diagrams (TDDs) are a commonly used variant of SCs. In time/distance diagrams, the vertical dimension reflects real time, not just sequence. In some cases, the horizontal dimension also models the real distance between the components. TDDs provide the right means for visualizing schedules of trains or buses.

Example 2.5: Figure 2.8 is an example of a TDD. This example refers to trains between Amsterdam, Cologne, Brussels, and Paris. Trains can run from either

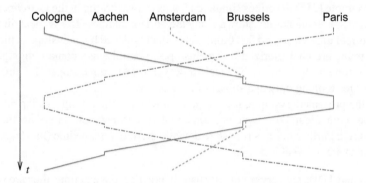

Fig. 2.8 Time/distance diagram

Amsterdam or Cologne to Paris via Brussels. Aachen is included as an intermediate stop between Cologne and Brussels. Vertical segments correspond to time spent at stations. For one of the trains, there is a timing overlap between the trains coming from Cologne and Amsterdam at Brussels. There is a second train which travels between Paris and Cologne which is not related to an Amsterdam train. This example and other examples can be simulated with the levi simulation software [473].   ∇

Example 2.6: A larger, more realistic example is shown in Fig. 2.9.

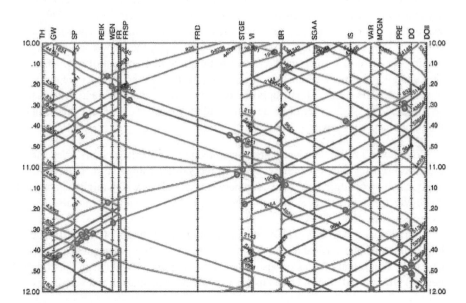

Fig. 2.9 Railway traffic displayed by a TDD (courtesy H. Brändli, IVT, ETH Zürich), © ETH Zürich

This example [215] describes simulated Swiss railway traffic in the Lötschberg area. Different station names are shown along the horizontal lines. The vertical dimension reflects real time. Slow and fast trains can be distinguished by their slope in the graph. Slow trains are characterized by steep slopes, possibly also containing significant waiting time at the stations (vertical slopes). For fast trains, slopes are almost flat. Trains are stopping only at a subset of the stations.

In the presented example, it is not known whether the timing overlap at stations happens coincidentally or whether some real synchronization for connecting trains is required. Furthermore, permissible deviations from the schedule (min/max timing behavior) are not visible.   ∇

SCs and TDDs are very frequently used in practice. For example, they are valuable for applications of the IoT. One of the key distinctions between SCs and TDDs is that SCs do not include any reference to real time. TDDs are appropriate means for representing typical schedules. However, SCs and TDDs fail to provide information about necessary synchronization.

UML was initially not designed for real-time applications. UML 2.0 includes **timing diagrams** as a special class of diagrams. Such diagrams enable referring to physical time, similar to TDDs. Also, certain UML "profiles" (see p. 117) allow additional annotations to refer to time [352].

2.3.3 Differential Equations

Differential equations can be written in the language of mathematics. Inputs for design tools typically require certain variants of this language. We exemplify such a variant with Modelica [382], a language aiming at modeling cyber-physical systems. Modelica has graphical as well as textual forms. Using the graphical form, systems can be described as sets of interconnected blocks. Each block can be described by equations. Connections between blocks denote common variables in the sense of mathematics. The information about each block together with information about connections can be transformed into a global set of equations. This process is called flattening of the hierarchy. Just like in mathematics, equations (and connections) have a bidirectional meaning (in contrast to programming languages).

Example 2.7: The following model[5] represents the bouncing ball example introduced on p. 10:

```
model StickyBall
    type Height = Real(unit ="m");
    type Velocity = Real(unit = "m/s");
    parameter Real s = 0.8 "Restitution";
    parameter Height h0=1.0 "Initial hight";
    constant Velocity eps=1e-3 "small velocity";
    Boolean stuck;
    Height h;
    Velocity v;
initial equation
    v = 0;
    h = h0;
    stuck = false;
equation
    v = der(h);
    der(v) = if stuck then 0 else -9.81;
    when h <= 0.0 then
        stuck = abs(v) < eps;
        reinit(v, if stuck then 0 else -s*v);
    end when;
end StickyBall;
```

In the equations part, velocity v is defined as the derivative of the height h. The derivative of velocity (the acceleration) is set to standard gravity (−9.81), unless the ball is already sticking to the surface. Equations have a bidirectional meaning. For this set of equations, there are boundary conditions defined in the initial equation part. Mathematical equations can be integrated numerically. This procedure is exploited in the description of the bouncing: when-clauses can be used to define *events* which happen while solving the equations. In the particular example, an event is generated

---

[5]This model has been derived from the model published by M. Tiller [517].

when the height becomes less or equal to zero. Whenever this event is generated while the velocity is still sufficiently large, the velocity is inverted and reduced by a factor of s, called **restitution**. s is the square root of the so-called rebound coefficient r [157]. The reinit-clause effectively defines another boundary condition.

However, if the velocity is smaller than eps, the ball is assumed to become sticky and the velocity is set to zero, suppressing all future activities. The resulting model can be simulated, for example, with OpenModelica[6].

The mathematical background is as follows: Let $v_0 = \sqrt{2gh_0}$ be the velocity just before the first bounce [157]. The time until the n-th bouncing is as follows [157]:

$$t_n = \frac{2v_0}{g} \sum_{k=0}^{n-1} s^k \tag{2.1}$$

As long as $s < 1$, this geometric series converges to

$$t_{final} = \lim_{n \to \infty} \frac{2v_0}{g} \sum_{k=0}^{n-1} s^k = \frac{2v_0}{g(1-s)} \tag{2.2}$$

This means that there is an upper bound on the time for the bounces, but not on the number of bounces. This corresponds to the fact that, mathematically speaking, infinite series may be converging to a finite value[7].

Using sets of equations involving derivatives in Modelica brings us close to the language of mathematics and physics. However, events introduce sequential behavior. The implicit numerical integration procedure also introduces the hazard of numerical precision problems. In fact, already the test $<= 0.0$ reflects the fact that we might miss the case of h being exactly 0. Another hazard is present in the published model for the non-sticky ball [517]: Numerical precision problems result in an OpenModelica solution for which the ball penetrates the floor for large times t. This problem is caused by not generating events if the time distance between bounces is too small.

This example demonstrates very nicely the advantages and limitations of Modelica: On the one hand, it is feasible to describe even the physical part of cyber-physical systems. On the other hand, we are not exactly using the language of mathematics, and in this way, we are introducing modeling hazards.   ▽

2.4 Communicating Finite State Machines (CFSMs)

In the following sections, we will consider the design of digital systems only. Compared to early design phases, we need more precise models of our SUD. We mentioned already on p. 15 and on p. 30 that we need to describe state-oriented behavior. State diagrams are a classical means of doing this. Figure 2.10 (the same as Fig. 2.1) shows an example of a classical state diagram, representing a **finite state machine (FSM)**.

---

[6]See https://openmodelica.org/.
[7]Note the link to the paradox of Achilles and the turtle [558].

Fig. 2.10 State diagram

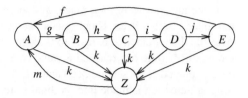

Circles denote states. We will consider FSMs for which only one of their states is active. Such FSMs are called **deterministic** FSMs. Edges denote state transitions. Edge labels represent events. Let us assume that a certain state of the FSM is active, and that an event happens which corresponds to one of the outgoing edges for the active state. Then, the FSM will change its state from the currently active state to the one indicated by the edge. FSMs may be implicitly clocked. Such FSMs are called **synchronous FSMs**. For synchronous FSMs, state changes will happen only at clock transitions. FSMs may also generate output (not shown in Fig. 2.10). For more information about classical FSMs, refer to, for example, Kohavi et al. [290].

2.4.1 Timed Automata

Classical FSMs do not provide information about time. In order to model time, classical automata have been extended to also include timing information. Timed automata are essentially automata extended with real-valued variables. *"The variables model the logical clocks in the system, that are initialized with zero when the system is started, and then increase synchronously with the same rate. Clock constraints, i.e., guards on edges, are used to restrict the behavior of the automaton. A transition represented by an edge can be taken when the clocks' values satisfy the guard labeled on the edge. Clocks may be reset to zero when a transition is taken"* [46].

Example 2.8: Figure 2.11 shows an example. The answering machine is usually in the initial state on the left. Whenever a *ring* signal is received, clock x is reset to 0 and a transition into a waiting state is made. If the called person lifts off the handset, talking can take place until the handset is returned. Otherwise, a transition to state *play text* can take place if time has reached a value of 4.

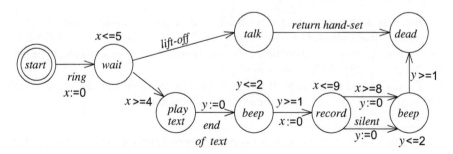

Fig. 2.11 Servicing an incoming line in an answering machine

Once the transition took place, a recorded message is played and this phase is terminated with a beep. Clock y ensures that this beep lasts at least one time unit. After the beep, clock x is reset to 0 again and the answering machine is ready for recording. If time has reached a value of 8 or if the caller remains silent, the next beep is played. This second beep again lasts at least one time unit. After the second beep, a transition is made into the final state. In this example, transitions are either caused by inputs (such as *lift-off*) or by so-called **clock constraints**.   ▽

Clock constraints describe transitions which **can** take place, but they do not have to. In order to make sure that transitions actually take place, additional **location invariants** can be defined. Location invariants $x <= 5$, $x <= 9$, and $y <= 2$ are used in the example such that transitions will take place no later than one time unit after the enabling condition became true. Using two clocks is for demonstration purposes only; a single clock would be sufficient.

Formally speaking, timed automata can be defined as follows [46]: Let C be a set of real-valued, non-negative variables representing clocks. Let Σ be a finite alphabet of possible inputs.

Definition 2.4: A **clock constraint** is a conjunctive formula of atomic constraints of the form $x \circ n$ or $(x - y) \circ n$ for $x, y \in C$, $\circ \in \{\le, <, =, >, \ge\}$ and $n \in \mathbb{N}$.

Note that constants n used in the constraints must be integers, even though clocks are real-valued. An extension to rational constants would be easy, since they could be turned into integers with simple multiplications. Let $B(C)$ be the set of clock constraints.

Definition 2.5 (Bengtson [46]): A **timed automaton** is a tuple (S, s_0, E, I) where

- S is a finite set of states.
- s_0 is the initial state.
- $E \subseteq S \times B(C) \times \Sigma \times 2^C \times S$ is the set of edges. $B(C)$ models the conjunctive condition which must hold and Σ models the input which is required for a transition to be enabled. 2^C reflects the set of clock variables which are reset whenever the transition takes place.
- $I : S \to B(C)$ is the set of invariants for each of the states. $B(C)$ represents the invariant which must hold for a particular state S. This invariant is described as a conjunctive formula.

This first definition is usually extended to allow parallel compositions of timed automata. Timed automata having a large number of clocks tend to be difficult to understand. More details about timed automata can be found, for example, in papers by Dill et al. [128] and Bengtsson et al. [46].

Simulation and verification of timed automata is possible with the popular tool UPPAAL[8]. UPPAAL supports concurrency and data variables.

---

[8]See http://www.uppaal.org for the academic and http://www.uppaal.com for the commercial version.

Timed automata extend classical automata with timing information. However, many of our requirements for specification techniques are not met by timed automata. In particular, in their standard form, they do no provide hierarchy and concurrency.

2.4.2   StateCharts: Implicit Shared Memory Communication

The StateCharts language is presented here as a very prominent example of a language based on automata and supporting hierarchical models as well as concurrency. It does include a limited way of specifying timing.

The StateCharts language was introduced by David Harel [197] in 1987 and later described more precisely in [135]. According to Harel, the name was chosen since it was *"the only unused combination of **flow** or **state** with **diagram** or **chart**."*

2.4.2.1   Modeling of Hierarchy

The StateCharts language describes extended FSMs. Due to this, they can be used for modeling state-oriented behavior. The key extension is **hierarchy**. Hierarchy is introduced by means of **super-states**.

Definition 2.6: States comprising other states are called **super-states**.

Definition 2.7: States included in super-states are called **sub-states** of the super-states.

Example 2.9: Figure 2.12 shows a StateCharts example. It is a hierarchical version of Fig. 2.10.

Fig. 2.12 Hierarchical state diagram

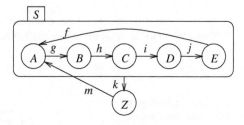

Super-state S includes states A, B, C, D, and E. Suppose the FSM is in state Z (we will also call Z to be an **active state**). Now, if input m is applied to the FSM, then A and S will be the new active states. If the FSM is in S and input k is applied, then Z will be the new active state, regardless of whether the FSM is in sub-states A, B, C, D, or E of S. In this example, all states contained in S are non-hierarchical states.   ∇

In general, sub-states of S could again be super-states consisting of sub-states themselves. Also, **whenever a sub-state of some super-state is active, the super-state is active as well**.

Definition 2.8: Each state which is not composed of other states is called a **basic state**.

Definition 2.9: For each basic state s, the super-states containing s are called **ancestor states**.

The FSM of Fig. 2.12 can only be in one of the sub-states of super-state S at any time. Super-states of this type are called **OR super-states**[9].

In Fig. 2.12, k might correspond to an exception for which state S has to be left. The example already shows that the hierarchy introduced in StateCharts enables a compact representation of exceptions.

StateCharts allows hierarchical descriptions of systems in which a system description comprises descriptions of subsystems which, in turn, may contain descriptions of subsystems. The **hierarchy** of the entire system can be represented by a **tree**. The root of the tree corresponds to the system as a whole, and all inner nodes correspond to hierarchical descriptions (in the case of StateCharts called super-nodes). The leaves of the hierarchy are non-hierarchical descriptions (in the case of StateCharts called basic states).

So far, we have used explicit, direct edges to basic states to indicate the next state. The disadvantage of that approach is that the internal structure of super-states cannot be hidden from the environment. However, in a true hierarchical environment, we should be able to hide the internal structure so that it can be described later or changed later without affecting the environment. This is possible with other mechanisms for describing the next state.

The first additional mechanism is the **default state mechanism**. It can be used in super-states to indicate the particular sub-states that will become active if the super-states become active. In diagrams, default states are identified by edges starting at small filled circles.

Example 2.10: Figure 2.13 shows a state diagram using the default state mechanism. It is equivalent to the diagram in Fig. 2.12. Note that the filled circle does not constitute a state itself.

Fig. 2.13 State diagram using the default state mechanism

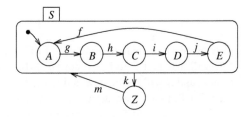

                                                                                              ▽

---

[9]More precisely, they should be called XOR super-states, since the FSM is in **either** A, B, C, D, or E. However, this name is not commonly used in the literature.

Another mechanism for specifying next states is the **history mechanism**. With this mechanism, it is possible to return to the last sub-state that was active before a super-state was left. The history mechanism is symbolized by a circle containing the letter H. Please do not confuse circles comprising this letter with states! We will be using a different font for states and the history mechanism in order to reduce the risk of confusion. In order to define the next state for the very initial transition into a super-state, the history mechanism is frequently combined with the default mechanism.

Example 2.11: Figure 2.14 shows an example. The behavior of the FSM is now somewhat different. If we input m while the system is in Z, then the FSM will enter A if this is the very first time we enter S, and otherwise, it will enter the last state that we were in before leaving S. This mechanism has many applications. For example, if k denotes an exception, we could use input m to return to the state we were in before the exception. States A, B, C, D, and E could also call Z like a procedure. After completing "procedure" Z, we would return to the calling state. In this way, we are adding elements of programming languages to StateCharts.

Fig. 2.14 State diagram using the history and the default state mechanism

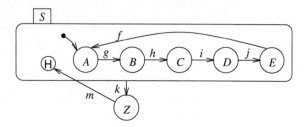

Fig. 2.15 Combining the symbols for the history and the default state mechanism

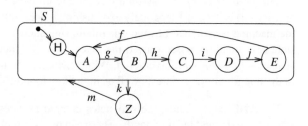

Figure 2.14 can also be redrawn as shown in Fig. 2.15. In this case, the symbols for the default and the history mechanism are combined.   ∇

Specification techniques must also be able to describe concurrency conveniently. Toward this end, the StateCharts language provides a second class of super-states, so-called **AND** states.

Definition 2.10: Super-states S are called **AND super-states** if the system containing S will be in **all** of the sub-states of S whenever it is in S.

Example 2.12: An AND super-state is included in the answering machine example shown in Fig. 2.16. An answering machine normally performs two tasks concurrently: It is monitoring the line for incoming calls and the keys for user input. In Fig. 2.16, the corresponding states are called *Lwait* and *Kwait*. Incoming calls are processed in state *Lproc* while the response to pressed keys is generated in state *Kproc*. State *Lproc* is left whenever the caller hangs up the phone. Returning to state *Lwait* due to call termination by the owner is not modeled. Hence, this model provides no protection against stalking.

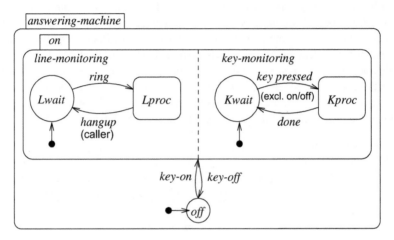

Fig. 2.16 Answering machine

For the time being, we assume that the on/off switch (generating events *key-off* and *key-on*) is decoded separately and pushing it does not result in entering *Kproc*. If the machine is switched off, the line monitoring state as well as the key monitoring state is left and reentered only if the machine is switched on. At that time, default states *Lwait* and *Kwait* are entered. While switched on, the machine will always be in the line monitoring state as well as in the key monitoring state.   ∇

For AND super-states, the sub-states entered as a result of entering the super-state can be defined independently. There can be any combination of history, default, and explicit transitions. It is crucial to understand that **all** sub-states will always be entered, even if there is just one explicit transition to one of the sub-states. Accordingly, transitions out of an AND super-state will always result in leaving **all** the sub-states.

Example 2.13: For example, let us modify our answering machine such that the on/off switch, like all other switches, is decoded in state *Kproc* (see Fig. 2.17).

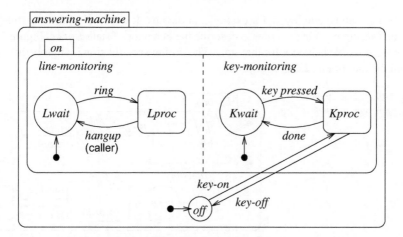

Fig. 2.17 Answering machine with modified on/off switch processing

If pushing that key is detected in *Kwait*, transitions are assumed first into state *Kproc* and then into the *off* state. The second transition results in leaving the line monitoring state as well. Switching the machine on again results in also entering the line monitoring state. ∇

AND super-states provide the key mechanism for describing concurrency inbreak StateCharts. Each sub-state can be considered a state machine by itself. These machines are communicating with each other, forming **communicating finite state machines** (CFSMs). This term has been used as the title of this section.

Summarizing, we can state the following: **States in StateCharts diagrams are either AND states, OR states, or basic states.**

2.4.2.2 Timers

Due to the requirement to model time in embedded systems, StateCharts also provides timers. Timers are denoted by the symbol shown in Fig. 2.18 on the left.

Fig. 2.18 Timer in StateCharts

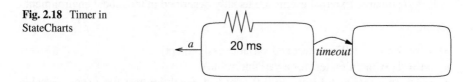

After the system has been in the state containing the timer for the specified time, a time-out will occur and the system will leave the specified state. Timers can also be used hierarchically.

Timers can be employed, for example, at the next lower level of the hierarchy of the answering machine in order to describe the behavior of state *Lproc*. Figure 2.19 shows a possible behavior for that state. The timing specification is slightly different from the one in Fig. 2.11.

Fig. 2.19 Servicing the incoming line in *Lproc*

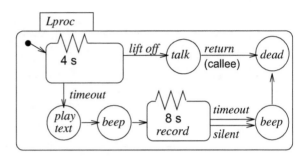

Due to the exception-like transition for hang-ups by the caller in Fig. 2.16, state *Lproc* is terminated whenever the caller hangs up. For hang-ups (returns) by the callee, the design of state *Lproc* results in an inconvenience: If the callee hangs up the phone first, the telephone will be dead (and quiet) until the caller has also hung up the phone.

The StateCharts language includes a number of other language elements. For a full description, refer to Harel [197]. A more detailed description of the semantics of StateCharts is described by Drusinsky and Harel [135].

2.4.2.3 Edge Labels and StateMate Semantics

Until now, we have not considered outputs generated by our extended FSMs. Generated outputs can be specified using edge labels. The general form of an edge label is "*event*[*condition*]/*reaction*." All three label parts are optional. The *reaction* part describes the reaction of the FSM to a state transition. Possible reactions include the generation of events and assignments to variables. The *condition* part implies a test of the values of variables or a test of the current state of the system. The *event* part refers to a test of current events. Events can be generated either internally or externally. Internal events are generated as a result of some transition and are described in *reaction* parts. External events are usually described in the model environment.

Examples:

- *on-key* / *on* := 1 (Event test and variable assignment),
- [*on* = 1] (Condition test for a variable value),
- *off-key* [not in *Lproc*] / *on* := 0 (Event test, condition test for a state, variable assignment. The assignment is performed if the event has occurred and the condition is true).

The semantics of edge labels can only be explained in the context of the semantics of StateMate [135], a commercial implementation of StateCharts. StateMate assumes a step-based execution of StateMate descriptions. Each step consists of three phases:

1. In the first phase, the impact of external changes on conditions and events is evaluated. This includes the evaluation of functions which depend on external events. This phase does not include any state changes. In our simple examples, this phase is not actually needed.
2. The next phase is to calculate the set of transitions that should be made in the current step. Variable assignments are evaluated, but the new values are only assigned to temporary variables.
3. In the third phase, state transitions become effective and variables obtain their new values.

The separation into phases 2 and 3 is important in order to guarantee a reproducible behavior of StateMate models.

Example 2.14: Consider the StateMate model of Fig. 2.20.

Fig. 2.20 Mutually dependent assignments

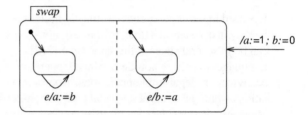

In the second phase, new values for a and b are stored in temporary variables, say a' and b'. In the final phase, temporary variables are copied into the user-defined variables:

```
phase 2: a':=b; b':=a;
phase 3: a:=a'; b:=b'
```

As a result, the values of the two variables will be swapped each time an event e happens. This behavior corresponds to that of two cross-coupled registers (one for each variable) connected to the same clock (see Fig. 2.21) and reflects the operation of a synchronous (clocked) finite state machine including those two registers[10].

Fig. 2.21 Cross-coupled D-type registers

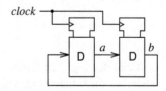

---

[10]We adopt IEEE standard schematic symbols [230] for gates and registers for all the schematics in this book. The symbols in Fig. 2.21 denote clocked D-type registers.

Without the separation into phases, the same value would be assigned to both variables. The result would depend on the sequence in which the assignments were performed.   ▽

The separation into (at least) two phases is quite typical for languages that try to reflect the operation of synchronous hardware. We will find the same separation in VHDL (see p. 102). Due to the separation, the results do not depend on the order in which parts of the model are executed by the simulation. This property is extremely important. Otherwise, there could be simulation runs generating different results, all of which would be considered correct. This is not what we expect from the simulation of a real circuit with a fixed behavior and it could be very confusing in design procedures. There are different names for this property:

- Kahn [266] calls this property **determinate**.
- In other papers, this property is called **deterministic**. However, the term "deterministic" is employed with different meanings:

  - It is used in the context of deterministic finite state machines, FSMs, which can be only in one state at a time. In contrast, non-deterministic finite state machines can be in several states at the same time [212].
  - Languages may have non-deterministic operators. For these operators, different behaviors are legal implementations. Approximate, non-deterministic computations would be a relevant special case of non-deterministic operators.
  - Many authors consider systems to be non-deterministic if their behavior depends on some input not known before run-time.
  - The term "deterministic" has also been used in the sense of "determinate," as introduced by Kahn.

In this book, we prefer to reduce possible confusion by following Kahn[11]. Note that StateMate models can be determinate only if there are no other reasons for an undefined behavior. For example, conflicts between transitions may be allowed (see Fig. 2.22).

Fig. 2.22 Left: conflict between different nesting levels; **right**: conflict at the same nesting level

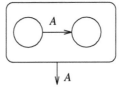 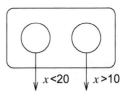

Consider Fig. 2.22 (**left**). If event A takes place while the system is in the left state, we must figure out which transition will take place. If these conflicts would be resolved arbitrarily, then we would have a non-determinate behavior. Typically,

---

[11] In the first edition of the book, we used the term "deterministic" together with an additional explanation.

priorities are defined such that this type of a conflict is eliminated. Now, consider Fig. 2.22 (**right**). There will be a conflict for $x=15$. Such conflicts are difficult to detect. Achieving a determinate behavior requires the absence of conflicts that are resolved in an arbitrary manner.

Note that there may be cases in which we would like to describe non-determinate behavior (e.g., if we have a choice to read from two inputs). In such a case, we would typically like to explicitly indicate that this choice can be taken at run-time (see the **select** statement of Ada on p. 107).

Implementations of hierarchical state charts other than StateMate typically do not exhibit determinate behavior. These implementations correspond to a software-oriented view onto hierarchical state charts. In such implementations, choices are usually not explicitly described.

The three phases described on p. 52 have to be repeatedly executed. Each execution is called a **step** (see Fig. 2.23).

Fig. 2.23 Steps during the execution of a StateMate model

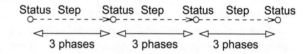

Steps are assumed to be executed each time events or variables have changed. The set of all values of variables, together with the set of events generated (and the current time), is defined as the **status**[12] of a StateMate model. After executing the third phase, a new status is obtained. The notion of steps allows us to define the semantics of **events** more precisely. Events are generated, as mentioned, either internally or externally. **The visibility of events is limited to the step following the one in which they are generated.** Thus, events behave like single bit values which are stored in permanently enabled registers at one clock transition and have an effect on the values stored at the next clock transition. They do not live forever.

Variables, in contrast, retain their values until they are reassigned. According to StateMate semantics, new values of variables are visible to all parts of the model from the step following the step in which the assignment was made onward. That means that StateMate semantics implies that new values of variables are propagated to all parts of a model between two steps. StateMate implicitly assumes a **broadcast mechanism for updates on variables**. Hence, StateCharts or StateMate can be implemented easily for shared memory-based platforms but is less appropriate for message passing and distributed systems. These languages essentially assume shared memory-based communication, even though this is not explicitly stated. For distributed systems, it will be very difficult to update all variables between two steps. Due to this broadcast mechanism, StateMate is not an appropriate language for modeling distributed systems.

---

[12]We would normally use the term "state" instead of "status." However, the term "state" has a different meaning in StateMate.

2.4.2.4   Evaluation and Extensions

StateCharts' main application domain is that of local, control-dominated systems. The capability of nesting hierarchies at arbitrary levels, with a free choice of AND and OR states, is a key advantage of StateCharts. Another advantage is that the semantics of StateMate is defined at a sufficient level of detail [135]. Furthermore, there are quite a number of commercial tools based on StateCharts. StateMate [220] and StateFlow [365] are examples of commercial tools based on StateCharts. Many of them are capable of translating StateCharts into equivalent descriptions in C or VHDL (see p. 94). From VHDL, hardware can be generated using synthesis tools. Therefore, StateCharts-based tools provide a complete path from StateCharts-based specifications down to hardware. Generated C programs can be compiled and executed. Hence, a path to software-based realizations exists as well.

Unfortunately, the efficiency of the automatic translation is sometimes a concern. For example, we could map sub-states of AND states to processes at the operating system level. This would hardly lead to efficient implementations on small processors. The productivity gain from object-oriented programming is not available in StateCharts, since it is not object-oriented. Furthermore, the broadcast mechanism makes it less appropriate for distributed systems. StateCharts do not comprise program constructs for describing complex computation and cannot describe hardware structures or non-functional behavior.

Commercial implementations of StateCharts typically provide some mechanisms for removing the limitations of the model. For example, C code can be used to represent program constructs and **module charts** of StateMate can represent hardware structures.

StateCharts allows timeouts. There is no straightforward way of specifying other timing requirements.

UML includes a variation of StateCharts and hence allows modeling state machines. In UML, these diagrams are called **state diagrams** in version 1 of UML and **state machine diagrams** from version 2.0 onward. Unfortunately, the semantics of state machine diagrams in UML is different from StateMate: The three simulation phases are not included.

2.4.3   Synchronous Languages

2.4.3.1   Motivation

Describing complex SUDs in terms of state machine diagrams is difficult. Such diagrams cannot express complex computations. Standard programming languages can express complex computations, but the sequence of executing several threads may be unpredictable. In a multi-threaded environment with preemptive scheduling, there can be many different interleavings of the different computations. Understanding all possible behaviors of such concurrent systems is difficult. A key reason for this is that, in general, many different execution orders are feasible, i.e., the execution order is not specified. The order of execution may well affect the result. The resulting non-determinate behavior can have a number of negative consequences, such as problems with verifying a certain design. For distributed systems with independent clocks, determinate behavior is difficult to achieve. However, for non-distributed systems, we can try to avoid the problems of unnecessary non-determinate semantics.

For synchronous languages, finite state machines and programming languages are merged into one model. Synchronous languages can express complex computations, but the underlying execution model is that of finite automata. They describe concurrently operating automata. Determinate behavior is achieved by the following key feature: "... *when automata are composed in parallel, a transition of the product is made of the 'simultaneous' transitions of all of them*" [191]. This means that we do not have to consider all the different sequences of state changes of the automata that would be possible if each of them had its own clock. Instead, we can assume the presence of a single global clock. Each clock tick, all inputs are considered, new outputs and states are calculated, and then the transitions are made. This requires a fast broadcast mechanism for all parts of the model. This idealistic view of concurrency has the advantage of guaranteeing **determinate behavior**. This is a restriction if compared to the general communicating finite state machines (CFSM) model, in which each FSM can have its own clock. Synchronous languages reflect the principles of operation in synchronous hardware and also the semantics found in control languages such as IEC 60848 [223] and STEP 7 [463]. See Potop-Butucaru et al. [433] for a survey on synchronous languages.

2.4.3.2   Examples of Synchronous Languages: Esterel, Lustre, and SCADE

Guaranteeing a determinate behavior for all language features has been a design goal for the synchronous languages such as Esterel [63, 148], Lustre [193], and Quartz [455].

Esterel is a reactive language: When activated with an input event, Esterel models react by producing an output event. Esterel is a synchronous language: All reactions are assumed to be completed in zero time and it is sufficient to analyze the behavior at discrete moments in time. This idealized model avoids all discussions about overlapping time ranges and about events that arrive while the previous reaction has not been completed. Like other concurrent languages, Esterel has a parallelism operator, written ||. Similar to StateCharts, communication is based on a broadcast mechanism. In contrast to StateCharts, however, communication is instantaneous. Instantaneous in this context means "within the same clock cycle." This means that all signals generated in a particular clock cycle are also seen by the others parts of the model in the same clock cycle and these other parts, if sensitive to the generated signals, react in the same clock cycle. Several rounds of evaluations may be required until a stable state is reached. The computation of resulting worst-case reaction times is performed, for example, by Boldt et al. [59]. The propagation of values during

the same macroscopic instant of time corresponds to the generation of a next status
for the same moment in time in StateMate, except that the broadcast is now instantaneous and not delayed until the next round of evaluations like in StateMate. For
more and updated information about Esterel, refer to the Esterel home page [148].

Esterel and Lustre use different syntactic techniques to denote CFSMs. Esterel
appears as a kind of imperative language, whereas Lustre looks more like a data
flow language (see p. 64 for a description of data flow). SyncCharts is a graphical
version of Esterel. In all three cases, semantics are explained by the closely related
underlying CFSMs. The commercial graphical language SCADE [147] combines
elements of all three languages. The so-called SCADE suite® is used for a number
of safety-critical software components, for example, by Airbus.

Due to the three simulation phases in StateMate, this tool has the key attributes
of synchronous languages and it is determinate if conflicts are resolved. According
to Halbwachs, "*StateMate is almost a synchronous language and the only feature
missing in StateMate is the instantaneous broadcast*" [192].

2.4.4 Message Passing: SDL as an Example

2.4.4.1 Features of the Language

StateCharts is not appropriate for modeling distributed communicating finite state
machines. For distributed systems, message passing is the better communication
paradigm. Therefore, we present a case of communicating finite state machines with
asynchronous message passing.

We use SDL (specification and description language) as an example. SDL was
designed for distributed applications. It dates back to the 1970s. Formal semantics
have been available since the 1980s. The language was standardized by the ITU
(International Telecommunication Union). The first standards document is the Z.100
Recommendation published in 1980, with updates in 1984, 1988, 1992 (SDL-92),
1996, and 1999. Relevant versions of the standard include SDL-88, SDL-92, SDL-
2000, and SDL-2010 [444, 457].

Many users prefer graphical specification languages, while others prefer textual
ones. SDL pleases both types of users since it provides textual as well as graphical
formats. Processes are the basic elements of SDL. Processes represent components
modeled as extended finite state machines. Extensions include operations on data.
Figure 2.24 shows the graphical symbols used in the graphical representation of SDL.

Fig. 2.24 Symbols used in
the graphical form of SDL

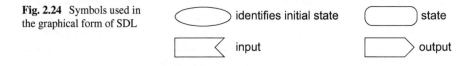

identifies initial state   state

input   output

Example 2.15: As an example, we will consider how the state diagram in Fig. 2.25 can be represented in SDL. Figure 2.25 is the same as Fig. 2.13, except that output has been added, state Z has been deleted, and the effect of signal k has been changed.

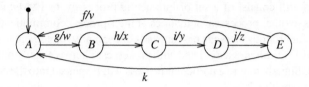

Fig. 2.25 FSM to be described in SDL

Figure 2.26 contains the corresponding graphical SDL representation.

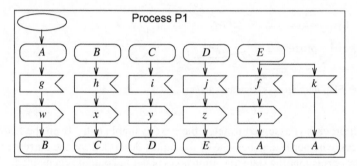

Fig. 2.26 SDL representation of Fig. 2.25

Obviously, the representation in Fig. 2.26 is equivalent to the state diagram of Fig. 2.25. ▽

As an extension to FSMs, SDL processes can perform operations on data. Variables can be declared locally for processes. Their type can either be predefined or defined in the SDL description itself. SDL supports abstract data types (ADTs). The syntax for declarations and operations is similar to that in other languages. Figure 2.27 shows how declarations, assignments, and decisions can be represented in SDL.

Fig. 2.27 Declarations, assignments, and decisions in SDL

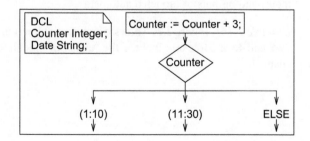

SDL also contains programming language elements such as procedures. Procedure calls can also be represented graphically. Object-oriented features became available with version SDL-1992 of the language and were extended with SDL-2000.

Extended FSMs are just the basic elements of SDL descriptions. In general, SDL descriptions will consist of a set of interacting processes, or FSMs. Processes can send signals to other processes. Semantics of inter-process communication in SDL is based on asynchronous message passing and conceptually implemented through *first-in first-out* (FIFO)-*queues* associated with processes. There is exactly one input queue per process. Signals sent to a particular process will be placed into the corresponding FIFO queue (see Fig. 2.28).

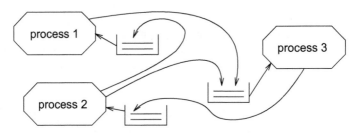

Fig. 2.28 SDL inter-process communication

Each process is assumed to fetch the next available entry from the FIFO queue and check whether it matches one of the inputs described for the current state. If it does, the corresponding state transition takes place and output is generated. The entry from the FIFO queue is ignored if it does not match any of the listed inputs (unless the so-called SAVE mechanism is used). FIFO queues are conceptually thought of as being of infinite length. This means that in the description of the semantics of SDL models, FIFO overflow is never taken into account. In actual systems, however, infinite FIFO queues cannot be implemented. They must be of finite length. This is one of the problems of SDL: In order to derive realizations from specifications, safe upper bounds on the length of the FIFO queues must be proven.

Process interaction diagrams can be used for visualizing which of the processes are communicating with each other. Process interaction diagrams include **channels** used for sending and receiving signals. In the case of SDL, the term "signal" denotes inputs and outputs of modeled automata.

Example 2.16: Figure 2.29 shows a process interaction diagram B1 with channels Sw1 and Sw2. Brackets include the names of signals propagated along a certain channel.

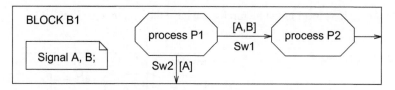

Fig. 2.29 Process interaction diagram   ▽

There are three ways of indicating the recipient of signals:

1. **Through process identifiers**: by using identifiers of recipient processes in the graphical output symbol (see Fig. 2.30 (**left**)).

Fig. 2.30 **Left**: process name identifies recipient; **right**: channel identifies recipient

   The number of processes does not need to be fixed at compile time, since processes can be generated dynamically at run-time. OFFSPRING represents identifiers of child processes generated dynamically by a process.
2. **Explicitly**: by indicating the channel name (see Fig. 2.30 (**right**)). Sw1 is the name of a channel.
3. **Implicitly**: If signal names imply the channel names, those channels are used. Example: For Fig. 2.29, signal B will implicitly always be communicated via channel Sw1.

No process can be defined within any other (processes cannot be nested). However, they can be grouped hierarchically into so-called **blocks**. Blocks at the highest hierarchy level are called **systems**. A system will not have any channels at its boundary if the environment is also modeled as a block. Process interaction diagrams are special cases of block diagrams. Process interaction diagrams are one level above the leaves of the hierarchical description.

Example 2.17: Block B1 of Example 2.16 can be used within intermediate-level blocks (such as within B in Fig. 2.31).

Fig. 2.31 SDL block

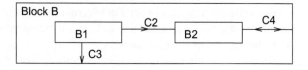

At the highest level in the hierarchy, we have the system (see Fig. 2.32).

Fig. 2.32 SDL system

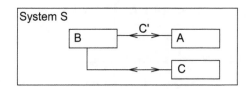

Figure 2.33 shows the hierarchy modeled by block diagrams Figs. 2.29, 2.31, and 2.32.

Fig. 2.33 SDL hierarchy

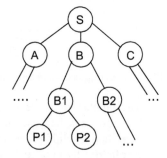

This example demonstrates that process interaction diagrams are next to the *leaves* of the hierarchical description, while system descriptions represent their *root*. ∇

Some of the restrictions of modeling hierarchy are removed in version SDL-2000 of the language. With SDL-2000, the descriptive power of blocks and processes is harmonized and replaced by a general *agent* concept.

In order to support the modeling of time, SDL includes **timers**. Timers can be declared locally for processes. They can be set using the SET primitive. This primitive has two parameters: an absolute time and a timer name. The absolute time defines a time at which the timer elapses. The built-in function now can be used to refer to the time at which the SET primitive is executed. Once a timer is elapsed, a signal is stored in the input queue. The name of this signal is obtained from the second parameter of the SET call. The signal will then typically cause a certain transition to take place in the FSM. However, this transition may be delayed by other entries in the input queue which have to be processed first. Hence, this timer concept is designed for soft timing constraints typically found in telecommunications and inappropriate for hard timing constraints. A second built-in function expirytime can be used to avoid some of the limitations of the now function.

Timers can be reset using the RESET primitive. This primitive will stop the count-ing process and—in case the signal has already been stored in the input queue—removes the signal from it. An implicit RESET is executed at the very beginning of executing a SET.

Example 2.18: Figure 2.34 shows the use of a timer T. The diagram corresponds to that of Fig. 2.26, with the exception that timer T is set to the current time now plus p during the transition from state *D* to *E*. For the transition from *E* to *A*, we now

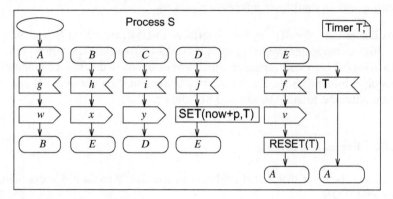

Fig. 2.34 Using timer T

have a timeout of p time units. If these time units have elapsed before signal *f* has arrived, a transition to state *A* is taken without generating output signal *v*. Strictly, periodic processing with a period of p is difficult to achieve this way, due to the possible delays by other entries in the input queue.   ∇

Example 2.19: SDL can be used to describe protocol stacks found in computer networks, and SDL is very appropriate for this. Figure 2.35 shows three processors connected through a router. Communication between processors and the router is based on FIFOs. The processors as well as the router implement layered protocols

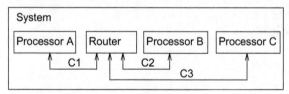

Fig. 2.35 Small computer network described in SDL

(see Fig. 2.36). Each layer describes communication at a more abstract level. The

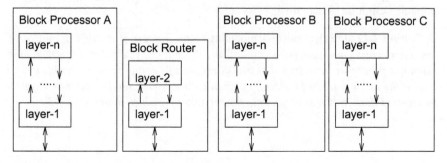

Fig. 2.36 Protocol stacks represented in SDL

behavior of each layer is typically modeled as a finite state machine. The detailed description of these FSMs depends on the network protocol and can be quite complex. Typically, this behavior includes checking and handling of error conditions, as well as sorting and forwarding of information packets.   ∇

Available tools for SDL include interfaces to UML (see p. 117), and SCs (see page 40). A comprehensive list of tools is available from the SDL forum [458].

Estelle [76] is another language which was designed to describe communication protocols. Similar to SDL, Estelle assumes communication via channels and FIFO buffers. Attempts to unify Estelle and SDL failed.

2.4.4.2 Evaluation of SDL

SDL is excellent for distributed applications and has been used, for example, for specifying ISDN.

SDL is not necessarily determinate (the order, in which signals arriving at some FIFO at the same time are processed, is not specified).

Reliable implementations require the knowledge of a upper bound on the length of the FIFOs. This upper bound may be difficult to compute. The timer concept is sufficient for soft deadlines, but not for hard ones.

Hierarchies are not supported in the same way as in StateCharts.

There is no full programming support (but revisions of the standard changed this) and no description of non-functional properties.

SDL is very useful as a reference model for asynchronous message passing, but the interest in SDL is decreasing.

2.5 Data Flow

2.5.1 Scope

Data flow is a very "natural" way of describing real-life applications. Data flow models reflect the way in which data flows from component to component [140]. Each component transforms the data in one way or the other. The following is a possible definition of data flow :

Definition 2.11 ([554]): Data flow modeling *"is the process of identifying, modeling, and documenting how data moves around an information system. Data flow modeling examines processes (activities that transform data from one form to another), data stores (the holding areas for data), external entities (what sends data into a system or receives data from a system), data flows (routes by which data can flow)"*.

A **data flow program** is specified by a directed graph where the nodes (vertices), called **actors**, represent computations and the arcs represent communication channels. The computation performed by each actor is assumed to be functional, that is, based on the input values only. Each process in a data flow graph is decomposed into a sequence of firings, which are atomic actions. Each firing produces and consumes tokens.

Example 2.20: Figure 2.37 describes, as an example, the flow of data in a video-on-demand (VOD) system [287]. Viewers are entering the system via the network

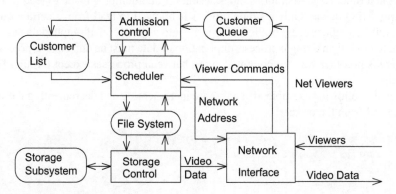

Fig. 2.37 Video-on-demand system

interface. Their admission request is added to the customer queue. Once they are admitted, their requests are scheduled for the file system. The file system, in cooperation with storage control, makes videos available to the customer.   ∇

For unrestricted data flow, it is difficult to prove requested system properties. Therefore, restricted models are commonly used.

A special type of data flow is used for implementing out-of-order execution of instructions in computer architectures. This type of execution is also known as *dynamic scheduling* of instructions. Two algorithms for dynamic scheduling are well-known: scoreboarding and the Tomasulo algorithm [520]. Both algorithms are covered in detail in books on computer architecture (see, for example, Hennessy et al. [205]). Therefore, they are not included in this book. However, there are variants of these algorithms which are applied at task level (for example, see Wang et al. [537]).

2.5.2 Kahn Process Networks

Kahn process networks (KPN) [266] are a special case of data flow models. Like other data flow models, KPNs consist of nodes and edges. Nodes correspond to computations performed by some program or task. KPN graphs, like all data flow graphs, show computations to be performed and their dependencies, but not the order in which the computations must be performed (in contrast to specifications in von-Neumann languages such as C). Edges imply communication via channels containing potentially infinite FIFOs. Computation times and communication times may vary, but communication is guaranteed to happen within a finite amount of time. Writes are non-blocking, since the FIFOs are assumed to be as large as needed. Reads must specify a single channel to be read from. A node cannot check whether data is available before attempting a read. A process cannot wait for data on more than one port at a time. Read operations block whenever an attempt is made to read from an empty FIFO queue. Only a single process is allowed to read from a certain queue, and only a single process is allowed to write into a queue. So, if output data has to be sent to more than a single process, duplication of data must be done inside processes. There is no other way for communication between processes except through FIFO queues.

In the following example, p1 and p2 are incrementing and decrementing the value received from the partner:

```
process p1(in int u, out int v){
    int i;
    i = 0;
    for (;;)    {
        send(i,v);                          /* send i via channel v */
        i = wait(u);                        /* read i from channel u */
        i = i-1;    }
    }
process p2(in int v, out int u){
    int i;
    for (;;)    {
        i = wait(v);
        i = i+1;
        send(i,u);    }
    }
```

Figure 2.38 shows a graphical representation of this KPN.

Fig. 2.38 Graphical representation of KPN

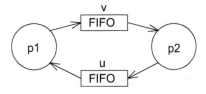

Obviously, we do not really need the FIFOs in this example, since messages cannot accumulate in the channels. This example and other examples can be simulated with the levi simulation software [471].

The restrictions are resulting in the **key beauty of KPNs**: The order in which a node is reading data from its channels is fixed by the sequence of read operations and does not depend on the order in which producers are transmitting data over the channels. This means that the sequence of operations is independent of the speed of the nodes producing data. **For a given set of input data, KPNs will always generate the same results, independently of the speed of the nodes.** This property is important, for example, for simulations: It does not matter how fast we are simulating the KPN, the result will always be the same. In particular, the result does not depend on using hardware accelerators for some of the nodes and a distributed execution will give the same result as a centralized one. This property has been called "determinate," and we are following this use. SDL-like conflicts at FIFOs do not exist. Due to this nice property, KPNs are frequently used as an internal representation within a design flow.

Sometimes, KPNs are extended with a "merge" operator (corresponding to Ada's `select` statement, see p. 107). This operator allows for queuing read commands containing a list of channels. The operator completes execution after the first of these channels has generated data. Such an operator introduces a non-determinate behavior: The order of processing inputs is not specified if both inputs arrive at the same time. This extension is useful in practice, but it destroys the key beauty of KPNs.

In general, Kahn processes require scheduling at run-time, since it is difficult to predict their precise behavior over time. These problems result from the fact that we do not make any assumptions regarding the speed of the channels and the nodes. Nevertheless, execution times are actually unknown during early design phases, and therefore, this model is very adequate.

KPNs are Turing complete, which means whatever can be computed by a Turing machine (the standard model for computability) can also be computed by a KPN. The proof is based on the fact that KPNs are a superset of so-called Boolean Dataflow (BDF), and according to Buck [75], BDF can simulate Turing machines. However, the number of processes has to be fixed at design time, which is an important limitation for many applications.

The question of whether or not finite-length FIFOs are sufficient for an actual KPN model is undecidable in the general case. However, useful scheduling algorithms [281] or proofs of the boundedness of the FIFOs [100] exist for some special cases. For example, these bounds can be derived for Polyhedral Process Networks (PPNs). For PPNs, the code for each of the nodes includes loops with bounds known at compile time. Derin [120] exploits knowledge about the code of the nodes for dynamic task migration.

2.5.3 Synchronous Data Flow

Scheduling becomes significantly easier, and questions regarding buffer sizes can decidably be answered if we impose restrictions on the timing of nodes and channels. Synchronous data flow (SDF) [319] is such a model. SDF can best be introduced by referring to its graphical notation. SDF models include a directed graph, i.e., SDF models contain nodes and directed edges. Nodes are also called **actors**. Edges can store tokens, by default an unlimited number of them. In general, some of the edges will initially contain some tokens. Each edge has an incoming and an outgoing weight. The execution of an SDF model assumes a clock. For an actor to be enabled, it is necessary that for each of the edges leading to that actor the number of tokens on that edge is at least equal to the outgoing weight for that edge.

Example 2.21: Figure 2.39 (**left**) shows a synchronous data flow graph. Actor B is enabled since there is a sufficient number of tokens on the edges leading to B. Actor A is not enabled. Input edges like the one shown at the top for actor A are assumed to supply an infinite stream of tokens. Each clock tick, all enabled actors **fire**. As a

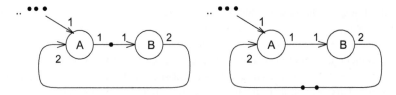

Fig. 2.39 Graphical representation of SDF: **left**: initial situation; **right**: after firing B

result, the number of tokens on the incoming edges get decreased by the incoming weight and the number of tokens on the outgoing edges are increased by the outgoing weight. Obviously, the number of tokens produced or consumed in a particular firing is static (does not vary during the execution of the model). For our example, the resulting number of tokens is shown in Fig. 2.39 on the right.   ∇

In practice, tokens will represent data, actors will represent computations, and edges should correspond to FIFO buffers.

Buffers on the edges imply that SDF uses **asynchronous** message passing. Instead of using the default unlimited buffer capacities, we can express limited buffer capacities with backward edges. The initial number of tokens on these backward edges corresponds to the capacity of the FIFO buffer. This is shown in Fig. 2.40. The two models shown in Fig. 2.40 are equivalent.

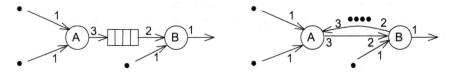

Fig. 2.40 Replacing explicit FIFO buffers by backward edges

For example, the first firing of A will consume three tokens from the backward edge, leaving only one token on the backward edge, corresponding to the one empty FIFO slot after the first firing of A on the left.

The property of producing and consuming a static number of tokens makes it possible to determine execution order and memory requirements at compile time. Hence, complex run-time scheduling of executions is avoided. SDF graphs can be translated into periodic schedules.

Example 2.22: Let us have a closer look at schedules of SDF models. Consider the example shown in Fig. 2.41. Suppose that initially there are six tokens for edge e_1.

Fig. 2.41 SDF loop

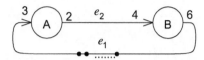

Then, Table 2.2 (**left**) shows the resulting schedule for firings. Due to the limited number of initial tokens, only sequential firings are feasible.

Table 2.2 Schedules for loop in SDF: **left**: six initial tokens on e_1, **right**: nine initial tokens on e_1

| Clock | Tokens on edges | | Next actor action | Clock | Tokens on edges | | Next actor action |
	e_1	e_2	A or B		e_1	e_2	A or B or (A and B)
0	6	0	A	0	9	0	A
1	3	2	A	1	6	2	A
2	0	4	B	2	3	4	A and B
3	6	0	A	3	6	2	A
4	3	2	A	4	3	4	A and B

Now, let us assume that there are nine initial tokens for edge e_1. Then, the schedule of Table 2.2 (**right**) is produced. Under this assumption, A and B fire synchronously.   ∇

During the generation of schedules, we could also consider constraints and objectives such as a limited number of available processors [60].

In this example, using edge labels 2, 3, 4, and 6 resulted in different execution rates of actors A and B. In general, edge labels facilitate the modeling of **multi-rate** signal processing applications, applications for which certain signals are generated at frequencies that are multiples of other frequencies. For example, in a TV set, some computations might be performed at a rate of 100 Hz while others are performed at a rate of 50 Hz. Ignoring some initial transient phase and considering longer periods, the number of tokens sent to an edge must be equal to the number of tokens consumed. Otherwise, tokens would accumulate in the FIFO buffers and no finite FIFO capacity would be sufficient. Let n_s be the number of tokens produced by some sender per firing, and let f_s be the corresponding rate. Let n_r be the corresponding number of tokens consumed per firing at the receiver, and let f_r be the corresponding rate. Then, we must have

$$n_s * f_s = n_r * f_r \tag{2.3}$$

This condition is met in the steady state for the example shown in Table 2.2.

SDF graphs may include delays, denoted by the symbol D on an edge (see Fig. 2.42).

Fig. 2.42 SDF delay

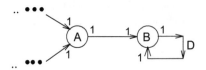

The observer pattern, mentioned as a problem for modeling with von-Neumann languages on p. 31, can be easily implemented correctly in SDF (see Fig. 2.43). There is no risk of deadlocks. However, SDF does not allow adding new observers at run-time.

Fig. 2.43 Observer pattern in SDF

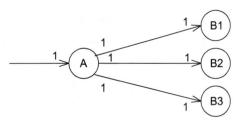

The letter S in SDF initially was meant to stand for the term **synchronous**, since enabled nodes fire synchronously. However, the two schedules in Table 2.2 demonstrate that cases of firing all actors synchronously may indeed be very rare. Therefore, the "S" in SDF has also been reinterpreted to denote the term "static" instead of "synchronous."

SDF models are determinate, but they are not appropriate for modeling control flow, such as branches. Several extensions and variations of SDF models have been proposed (see, for example, Stuijk [491]):

- For example, we can have **modes** corresponding to states of an associated finite state machine. For each of the modes, a different SDF graph could be relevant. Certain events could then cause transitions between these modes.
- **Homogeneous synchronous data flow** (HSDF) graphs are a special case of SDF graphs. For HSDF graphs, the number of tokens consumed and produced per firing is always 1.
- For **cyclo-static data flow** (CSDF), the number of tokens produced and consumed per firing can vary over time, but has to be periodic.

Complex SUDs including control flow must be modeled using more general computational graph structures.

2.5.4 Simulink

Computational graph structures are also frequently used in control engineering. For this domain, the Simulink® toolbox of MATLAB® [506, 510] is very popular. MATLAB is a modeling and simulation tool based on mathematical models including partial differential equations. Figure 2.44 shows an example of a Simulink model [349].

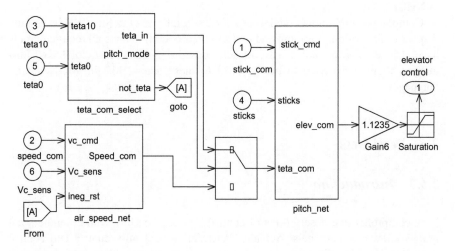

Fig. 2.44 Simulink model

The amplifier and the saturation component on the right demonstrate the inclusion of analog modeling. In the general case, the "schematic" could contain symbols denoting analog components such as integrators, differentiators. The switch in the center indicates that Simulink also allows some control flow modeling.

The graphical representation is intuitive and allows control engineers to focus on the control function, without caring about the code necessary to implement the function. The graphical symbols suggest that analog circuits are used as traditional components in control designs. A key goal is to synthesize software from such models. This approach is typically associated with the term **model-based design**.

Semantics of Simulink models reflect the simulation on a digital computer, and the behavior may be similar to that of analog circuits, but possibly not quite the same. What is actually the semantics of a Simulink model? Marian and Ma [349] describe the semantics as follows: "*Simulink uses an idealized timing model for block (node) execution and communication. Both happen infinitely fast at exact points in simulated time. Thereafter, simulated time is advanced by exact time steps. All values on edges are constant in between time steps.*" This means that we execute the model time step after time step. For each step, we compute the function of the nodes (in zero time) and propagate the new values to connected inputs. This explanation does not specify the distance between time steps. Also, it does not immediately tell us how to implement the system in software, since even slowly varying outputs may be recomputed frequently.

This approach is appropriate for modeling physical systems such as cars or trains at a high level and then simulating the behavior of these systems. Also, digital signal processing systems can be conveniently modeled with MATLAB® and Simulink®. In order to generate implementations, MATLAB/Simulink models first must be translated into a language supported by software or hardware design systems, such as C or VHDL.

Components in Simulink models provide a special case of **actors**. We can assume that actors are waiting for input and perform their operation once all required inputs have arrived. SDF is another case of actor-based languages. In **actor-based languages**, there is no need to pass control to these actors, like in von-Neumann languages.

2.6 Petri Nets

2.6.1 Introduction

Very comprehensive descriptions of control flow are feasible with computational graphs known as Petri nets. Actually, Petri nets model **only** control and control dependencies. Modeling data as well requires extensions of Petri nets. Petri nets focus on the modeling of causal dependencies.

In 1962, Carl Adam Petri published his method for modeling causal dependencies, which became known as Petri nets [427]. Petri nets do not assume any global synchronization and are therefore especially suited for modeling distributed systems.

Conditions, **events**, and a **flow relation** are the key elements of Petri nets. Conditions are either satisfied or not satisfied. Events can happen. The flow relation describes the conditions that must be met before events can happen and it also describes the conditions that become true if events happen. Graphical notations for Petri nets typically use circles to denote conditions and boxes to denote events. Arrows represent flow relations.

Example 2.23: Figure 2.45 shows a first example. This example describes mutual exclusion for trains on a railroad track that must be used in both directions. A token is used to prevent collisions of trains going into opposite directions. In the Petri net, that token is symbolized by a condition in the center of the model. A partially filled circle (a circle containing a second, filled circle) denotes that a condition is met (this means that the track is available). When a train wants to travel to the right (also denoted by a partially filled circle in Fig. 2.45), the two conditions that are necessary for the event "train entering track from the left" are met. We call these two conditions **preconditions**. If the preconditions of an event are met, it can happen.

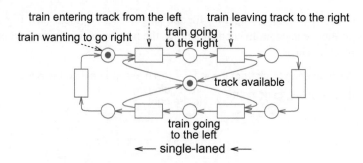

Fig. 2.45 Single-track railroad segment

As a result of that event happening, the token is no longer available and there is no train waiting to enter the track. Hence, the preconditions are no longer met and the partially filled circles disappear (see Fig. 2.46).

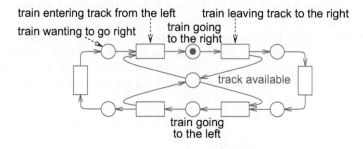

Fig. 2.46 Using resource "track"

However, there is now a train going on that track from the left to the right, and thus, the corresponding condition is met (see Fig. 2.46). A condition which is met after an event happened is called a **postcondition**. In general, an event can happen only if all its preconditions are true (or met). If it happens, the preconditions are no longer met and the postconditions become valid. Arrows identify those conditions which are preconditions of an event and those that are postconditions of an event. Continuing with our example, we see that a train leaving the track will return the token to the condition at the center of the model (see Fig. 2.47).

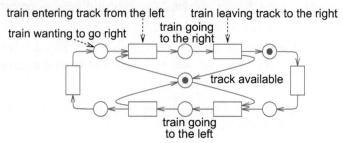

Fig. 2.47 Freeing resource "track"

If there are two trains competing for the single-track segment (see Fig. 2.48), only one of them can enter. In such situations, the next transition to be fired is non-deterministically chosen.

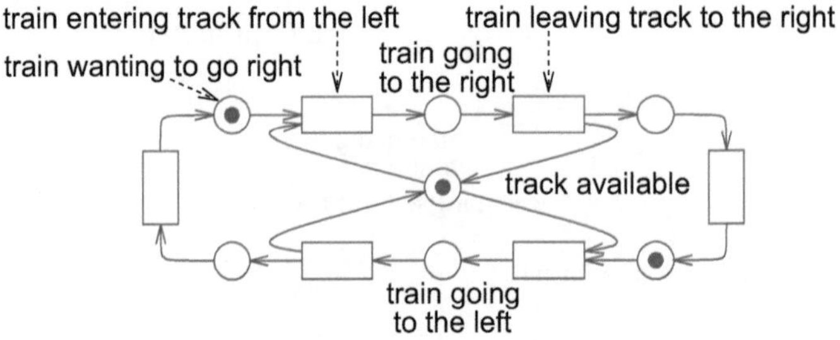

Fig. 2.48 Conflict for resource "track"   ▽

Analyses of the net must consider all possible firing sequences. For Petri nets, we are intentionally modeling non-determinism.

A key advantage of Petri nets is that they can be the basis for formal proofs about system properties and that there are standardized ways of generating such proofs. In order to enable such proofs, we need a more formal definition of Petri nets. We will consider three classes of Petri nets: condition/event nets, place/transitions nets, and predicate transition nets.

2.6.2 Condition/Event Nets

Condition/event nets are the first class of Petri nets that we will define more formally.

Definition 2.12: $N = (C, E, F)$ is called a **net** iff the following holds:

1. C and E are disjoint sets.
2. $F \subseteq (E \times C) \cup (C \times E)$ is a binary relation, called flow relation.

The set C is called conditions and the set E is called events.

Definition 2.13: Let N be a net and let $x \in (C \cup E)$. $^\bullet x := \{y | y F x, y \in (C \cup E)\}$ is called the **preset** of x. If x denotes an event, $^\bullet x$ is also called the set of **preconditions** of x.

Definition 2.14: Let N be a net and let $x \in (C \cup E)$. $x^\bullet := \{y | x F y, y \in (C \cup E)\}$ is called the **postset** of x. If x denotes an event, $x^\bullet$ is also called the set of **postconditions** of x.

The terms preconditions and postconditions are preferred if these sets actually denote conditions $\in C$, that is, if $x \in E$.

Definition 2.15: Let $(c, e) \in C \times E$. (c, e) is called a **loop** if $c F e \wedge e F c$.

Definition 2.16: Let $(c, e) \in C \times E$. N is called **pure** if F does not contain any loops (see Fig. 2.49, **left**).

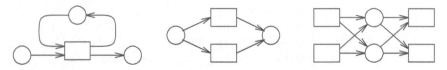

Fig. 2.49 Nets which are not pure (**left**) and not simple (**center** and **right**)

Definition 2.17: A net is called **simple** if no two transitions t_1 and t_2 have the same set of pre- and postconditions (see Fig. 2.49 (**center** and **right**)).

Simple nets with no isolated elements meeting some additional restrictions are called **condition/event nets**. Condition/event nets are a special case of bipartite graphs (graphs with two disjoint sets of nodes). We will not discuss those additional restrictions in detail since we will consider more general classes of nets in the following.

2.6.3 Place/Transition Nets

For condition/event nets, there is at most one token per condition. For many applications, it is useful to remove this restriction and to allow more tokens per condition. Nets allowing more than one token per condition are called place/transition nets. Places correspond to what we so far called conditions and transitions correspond to what we so far called events. The number of tokens per place is called a **marking**. Mathematically, a marking is a mapping from the set of places to the set of natural numbers extended by a special symbol ω denoting infinity.

Let $\mathbb{N}_0$ denote the natural numbers including 0. Then, formally speaking, place/transition nets can be defined as follows:

Definition 2.18: (P, T, F, K, W, M_0) is called a place/transition net $\iff$

1. $N = (P, T, F)$ is a net with places $p \in P$, transitions $t \in T$, and flow relation F.
2. Mapping $K : P \to (\mathbb{N}_0 \cup \{\omega\}) \setminus \{0\}$ denotes the capacity of places (ω symbolizes infinite capacity).
3. Mapping $W : F \to (\mathbb{N}_0 \setminus \{0\})$ denotes the weight of graph edges.
4. Mapping $M_0 : P \to \mathbb{N}_0 \cup \{\omega\}$ represents the initial marking of places.

Edge weights affect the number of tokens that are required before transitions can happen and also identify the number of tokens that are generated if a certain transition takes place. Let $M(p)$ denote a current marking of place $p \in P$, and let $M'(p)$ denote a marking after some transition $t \in T$ took place. The weight of edges belonging to preconditions represents the number of tokens that are removed from places in the preset. Accordingly, the weight of edges belonging to the postconditions represents the number of tokens that are added to the places in the postset. Formally, marking M' is computed as follows:

$$M'(p) = \begin{cases} M(p) - W(p,t), & \text{if } p \in {}^{\bullet}t \setminus t^{\bullet} \\ M(p) + W(t,p), & \text{if } p \in t^{\bullet} \setminus {}^{\bullet}t \\ M(p) - W(p,t) + W(t,p), & \text{if } p \in {}^{\bullet}t \cap t^{\bullet} \\ M(p) & \text{otherwise} \end{cases}$$

Figure 2.50 demonstrates how transition t_j affects the current marking.

Fig. 2.50 Generation of a new marking

By default, unlabeled edges are considered to have a weight of 1 and unlabeled places are considered to have unlimited capacity ω.

We now need to explain the two conditions that must be met before a transition $t \in T$ can take place:

- for all places p in the preset, the number of tokens must at least be equal to the weight of the edge from p to t and
- for all places p in the postset, the capacity must be large enough to accommodate the new tokens which t will generate.

Transitions meeting these two conditions are called **M-activated**. Formally, this can be defined as follows:

Definition 2.19: Transition $t \in T$ is said to be M-activated $\iff$

$$(\forall p \in {}^{\bullet}t : M(p) \geq W(p,t)) \wedge (\forall p' \in t^{\bullet} : M(p') + W(t,p') \leq K(p'))$$

Activated transitions can happen, but they do not need to. If several transitions are activated, the sequence in which they happen is not deterministically defined.

The impact of a firing transition t on the number of tokens can be represented conveniently by a vector $\underline{t}$ associated with t. $\underline{t}$ is defined as follows:

$$\underline{t}(p) = \begin{cases} -W(p,t), & \text{if } p \in {}^{\bullet}t \setminus t^{\bullet} \\ +W(t,p), & \text{if } p \in t^{\bullet} \setminus {}^{\bullet}t \\ -W(p,t) + W(t,p), & \text{if } p \in {}^{\bullet}t \cap t^{\bullet} \\ 0 & \text{otherwise} \end{cases}$$

The new number M' of tokens, resulting from the firing of transition t, can be computed for all places p as follows:

$$M'(p) = M(p) + \underline{t}(p)$$

Using "+" to denote vector addition, we can rewrite this equation as follows:

$$M' = M + \underline{t}$$

The set of all vectors $\underline{t}$ form an incidence matrix $\underline{N}$. $\underline{N}$ contains vectors $\underline{t}$ as columns.

$$\underline{N} : P \times T \to \mathbb{Z}; \quad \forall t \in T : \underline{N}(p, t) = \underline{t}(p)$$

It is possible to formally prove system properties by using matrix $\underline{N}$. For example, we are able to compute sets of places, for which firing transitions will not change the overall number of tokens [445]. Such sets are called **place invariants**. Let us initially consider a single transition t_j in order to find such invariants. Let us search for sets $R \subseteq P$ of places such that the total number of tokens does not change if t_j fires. The following must hold for such sets:

$$\sum_{p \in R} \underline{t}_j(p) = 0 \qquad (2.4)$$

Figure 2.51 shows a transition for which the total number of tokens does not change if it fires.

Fig. 2.51 Transition with a constant number of tokens

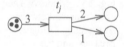

We are now introducing the characteristic vector $\underline{c}_R$ of some set R of places:

$$\underline{c}_R(p) = \begin{cases} 1 \text{ iff } p \in R \\ 0 \text{ iff } p \notin R \end{cases}$$

With this definition, we can rewrite Eq. (2.4) as

$$\sum_{p \in R} \underline{t}_j(p) = \sum_{p \in P} \underline{t}_j(p) \cdot \underline{c}_R(p) = \underline{t}_j \cdot \underline{c}_R = 0. \qquad (2.5)$$

which denotes the scalar product. Now, we search for sets of places such that firings of **any** transition will not change the total number of tokens. This means that Eq. (2.5) must hold for all transitions t_j:

$$\underline{t}_1 \cdot \underline{c}_R = 0$$
$$\underline{t}_2 \cdot \underline{c}_R = 0$$
$$\dots$$
$$\underline{t}_n \cdot \underline{c}_R = 0$$

$$(2.6)$$

Equation (2.6) can be combined into the following equation by using the transposed incidence matrix $\underline{N}^T$:

$$\underline{N}^T \cdot \underline{c}_R = 0 \qquad (2.7)$$

Equation (2.7) represents a system of linear, homogeneous equations. Matrix $\underline{N}$ represents edge weights of our Petri nets. We are looking for solution vectors $\underline{c}_R$ for this system of equations. Solutions must be characteristic vectors. Therefore, their components must be 1 or 0 (integer weights can be accepted if we use weighted sums of tokens). This is more complex than solving systems of linear equations with real-valued solution vectors. Nevertheless, it is possible to obtain information by solving equation (2.7). Using this proof technique, we can, for example, show that we are correctly implementing mutually exclusive access to shared resources.

Example 2.24: Let us now consider a larger example: We are again considering the synchronization of trains. In particular, we are trying to model high-speed Thalys trains traveling between Amsterdam, Cologne, Brussels, and Paris. Segments of the train run independently from Amsterdam and Cologne to Brussels. There, the segments get connected and then they run to Paris. On the way back from Paris, they get disconnected at Brussels again. We assume that Thalys trains must synchronize with some other train at Paris. The corresponding Petri net is shown in Fig. 2.52.

Fig. 2.52 Model of Thalys trains running between Amsterdam, Cologne, Brussels, and Paris

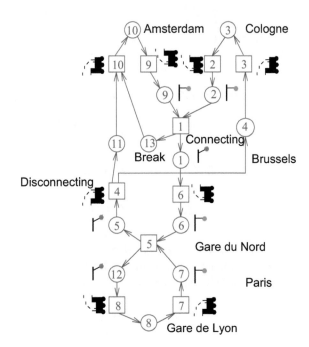

Places 3 and 10 model trains waiting at Cologne and Amsterdam, respectively. Transitions 2 and 9 model trains driving from these cities to Brussels. After their arrival at Brussels, places 2 and 9 contain tokens. Transition 1 denotes connecting the two trains. The cup symbolizes the driver of one of the trains, who will have a break at Brussels while the other driver is continuing on to Paris. Transition 5 models synchronization with other trains at the Gare du Nord station of Paris. These other trains connect Gare du Nord with some other stations (we have used Gare de Lyon as an example, even though the situation at Paris is somewhat more complex). Of course, Thalys trains do not use steam engines; they are just easier to visualize than modern high-speed trains. Table 2.3 shows matrix N^T for this example.

Table 2.3 N^T for the Thalys example

	p_1	p_2	p_3	p_4	p_5	p_6	p_7	p_8	p_9	p_{10}	p_{11}	p_{12}	p_{13}
t_1	1	-1							-1				1
t_2		1	-1										
t_3			1	-1									
t_4				1	-1						1		
t_5					1	-1	-1				1		
t_6	-1				1								
t_7							1	-1					
t_8								1				-1	
t_9									1	-1			
t_{10}											1	-1	-1

For example, row 2 indicates that firing t_2 will increase the number of tokens on p_2 by 1 and decrease the number of tokens on p_3 by 1. Using techniques from linear algebra, we are able to show that the following four vectors are solutions for this system of linear equations:

$$c_{R,1} = (1, 1, 1, 1, 1, 1, 0, 0, 0, 0, 0, 0, 0)$$
$$c_{R,2} = (1, 0, 0, 0, 1, 1, 0, 0, 1, 1, 1, 0, 0)$$
$$c_{R,3} = (0, 0, 0, 0, 0, 0, 0, 0, 1, 1, 0, 0, 1)$$
$$c_{R,4} = (0, 0, 0, 0, 0, 0, 1, 1, 0, 0, 0, 1, 0)$$

These vectors correspond to the places along the track for trains from Cologne, to the places along the track for trains from Amsterdam, to the places along the path for drivers of trains from Amsterdam, and to the places along the track within Paris, respectively. Therefore, we are able to show that the number of trains and drivers along these tracks is constant (something which we actually expect). This example demonstrates that place invariants provide us with a standardized technique for proving properties about systems.   ∇

2.6.4 Predicate/Transition Nets

Condition/event nets as well as place/transition nets can quickly become very large for large examples. A reduction of the size of the nets is frequently possible with predicate/transition nets.

Example 2.25: We will demonstrate this, using the so-called dining philosophers problem as an example. The problem is based on the assumption that a set of philosophers is dining at a round table. In front of each philosopher, there is a plate containing spaghetti (see Fig. 2.53). Between each of the plates, there is just one

Fig. 2.53 The dining philosophers problem

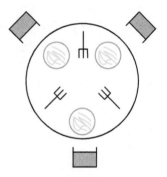

fork. Each philosopher is either eating or thinking. Eating philosophers need their two adjacent forks for that, so they can only eat if their neighbors are not eating.

This situation can be modeled as a condition/event net, as shown in Fig. 2.54.

Fig. 2.54 Place/transition net model of the dining philosophers problem

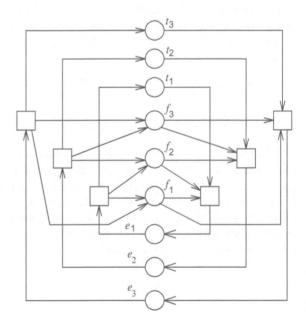

Conditions t_j correspond to the thinking states, conditions e_j correspond to the eating states, and conditions f_j represent available forks. Considering the small size of the problem, this net is already very large. The size of this net can be reduced by using predicate/transition nets. Figure 2.55 is a model of the same problem as a predicate/transition net. With predicate/transition nets, tokens have an identity

Fig. 2.55 Predicate/transition net model of the dining philosophers problem

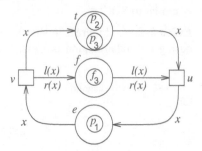

and can be distinguished from each other. Predicate/transition nets have also been called **Colored Petri Nets** (CPN). See Jensen [260] for a survey of applications of CPNs for modeling of IT systems, including communication protocols. We use this in Fig. 2.55 in order to distinguish between the three different philosophers p_1 to p_3 and to identify fork f_3. Furthermore, edges can be labeled with variables and functions. In the example, we use variables to represent the identity of philosophers and functions $l(x)$ and $r(x)$ to denote the left and right forks of philosopher x, respectively. These two forks are required as a precondition for transition u and returned as a postcondition by transition v. This model can be easily extended to $n > 3$ philosophers. We just need to add more tokens. In contrast to the net in Fig. 2.54, the structure of the net does not have to be changed. ▽

2.6.5 Evaluation

The key advantage of Petri nets is their power for modeling causal dependencies. Standard Petri nets have no notion of time, and all decisions can be taken locally by just analyzing transitions and their pre- and postconditions. Therefore, they can be used for modeling geographically distributed systems. Furthermore, there is a strong theoretical foundation for Petri nets, simplifying formal proofs of system properties. Petri nets are not necessarily determinate: Different firing sequences can lead to different results. The descriptive power of Petri nets encompasses that of other MoCs, including finite state machines.

In certain contexts, their strength is also their weakness. If time is to be modeled, standard Petri nets cannot be used. Furthermore, standard Petri nets have no notion of hierarchy and no programming language elements, let alone object-oriented features. In general, it is difficult to represent data.

There are extended versions of Petri nets avoiding the mentioned weaknesses. However, there is no universal extended version of Petri nets meeting all requirements mentioned at the beginning of this chapter. Nevertheless, due to the increasing amount of distributed computing, Petri nets became more popular.

UML includes extended Petri nets called **activity diagrams**. Extensions include symbols denoting decisions (like in ordinary flow charts). The placement of symbols is similar to SDL.

Example 2.26: Figure 2.56 shows an activity chart of the procedure to be followed during a standardization process. Forks and joins of control correspond to transi-

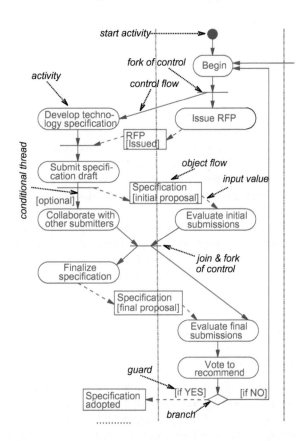

Fig. 2.56 Activity diagram [288]

tions in Petri nets, and they use the symbols (horizontal bars) that were initially used for Petri nets as well. The diamond at the bottom shows the symbol used for decisions. Activities can be organized into "swim lanes" (areas between vertical dotted lines) such that the different responsibilities and the documents exchanged can be visualized.   ▽

Interestingly, Petri nets were initially not a mainstream technique. Decades after their invention, they have become a popular technique due to their inclusion in UML.

2.7 Discrete Event-Based Languages

2.7.1 Basic Discrete Event Simulation Cycle

The discrete event-based model of computation is based on simulating the generation of events and processing them over time. We use a queue of future events, and these are sorted by the time at which they should be processed. Semantics is defined by fetching the event at the head of the queue, performing the corresponding actions, and possibly entering new events into the queue. Time is advanced whenever no action exists which should be performed at the current time. This is the basic algorithm:

```
loop
    fetch next entry from queue;
    perform function (e.g., assignment of variables as listed in the entry)
        (this may include the generation of new events);
until termination criterion is met;
```

Hardware description languages (HDLs) are typically based on the discrete event model. We will use HDLs as a prominent example of discrete event modeling.

Example 2.27: We demonstrate the application of this general scheme to simulate an RS latch (see Fig. 2.57). The latch consists of two cross-coupled NOR gates.

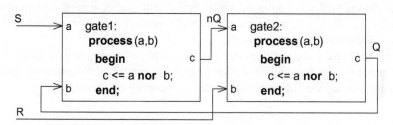

Fig. 2.57 Two cross-connected NOR gates forming an RS latch

The corresponding code in a hardware description language, in this case VHDL, is included in Fig. 2.57 as well. A representative sequence of values at the inputs and outputs is shown in Table 2.4.

Table 2.4 Sequence of values at inputs and outputs of RS latch

	$t < 0$	$t = 0$	$t > 0$		
R	0	1	1	1	1
S	0	0	0	0	0
Q	1	1	0	0	0
nQ	0	0	0	1	1

Let us assume that initially, the latch is set and this state is maintained, i.e., output Q is 1 and R = S = 0. The operation of both NOR gates is described by processes gate1 and gate2. These processes are initially inactive, waiting for some event on their inputs a or b. This waiting is expressed by the lists (a, b). gate1 and gate2 are said to be **sensitive** to the entries in that list.

Now, suppose that at time 0, input R, the reset input, is changed to 1. We expect the latch to be reset. In terms of events, this works as follows: The change at input R is an event, which is stored in the queue of future events.

This event is immediately processed, since it is the only event in the queue. This event will wake up gate2, since this gate is sensitive to changes on its input b. gate2 will compute the NOR function, with a result of 0, and will then execute the assignment c <= expression. This notation indicates a **signal assignment**. This means that the new values will initially be stored only in the entries of future events. The actual assignment to the variable on the left becomes effective only when the time for processing this entry in the list of future events has been reached. In our example, an event requesting output c of gate2 to be set to 0 will be created and stored in the event queue.

This event will be immediately fetched, since it is the only event. The event will set output c to 0. This wakes up gate1, due to its sensitivity. gate1 will compute the NOR function as well. This computation results in an event, requesting output c of gate1 to be set to 1. This event will also be stored in the queue.

This event will also be immediately processed, setting the output as requested. This change will wake up gate2 again. gate 2 will again compute an output of 0. Further details will depend somewhat on the mechanism which is used to detect stable situations not requiring further events to be generated.

We could have added delays in terms of real physical units to each of the signal assignments, which would have allowed us to keep track of elapsed time. Overall, this event-based simulation approximates the behavior of a real latch.   ∇

2.7.2 Multi-valued Logic

Which values could we use for the signals in the above example? In this book, we are restricting ourselves to embedded systems implemented with binary logic. Nevertheless, it may be advisable or necessary to use more than two values for modeling such systems. For example, our systems might contain electrical signals of different strengths. It may be necessary to compute the strength and the logic level resulting from a connection of two or more sources of electrical signals. In the following, we will therefore distinguish between the **level** and the **strength** of a **signal**. While the former is an abstraction of the signal voltage, the latter is an abstraction of the impedance (resistance) of the voltage source. We will be using discrete sets of signal values representing the signal level and the strength. **Using discrete sets of strengths avoids the problems of having to solve Kirchhoff's network equations and enables us to avoid analog models used in electrical engineering.** We will also model unknown electrical signals by special signal values.

In practice, electronic design systems use a variety of value sets. Some systems allow only two, while others allow 9 or 46. The overall goal of developing discrete value sets is to avoid the problems of solving network equations and still model existing systems with sufficient precision.

In the following, we will present a systematic technique for building up value sets and relating these to each other. We will use the strength of electrical signals as the key parameter for distinguishing between various value sets. A systematic way of building up value sets, called CSA theory, was presented by Hayes [200]. CSA stands for "connector, switch, attenuator." These three elements are key elements of this theory. We will later show how the standard value set used for most cases of VHDL-based modeling can be derived as a special case.

One Signal Strength (Two Logic Values)

In the simplest case, we will start with just two logic values, called '0' and '1'. These two values are considered to be of the same strength. This means that if two wires **connect** values '0' and '1', we will not know anything about the resulting signal level.

A single signal strength may be sufficient if no two wires carrying values '0' and '1' are connected and no signals of different strength meet at a particular node of electronic circuits.

Two Signal Strengths (Three and Four Logic Values)

In many circuits, there may be instances in which a certain electrical signal is not actively driven by any output. This may be the case, when a certain wire is not connected to ground, the supply voltage, or any circuit node.

For example, systems may contain open collector outputs (see Fig. 2.58, (**left**))[13].

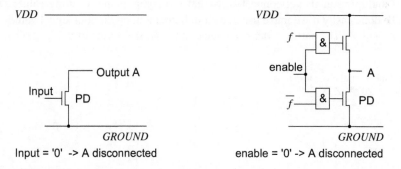

Fig. 2.58 Effectively disconnectable outputs: **left**: open collector output; **right**: tristate output

If the "pull-down" transistor PD is non-conducting, the output is effectively disconnected. For the tristate outputs (see Fig. 2.58, (**right**)), an enable signal of '0' will

---

[13]Schematics should **help** students to understand signal values, not make it more difficult. Students unfamiliar with schematics could just study logic values.

generate a '0' at the outputs of the AND gates (denoted by &), and will make both transistors non-conducting. As a result, output A will be disconnected[14]. Hence, using appropriate input signals, such outputs can be effectively disconnected from a wire.

The signal strength of disconnected outputs is the smallest strength that we can think of. We will denote the value at disconnected outputs as 'Z'. The signal strength of 'Z' is smaller than that of '0' and '1'. Furthermore, the signal level of 'Z' is unknown. If a signal of value 'Z' is connected to another signal, that other signal will always dominate. For example, if two tristate outputs are connected to the same bus and if one output contributes a value of 'Z', the resulting value on the bus will always be the value contributed by the second output (see Fig. 2.59).

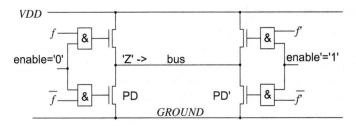

Fig. 2.59 Right output dominates bus

In most cases, three-valued logic sets {'0','1','Z'} are extended by a fourth value called 'X'. 'X' represents an unknown signal level of the same strength as '0' or '1'. More precisely, we are using 'X' to represent unknown values of signals that can be either '0' or '1' or some voltage representing neither '0' nor '1'[15].

If multiple signals get connected, we have to compute the resulting value. This can be done easily if we make use of a partial order among the four signal values '0', '1', 'Z', and 'X'. The partial order is depicted in the **Hasse diagram** in Fig. 2.60.

Fig. 2.60 Partial order for
value set {'0','1','Z','X'}

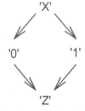

---

[14]Pull-up transistors may be depletion transistors, and the tristate outputs may be inverting.

[15]There are other interpretations of 'X' [67], but ours is the most useful in our context.

Edges in this figure reflect the domination of signal values. Edges define a relation $>$. If $a > b$, then a dominates b. '0' and '1' dominate 'Z'. 'X' dominates all other signal values. Based on the relation $>$, we define a relation $\geq$. $a \geq b$ holds iff $a > b$ or $a = b$.

We define an operation sup on two signals, which returns the **supremum** of the two signal values.

Definition 2.20: Let a and b be two signal values from a partially ordered set $(S, \geq)$. The **supremum** $c \in S$ of the two values a and b is the smallest value for which $c \geq a$ and $c \geq b$ hold.

For example, sup ('Z', '0') = '0' and sup('Z','1') = '1'.

Lemma 2.1: Let a and b be two signals having values from a partially ordered set, where the partial order has been selected as shown above. Then, **the sup function computes resulting signal values if the two signals get connected.**

The supremum corresponds to the **connect** element of the CSA theory.

Three Signal Strengths (Seven Signal Values)

In many circuits, two signal strengths are not sufficient. A common case that requires more values is the use of depletion transistors (see Fig. 2.61).

Fig. 2.61 Output using depletion transistor

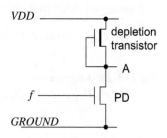

The effect of the depletion transistor is similar to that of a resistor providing a low conductance path to the supply voltage *VDD*. The depletion transistor as well as the "pull-down transistor" PD acts as drivers for node A of the circuit, and the signal value at node A can be computed using the supremum function. The pull-down transistor provides a driver value of '0' or 'Z', depending upon the input to PD. The depletion transistor provides a signal value, which is weaker than '0' and '1'. Its signal level corresponds to the signal level of '1'. We represent the value contributed by the depletion transistor by 'H', and we call it a "weak logic one." Similarly, there can be weak logic zeros, represented by 'L'. The value resulting from the possible connection between 'H' and 'L' is called a "weak logic undefined," denoted as 'W'. As a result, we have three signal strengths and seven logic values {'0','1','L','H','W','X','Z'}. Computing the resulting signal value can again be based on a partial order among these seven values. The corresponding partial order is shown in Fig. 2.62.

Fig. 2.62 Partial order for
value set
{'0','1','L','H','W','X','Z'}

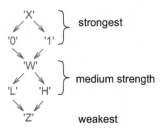

sup is also defined for this partially ordered set. For example, $sup('H','0') = '0'$, $sup('H','Z') = 'H'$ and $sup('H','L') = 'W'$.

'0' and 'L' represent the same signal levels, but a different strength. The same holds for the pairs '1' and 'H'. Devices increasing signal strengths are called **amplifiers**, devices reducing signal strengths are called **attenuators**.

Ten Signal Values (Four Signal Strengths)

In some cases, three signal strengths are not sufficient. For example, there are circuits using charges stored on wires. Such wires are charged to levels corresponding to '0' or '1' during some phases of the operation of the electronic circuit. This stored charge can control the (high impedance) inputs of some transistors. However, if these wires get connected to even the weakest signal source (except 'Z'), they lose their charge and the signal value from that source dominates.

Example 2.28: In Fig. 2.63, we are driving a bus from a specialized output.

Fig. 2.63 Precharging a bus

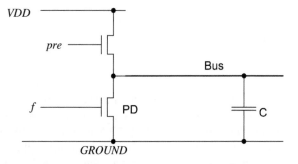

The bus has a high capacitive load C. While function f is still '0', we set *pre* to '1', charging capacitor C. Then, we set *pre* to '0'. If the real value of function f becomes known and it turns out to be '1', we discharge the bus.   ▽

The key reason for using precharging is that charging a bus using an output such as the one shown in Fig. 2.61 is a slow process, since the resistance of depletion transistors is large. Discharging through regular pull-down transistors PD is a much faster process.

In order to model such cases, we need signal values which are weaker than 'H' and 'L', but stronger than 'Z'. We call such values "very weak signal values" and denote them by 'h' and 'l'. The corresponding very weak unknown value is denoted by 'w'. As a result, we obtain ten signal values {'0','1','L','H','l','h','X','W','w','Z'}. Using signal strengths, we can again define a partial order among these values (see Fig. 2.64).

Fig. 2.64 Partially ordered set {'0','1','Z','X','H','L', 'W','h','l','w'}

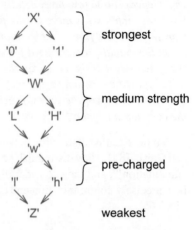

Note that precharging is not without risks. Once a precharged wire is discharged due to a transient signal, it cannot be recharged during the same clock period.

Five Signal Strengths

So far, we have ignored power supply signals. These are stronger than the strongest signals we have considered so far. Signal value sets taking power supply signals into account have resulted in the definition of initially popular 46-valued value sets [106]. However, such models are hardly used anymore.

2.7.3 Transaction-Level Modeling (TLM)

Discrete event simulation allows us to keep track of simulated time. However, it is not obvious how precisely we will be modeling time. A very precise model reflecting detailed timing of hardware signals will require long simulation times. In particular, very long simulation times are needed when we model electrical circuits. Faster simulation is feasible with cycle-accurate models reflecting the number of clock cycles in a clocked (synchronous) system implementation. More simulation speed can be gained from more coarse-grained timing models. In particular, transaction-level modeling (TLM) has received much attention. TLM has been defined as follows [184]:

Definition 2.21: *"**Transaction-level modeling** (TLM) is a high-level approach to modeling digital systems where details of communication among modules are separated from the details of the implementation of functional units or of the communication architecture. Communication mechanisms such as buses or FIFOs are modeled as channels, and are presented to modules using SystemC interface classes. Transaction requests take place by calling interface functions of these channel models, which encapsulate low-level details of the information exchange. At the transaction level, the emphasis is more on the functionality of the data transfers—what data is transferred to and from what locations—and less on their actual implementation, that is, on the actual protocol used for data transfer. This approach makes it easier for the system-level designer to experiment, for example, with different bus architectures (all supporting a common abstract interface) without having to recode models that interact with any of the buses, provided these models interact with the bus through the common interface."*

A more detailed distinction between different timing models was described by Cai and Gajski [84]. They distinguish between timing models for communication and for computation[16], and they consider different cases of timing models, depending upon how precisely communication and computation are modeled. Six cases are shown in Fig. 2.65.

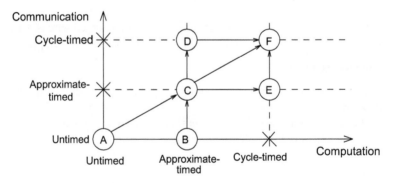

Fig. 2.65 Distinction between different timing models

For communication as well as for computations, we distinguish between untimed, approximately timed, and cycle-timed models. In diagram Fig. 2.65, crosses mark three unbalanced combinations of timing models, which have not been considered by Cai and Gajski. As a result, we consider six remaining cases [84]:

A **Untimed models**: In this case, we model only the functionality and do not consider timing at all. Such models are appropriate for early design phases. They can be called **specification model**.

---

[16]This is very much in line with the same distinction which we have made in Table 2.1 on p. 39.

B In the specification model, we can replace pure functionality descriptions by descriptions of components using rough timing models. For example, we might know the WCET of some code running on a processor. We would still model communication by abstract communication primitives. As a result, we obtain node B in Fig. 2.65. Such a model can be called **component assembly model**.

C In a model of type B, we could replace abstract communication primitives by communication models which are approximately timed. This means that we try to model access conflicts and their impact on the timing, but we do not model the impact of each and every signal, nor do we model any links to clock cycles. Such a model can be called **bus arbitration model**.

D In a model of type C, we could replace rough communication timing models with cycle-timed models. This implies that we keep track of elapsed clock cycles in our simulation. We might even consider real, physical time. The resulting model, denoted as node D in Fig. 2.65, can be called a **bus functional model** [84].

E In a model of type C, we could also replace rough computation timing models by cycle-accurate timing models of the computation. This allows us, for example, to capture memory references in detail. The resulting model can be called a **cycle-accurate computation model**.

F The node labeled F is obtained when communication and computation are modeled in a cycle-accurate way. Such a model can be called an **implementation model**.

Design procedures need to traverse the diagram in Fig. 2.65 from node A to node F.

2.7.4 SpecC

The SpecC language [167] provides us with an nice example for demonstrating TLMs and a clear separation between communication and computation. SpecC models systems as hierarchical networks of behaviors communicating through channels. SpecC descriptions consist of behaviors, channels, and interfaces. Behaviors include ports, locally instantiated components, private variables and functions, and a public main function. Channels encapsulate communication. They include variables and functions, which are used for the definition of a communication protocol. Interfaces are linking behaviors and channels together. They declare the communication protocols which are defined in a channel.

SpecC can model hierarchies with nested behaviors.

Example 2.29: Figure 2.66 [167] shows a component G including sub-components g1 and g2 as leaves in the hierarchy.

Fig. 2.66 Structural
hierarchy of SpecC example

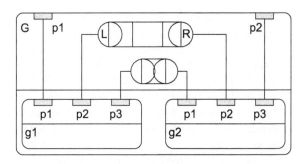

This structural hierarchy is described in the following SpecC model:

```
01: interface L {void Write(int x); };
02: interface R {int Read(void); };
03: channel H implements L,R
04:   {int Data; bool Valid;
05:    void Write(int x) {Data=x; Valid=true;}
06:    int Read (void)
07:       {while (!Valid) waitfor (10); return (Data);}
08:   }
09: behavior G1(in int p1, L p2, in int p3)
10:    {void main (void) {/* ...*/ p2.Write(p1);} };
11: behavior G2 (out int p1, R p2, out int p3)
12:    {void main(void) {/*...*/ p3=p2.Read(); } };
13: behavior G(in int p1, out int p2)
14:   {int h1;   H h2;   G1 g1(p1, h2, h1);   G2 g2(h1, h2, p2);
15:   void main (void)
16:     {par {g1.main(); g2.main();}}
17:   };
```

Concurrent execution of sub-components is denoted by the keyword **par** in line 16.
As indicated in line 14, sub-components are communicating through integer h1 and
through channel h2. Note that the interface protocol implemented in channel H (see
line 03), consisting of methods for read and write operations (lines 05 and 06), can be
changed without changing behaviors G1 and G2. For example, communication can
be bit serial or parallel and the choice does not affect the models of G1 and G2. This
is a necessary feature for reuse of hardware components or intellectual property (IP).
The presented SpecC model does not include any timing information. Hence, it is a
specification model (model of type A in Fig. 2.65).   ∇

The design flow for SpecC was already shown in Fig. 1.9 on p. 20. The path in
Fig. 2.65 is A, B, D, F [84]. At the specification level, SpecC can model any kind
of communication and typically uses message passing. The communication model
of SpecC has inspired the communication model in SystemC 2.0.

Note that SpecC is based on C and C++ syntax. The reason for this is the following:

There is the trend of implementing more and more functionality in software and using C for this purpose. For example, embedded systems implement standards such as MPEG 1/2/4 or decoders for mobile phone standards such as GSM, UMTS, or LTE. These standards are frequently available in the form of "reference implementations," consisting of C programs not optimized for speed but providing the required functionality. The disadvantage of design methodologies based on special hardware description languages (like VHDL or Verilog, see below) is that these standards must be rewritten in order to generate systems. Furthermore, simulating hardware and software together requires interfacing software and hardware simulators. Typically, this involves a loss of simulation efficiency and inconsistent user interfaces. Also, designers would need to learn several languages.

Therefore, there has been a search for techniques for representing hardware structures in software languages. Some fundamental problems had to be solved before hardware could be modeled with software languages:

- **Concurrency**, as it is found in hardware, has to be modeled in software.
- There has to be a representation of simulated **time**.
- **Multiple-valued logic** as described earlier must be supported.
- The **determinate behavior** of almost all useful hardware circuits should be guaranteed.

For the SpecC language, as well as for other hardware description languages, these problems were solved.

2.7.5 SystemC™

TLM modeling and the separation between communication and computation are also available in SystemC™. SystemC (like SpecC) is based on C and C++. Similar to SpecC, SystemC provides channels, ports, and interfaces as abstract components for communication. The introduction of these mechanisms facilitates transaction-level modeling.

SystemC™ [416, 498] is a C++ class library. With SystemC, specifications can be written in C or C++, making appropriate references to the class library.

SystemC comprises a notion of processes executed concurrently. The execution of these processes is controlled by calls to **wait** primitives and **sensitivity lists** (lists of signals for which value changes start a re-execution of code). The sensitivity list concept includes dynamic sensitivity lists, i.e., the list of relevant signals can change during the execution.

SystemC includes a model of time. Earlier, SystemC 1.0 used floating point numbers to denote time. In the current standard, an integer model of time is preferred. SystemC also supports physical units such as picoseconds, nanoseconds, and microseconds.

SystemC data types include all common hardware types: Four-valued logic ('0','1','X' and 'Z') and bitvectors of different lengths are supported. Writing digital signal processing applications is simplified due to the availability of fixed-point data types.

Determinate behavior (see p. 54) of SystemC is not guaranteed in general, unless a certain modeling style is used. Using a command line option, the simulator can be directed to run processes in different orders. This way, the user can check whether the simulation results depend on the sequence in which the processes are executed. However, for models of realistic complexity, only the presence of non-determinate behavior can be shown, not its absence.

Transaction-level modeling with SystemC has been described in a White Paper by Montoreano [384]. The White Paper distinguishes only between two types of TLM models:

- **Loosely timed models**: They are described as follows [384]: *"These models have a loose dependency between timing and data, and are able to provide timing information and the requested data at the point when a transaction is being initiated. These models do not depend on the advancement of time to be able to produce a response. Normally, resource contention and arbitration are not modeled using this style. Due to the limited dependencies and minimal context switches, these models can be made to run the fastest and are particularly useful for doing software development on a Virtual Platform."*
- **Approximately timed models**: They are described as follows [384]: *"These models can depend on internal/external events firing and/or time advancing before they can provide a response. Resource contention and arbitration can be modeled easily with this style. Since these models must synchronize/order the transactions before processing them, they are forced to trigger multiple context switches in the simulation, resulting in performance penalties."*

Hardware synthesis starting from SystemC has become available [207, 208]. A synthesizable subset of the language has been defined [7]. There are also commercial synthesis offerings. Commercial offerings are expected to support the synthesizable subset as a minimum. Methodology and applications for SystemC-based design are described in a book on that topic [390]. At the time of writing, the most recent version of SystemC is SystemC 2.3.1 [6].

2.7.6   VHDL

2.7.6.1   Introduction

VHDL is another language which is based on the discrete event paradigm. In contrast to SpecC and SystemC, it does not support a clear distinction between communication and computation, making reuse of components somewhat more difficult. However, VHDL is supported by many industrial and academic tools and is in widespread use. It is an example of a hardware description language (HDL). Having presented an initial example of event-based modeling already on p. 84, we would like to delve deeper into VHDL.

VHDL uses **processes** for modeling concurrency. Each process models one component of the potentially concurrent hardware. For simple hardware components, a single process may be sufficient. More complex components may need several processes for modeling their operations. Processes communicate through **signals**. Signals roughly correspond to physical connections (wires).

The origin of VHDL can be traced back to the 1980s. At that time, most design systems used graphical HDLs. The most common building block was the gate. However, in addition to using graphical HDLs, we can also use textual HDLs. The strength of textual languages is that they can easily represent complex computations including variables, loops, function parameters, and recursion. Accordingly, when digital systems became more complex in the 1980s, textual HDLs almost completely replaced graphical HDLs. Textual HDLs were initially a research topic at universities. See Mermet et al. [374] for a survey of languages designed in Europe at that time. MIMOLA was one of these languages, and the author of this book contributed to its design and applications [357, 362]. Textual languages became popular when VHDL and its competitor Verilog (see p. 104) were introduced.

VHDL was designed in the context of the VHSIC program of the Department of Defense (DoD) in the USA. VHSIC stands for *very high-speed integrated circuits*[17]. Initially, the design of VHDL (VHSIC hardware description language) was done by three companies: IBM, Intermetrics, and Texas Instruments. A first version of VHDL was published in 1984. Later, VHDL became an IEEE standard, called IEEE 1076. The first IEEE version was standardized in 1987; updates were published in 1993, 2000, 2002, and 2008 [229, 231–233, 235][18]. VHDL-AMS [236] allows modeling analog and mixed-signal systems by including differential equations in the language. The design of VHDL used Ada (see p. 106) as the starting point, since both languages were designed for the DoD. Since Ada is based on PASCAL, VHDL has some of the syntactical flavor of PASCAL. However, the syntax of VHDL is much more complex and it is necessary not to get distracted by the syntax. In the current book, we will just focus on some concepts of VHDL which are useful also in other languages. A full description of VHDL is beyond the scope of this book. The standard is available from IEEE (see, for example, [235]).

2.7.6.2 Entities and Architectures

VHDL, like all other HDLs, includes support for modeling concurrent operation of hardware components. Components are modeled by so-called **design entities** or **VHDL entities**. Entities contain **processes** used to model concurrency. According to the VHDL grammar, design entities are composed of two types of ingredients: an **entity declaration** and one (or several) **architectures** (see Fig. 2.67).

Fig. 2.67 Entity consisting of an entity declaration and architectures

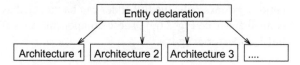

---

[17]The design of the Internet was also part of the VHSIC program.

[18]The next update can be expected for 2017.

For each entity, the most recently analyzed architecture will be used by default. The use of other architectures can be specified. Architectures may contain several processes.

Example 2.30: We will discuss a full adder as an example. Full adders have three input ports and two output ports (see Fig. 2.68).

Fig. 2.68 Full adder and its interface signals

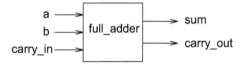

An entity declaration corresponding to Fig. 2.68 is the following:

```
entity full_adder is                            -- entity declaration
   port (a, b, carry_in: in Bit;                       -- input ports
         sum, carry_out: out Bit);                     -- output ports
end full_adder;
```

Two hyphens (--) are starting comments. They extend until the end of the line.   ▽

Architectures consist of architecture headers and architectural bodies. We can distinguish between different styles of bodies, in particular between structural and behavioral bodies. We will show how the two are different using the full adder as an example. Behavioral bodies include just enough information to compute output signals from input signals and the local state (if any), including the timing behavior of the outputs.

Example 2.31: The following is an example of this:

```
architecture behavior of full_adder is               -- architecture
begin
   sum       <= (a xor b) xor carry_in after 10 ns;
   carry_out <= (a and b) or (a and carry_in) or
                     (b and carry_in) after 10 ns;
end behavior;
```

VHDL-based simulators are capable of displaying output signal waveforms resulting from stimuli applied to the inputs of the full adder described above.

In contrast, structural bodies describe the way entities are composed of simpler entities. For example, the full adder can be modeled as an entity consisting of three components (see Fig. 2.69). These components are called i1 to i3 and are of type half_adder or or_gate.

Fig. 2.69 Schematic describing structural body of the full adder

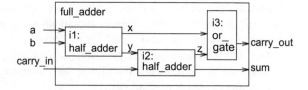

In the 1987 version of VHDL, these components must be declared in a so-called component declaration. This declaration is very similar to (and it serves the same purpose) as forward declarations in other languages. This declaration provides the necessary information about the component even if the full description of that component is not yet stored in the VHDL database (this may happen in the case of so-called top-down designs). From the 1992 version of VHDL onward, such declarations are not required if the relevant components are already stored in the component database.

Connections between local component and entity ports are described in **port maps**. The following VHDL code represents the structural body shown in Fig. 2.69:

```
architecture structure of full_adder is               -- architecture head
  component half_adder
    port (in1, in2: in Bit; carry: out Bit; sum: out Bit);
  end component;
  component or_gate
    port (in1, in2: in Bit; o: out Bit);
  end component;
  signal x, y, z: Bit;                                 --local signals
  begin                                                -- port map section
  i1: half_adder                          -- introduction of half_adder i1
     port map (a, b, x, y);                     --connections between ports
  i2: half_adder port map (y, carry_in, z, sum);
  i3: or_gate       port map (x, z, carry_out);
  end structure;
```

▽

2.7.6.3 Assignments

Example 2.31 contains several assignments. Let us look at assignments more closely! Assignments are special cases of statements. In VHDL, there are two kinds of assignments:

• **Variable assignments**: The syntax of variable assignments is

variable := expression

Whenever control reaches such an assignment, the expression is computed and assigned to the variable. Such assignments behave like assignments in common programming languages.

- **Signal assignments**: Signal assignments (as mentioned already on pp. 83 and 95) are evaluated concurrently. Signals and signal assignments are introduced in an attempt to model electrical signals in real hardware systems. Signals associate values with instances in time. In VHDL, such a mapping from time to values is represented by **waveforms**. Waveforms are computed from signal assignments. The syntax of signal assignments is

```
signal <= expression;
signal <= transport expression after delay;
signal <= expression after delay;
signal <= reject time inertial expression after delay;
```

Whenever control reaches such an assignment, the expression is computed and used to extend predicted future values of the waveform. In VHDL, each signal is associated with a so-called signal **driver**. Computing the value resulting from the contributions of multiple drivers to the same signal is called **resolution,** and resulting values are computed by functions called **resolution functions**. In this way, the *sup* function mentioned in the context of CSA theory is implemented if signals are connected.

In order to compute future values, **simulators are assumed to include a queue of events to happen later than the current simulated time**. This queue is sorted by the time at which future events (e.g., updates of signals) should happen. Executing a signal assignment results in the creation of entries in this queue. Each entry contains a time for executing the event, the affected signal, and the value to be assigned. For signal assignments not containing any **after** clause (first syntactical form), the entry will contain the current simulation time as the time at which this assignment has to be performed. In this case, the change will take place after an infinitesimally small amount of time, called δ-delay (see below). This allows us to update signals without changing macroscopic time.

For signal assignments containing a **transport** prefix (second syntactical form), the update of the signal will be delayed by the specified amount. This form of the assignment is following the so-called **transport delay model**. This model is based on the behavior of simple wires: Wires are (as a first order of approximation) delaying signals. Even short pulses propagate along wires. The transport delay model can be used for logic circuits, even though its main application is to model wires.

Example 2.32: Suppose that we model a simple OR gate using a transport delay signal assignment:

```
c <= transport a or b after 10 ns;
```

Such a model would propagate even short pulses (see Fig. 2.70).

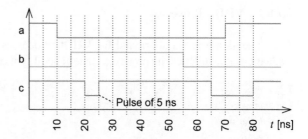

Fig. 2.70 Gate modeled with transport delay

Output signal c includes a short pulse of 5 ns, which would be suppressed for a transport delay model.   ∇

Transport delay signal assignments will delete all entries in the queue corresponding to the time of the computed update or later times (if we first execute an assignment with a rather large delay and then execute an assignment with a smaller delay, then the entry resulting from the first assignment will be deleted).

For signal assignments containing an `after` clause, but no `transport` clause, **inertial delay** is assumed. The inertial delay model reflects the fact that real circuits come with some "inertia." This means that short spikes will be suppressed. For the third syntactical form of the signal assignment, all signals changes which are shorter than the specified delay are suppressed. For the fourth form, all signal changes which are shorter than the indicated amount are removed from the predicted waveform. The subtle rules for removals are not repeated here.

Example 2.33: Suppose that we model a simple OR gate using inertial delay:

```
c <= a or b after 10 ns;
```

For such a model, short spikes would be suppressed (see Fig. 2.71).

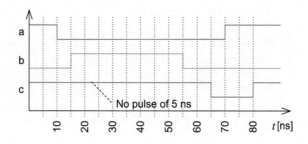

Fig. 2.71 Gate modeled with inertial delay

For output signal c, there is no short pulse of 5 ns, but the 15 ns pulse arrives at the output.   ∇

2.7.6.4   VHDL Processes

Assignments are just a shorthand for VHDL processes. More control over signal evaluations is available with processes. The general syntax for processes is as follows:

```
label:                                                              -- optional
process
   declarations                                                     --optional
   begin
      statements                                                    --optional
   end process;
```

In addition to assignments, processes may contain **wait** statements. Such statements can be used to explicitly suspend a process. There are the following kinds of **wait** statements:

```
wait on  signal list; --suspend until one of the signals in the list changes;
wait until  condition;          --suspend until condition is met, e.g.,  a ='1';
wait for  duration;             --suspend for a specified period of time;
wait;                                          --suspend indefinitely.
```

As an alternative to explicit **wait** statements, a list of signals can be added to the process header. In that case, the process is activated whenever one of the signals in that list changes its value.

Example 2.34: The following model of an AND gate will execute its body once and will restart from the beginning every time one of the inputs changes its value:

```
process(x, y) begin
   prod <= x  and y ;
end process;
```

This model is equivalent to

```
process begin
   prod <= x  and y ;
   wait on x,y;
end process;
```

where there is an explicit **wait** statement at the end.   ∇

2.7.6.5   The VHDL Simulation Cycle

According to the original standards document [229], the execution of a VHDL model is described as follows: *"The execution of a model consists of an **initialization phase** followed by the **repetitive execution of process statements** in the description of that model. Each such repetition is said to be a **simulation cycle**. In each cycle, the values* *of all signals in the description are computed. If as a result of this computation an event occurs on a given signal, process statements that are sensitive to that signal will resume and will be executed as part of the simulation cycle."*

The initialization phase takes signal initializations into account and executes each process once. It is described in the standards as follows[19]:

"At the beginning of initialization, the current time, T_c, is assumed to be 0 ns. The initialization phase consists of the following steps[20]:

- *The driving value and the effective value of each explicitly declared signal are computed, and the current value of the signal is set to the effective value. This value is assumed to have been the value of the signal for an infinite length of time prior to the start of the simulation. ...*
- *Each ... process in the model is executed until it suspends. ...*
- *The time of the next simulation cycle (which in this case is the first simulation cycle), T_n, is calculated according to ... step e of the simulation cycle, below."*

Each simulation cycle starts with setting the current time to the next time at which changes must be considered. This time T_n was either computed during the initialization or during the last execution of the simulation cycle. Simulation terminates when the current time reaches its maximum, $TIME'HIGH$. The standard describes the simulation cycle as follows: *"A simulation cycle consists of the following steps:*

(a) *The current time, T_c, is set equal to T_n. Simulation is complete when $T_n = TIME'HIGH$ and there are no active drivers or process resumptions at T_n.*

(b) *Each active explicit signal in the model is updated. (Events may occur as a result.)"* ...

In the cycle preceding the current cycle, future values for some signals have been computed. If T_c corresponds to the time at which these values become valid, they are now assigned. Values of newly computed signals are not assigned before the next simulation cycle, at the earliest. Signals that change their value generate events which, in-turn, may release processes that are sensitive to that signal.

(c) *"For each process P, if P is currently sensitive to a signal S and if an event has occurred on S in this simulation cycle, then P resumes.*

(d) *Each ... process that has resumed in the current simulation cycle is executed until it suspends.*

(e) *T_n (the time of the next simulation cycle) is set to the earliest of*

1. *TIME'HIGH (this is the end of simulation time).*
2. *The next time at which a driver becomes active (this is the next instance in time, at which a driver specifies a new value), or*
3. *The next time at which a process resumes (as computed from **wait for** statements).*

If $T_n = T_c$, then the next simulation cycle (if any) will be a delta cycle."

The iterative nature of simulation cycles is shown in Fig. 2.72.

---

[19] We leave out the discussion of implicitly declared signals and so-called postponed processes.

[20] Some sections of the standard are omitted in the citation (indicated by "...").

Fig. 2.72 VHDL simulation cycles

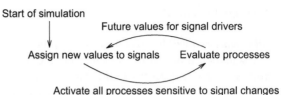

Delta (δ) simulation cycles have been the source of discussions. They introduce an infinitesimally small delay if the user did not specify any.

Example 2.35: Let us come back to our latch example and look more closely at timing. Figure 2.73 shows the latch again, this time using standard schematic symbols.

Fig. 2.73 RS Flip-flop

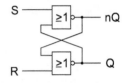

The flip-flop is modeled in VHDL as follows:

```
entity RS_Flipflop is
  port (R: in    BIT;                                    -- reset
        S: in    BIT;                                    -- set
        Q: inout BIT;                                    -- output
        nQ: inout BIT; );                                -- Q-bar
end RS_Flipflop;
architecture one of RS_Flipflop is
  begin
    process: (R,S,Q,nQ)
      begin
        Q  <= R nor nQ;  nQ <= S nor Q;
      end process;
  end one;
```

Ports Q and nQ must be of mode **inout** since they are also read internally, which would not be possible if they were of mode **out**. Table 2.5 shows the simulation times at which signals are updated for this model. During each cycle, updates are

Table 2.5 δ cycles for RS flip-flop

	<0ns	0ns	0ns+δ	0ns+2*δ	0ns+3*δ
R	0	1	1	1	1
S	0	0	0	0	0
Q	1	1	0	0	0
nQ	0	0	0	1	1

propagated through one of the gates. Simulation terminates after three δ cycles. The last cycle does not change anything, since Q is already '0'.   ∇

δ cycles correspond to an infinitesimally small unit of time, which will always exist in reality. δ cycles ensure that simulation respects causality.

The results do not depend on the order in which parts of the model are executed by the simulation. This feature is enabled by the separation between the computation of new values for signals and their actual assignment. In a model containing the lines

```
a <= b;
b <= a;
```

signals a and b will always be swapped. If the assignments were performed immediately, the result would depend on the order in which we execute the assignments (see also p. 53). **VHDL models are therefore determinate**. This is what we expect from the simulation of a real circuit with a fixed behavior.

There can be arbitrarily many δ cycles before the current time T_c is advanced. This possibility of infinite loops can be confusing. One of the options of avoiding this possibility would be to disallow zero delays, which we used in our model of the flip-flop.

The propagation of values using signals also allows an easy implementation of the observer pattern (see p. 31). In contrast to SDF, the number of observers can vary, depending on the number of processes waiting for changes on a signal.

What is the communication model behind VHDL? The description of the semantics of VHDL relies heavily on a **single, centralized** queue of future events, storing values of all signals in the future. The purpose of this queue is **not** to implement asynchronous message passing. Rather, this queue is supposed to be accessed by the simulation kernel, one entry at a time, in a non-distributed fashion. Attempts to perform distributed VHDL simulations are typically suffering from a poor performance. All modeled components can access values of signals and variables which are in their scope without any message-based communication. Therefore, we tend toward associating VHDL with a shared memory-based implementation of the communication. However, FIFO-based message passing could be implemented in VHDL on top of the VHDL simulator as well.

2.7.6.6   IEEE 1164

In VHDL, there is no predefined number of signal values, except for some basic support for two-valued logic. Instead, the used value sets can be defined in VHDL itself and different VHDL models can use different value sets.

However, portability of models would suffer in a very severe manner if this capability of VHDL was applied in this way. In order to simplify exchanging VHDL models, a standard value set was defined and standardized by the IEEE. This standard is called IEEE 1164 and is employed in many system models. IEEE 1164 has nine values: {'0','1','L','H','X','W','Z', 'U','-'}. The first seven values correspond to the seven signal values described from pp. 84 to 89. 'U' denotes an uninitialized value. It is used by simulators for signals that have not been explicitly initialized.

'-' denotes the **input don't care**. This value needs some explanation. Frequently, hardware description languages are used for describing Boolean functions. The VHDL **select** statement is a very convenient means for doing that. The **select** statement corresponds to **switch** and **case** statements found in other languages, and its meaning is different from the **select** statement in Ada (see p. 107).

Example 2.36: Suppose that we would like to represent the Boolean function

$$f(a, b, c) = a\overline{b} + bc$$

Furthermore, suppose that f should be undefined for the case of $a = b = c = '0'$. A very convenient way of specifying this function would be the following:

```
f <= select a & b & c                    -- & denotes concatenation
       '1' when "10-"                     -- corresponds to first term
       '1' when "-11"                     -- corresponds to second term
       'X' when "000"
```

This way, functions given above could be easily translated into VHDL. Unfortunately, the **select** statement denotes something completely different. Since IEEE 1164 is just one of a large number of possible value sets, it does not include any knowledge about the "meaning" of '-'. Whenever VHDL tools evaluate select statements such as the one above, they check if the selecting expression (a & b & c in the case above) is equal to the values in the when-clauses. In particular, they check if, e.g., a & b & c is equal to "10-". In this context, '-' behaves like any other value: VHDL systems check if c has a value of '-'. Since '-' is never assigned to any of the variables, these tests will never be true.   ∇

Therefore, '-' is of limited benefit. The non-availability of convenient input don't care values is the price that one has to pay for the flexibility of defining value sets in VHDL itself[21].

The nice property of the general discussion on pp. 84 to 89 is the following: It allows us to immediately draw conclusions about the modeling power of IEEE 1164. The IEEE standard is based on the seven-valued value set described on p. 85, and therefore, is capable of modeling circuits containing depletion transistors. It is, however, not capable of modeling charge storage[22].

2.7.7 Verilog and SystemVerilog

Verilog is another hardware description language. Initially, it was a proprietary language, but it was later standardized as IEEE standard 1364, with versions called IEEE

---

[21] This problem was corrected in VHDL 2006 [326].

[22] As an exception, if the capability of modeling depletion transistors or pull-up resistors is not needed, one could interpret weak values as stored charges. This is, however, not very practical since pull-up resistors are found in most actual systems.

standard 1364–1995 (Verilog version 1.0) and IEEE standard 1364–2001 (Verilog 2.0). Some features of Verilog are quite similar to VHDL. Just like in VHDL, designs are described as a set of connected design entities, and design entities can be described behaviorally. Also, processes are used to model concurrency of hardware components. Just like in VHDL, bitvectors and time units are supported. There are, however, some areas in which Verilog is less flexible and focuses more on comfortable built-in features. For example, standard Verilog does not include the flexible mechanisms for defining enumerated types such as the ones defined in the IEEE 1164 standard. However, support for four-valued logic is built into the Verilog language, and the standard IEEE 1364 also provides multiple-valued logic with eight different signal strengths. Multiple-valued logic is more tightly integrated into Verilog than into VHDL. The Verilog logic system also provides more features for transistor-level descriptions. However, VHDL is more flexible. For example, VHDL allows hardware entities to be instantiated in loops. This can be used to generate a structural description for, e.g., n-bit adders without having to specify n adders and their interconnections manually.

Verilog has a similar number of users as VHDL. While VHDL is more popular in Europe, Verilog is more popular in the USA.

Verilog versions 3.0 and 3.1 are also known as SystemVerilog. They include numerous extensions to Verilog 2.0. These extensions include [237, 494]:

- additional language elements for modeling behavior,
- C data types such as `int` and type definition facilities such as `typedef` and `struct`,
- definition of interfaces of hardware components as separate entities,
- standardized mechanism for calling C/C++ functions and, to some extent, to call built-in Verilog functions from C,
- significantly enhanced features for describing an environment (called test bench) for the hardware circuit under design (called CUD), and for using the test bench to validate the CUD by simulation,
- classes known from object-oriented programming for use within test benches,
- dynamic process creation,
- standardized inter-process communication and synchronization, including semaphores,
- automatic memory allocation and deallocation,
- language features that provide a standardized interface to formal verification (see p. 2 31).

Due to the capability of interfacing with C and C++, interfacing to SystemC models is also possible. Improved facilities for simulation- as well as for formal verification-based design validation and the possible interfacing to SystemC will potentially create a very good acceptance. Verilog and SystemVerilog have been merged into one standard, IEEE 1800–2009 [234].

2.8 Von-Neumann Languages

The sequential execution and explicit control flow of von-Neumann languages are their common characteristics. Also, such languages allow an almost unrestricted access to global variables and we may need explicit communication and synchronization. Model-based design using CFSMs and computational graphs is very appropriate for embedded system design. Nevertheless, the use of standard von-Neumann languages is still widespread. Therefore, we cannot ignore these languages.

Also, the distinction between models such as KPNs and properly restricted von-Neumann languages is blurring. For KPNs, we do also have sequential execution of the code for each of the nodes. We are still keeping the distinction between KPN and von-Neumann languages since the KPN style of modeling has its advantages like determinate execution.

For the first two languages covered next, communication is built into the languages. For the remaining languages, focus is on the computations and communication can be replaced by selecting different libraries.

2.8.1 CSP

CSP (*communicating sequential processes*) [209] is one of the first languages comprising mechanisms for inter-process communication. Communication is based on channels.

Example 2.37: Consider input/output for channel c in this example:

```
process A                          process B

. . . . .                          . . . . . .

var a ..                           var b ...
  a := 3;                            . . .
  c!a; -- output to channel c        c?b; -- input from channel c
end;                               end;
```

Both processes will wait for the other process to arrive at the input or output statement. This is a case of **rendez-vous**-based, **blocking**, or **synchronous message passing**.   ∇

CSP is determinate, since it relies on the commitment to wait for input from a particular channel, like in Kahn process networks.

CSP has laid the foundation for the OCCAM language that was proposed as a programming language of the **transputer** [378]. The focus on communication channels has been picked up again in the design of the XS1 processor [575].

2.8.2 Ada

During the 1980s, the Department of Defense (DoD) in the USA realized that the dependability and maintainability of the software in its military equipment could soon become a major source of problems, unless some strict policy was enforced. It was decided that all software should be written in the same real-time language. Requirements for such a language were formulated.

No existing language met the requirements, and, consequently, the design of a new one was started. The language which was finally accepted was based on PASCAL. It was called Ada (after Ada Lovelace, regarded as being the first (female) programmer). Ada'95 [81, 274] is an object-oriented extension of the original standard.

One of the interesting features of Ada is the ability to have nested declarations of processes (called tasks in Ada). Tasks are started whenever control passes into the scope in which they are declared.

Example 2.38: The following code has been adopted from Burns et al. [81]:

```
procedure example1 is
   task a;
   task b;
   task body a is
      -- local declarations for a
      begin
        -- statements for a
      end a;
   task body b is
      -- local declarations for b
      begin
        -- statements for b
      end b;
  begin
   -- body of procedure example1
  end;
```

Tasks a and b will start before the first statement of the code of example1.   ∇

The communication concept of Ada is another key concept. It is based on the synchronous ***rendez-vous*** paradigm. Whenever two tasks want to exchange information, the task reaching the "meeting point" first has to wait until its partner has also reached a corresponding point of control. Syntactically, procedures are used for describing communication. Procedures which can be called from other tasks must be identified by the keyword **entry**.

Example 2.39: This code has also been adopted from Burns et al. [80]:

```
task screen_out is
  entry call (val : character; x, y : integer);
end screen_out;
```

Task `screen_out` includes a procedure named `call` which can be called from other processes. Some other task can call this procedure by prefixing it with the name of the task:

```
screen_out.call('Z',10,20);
```

The calling task has to wait until the called task has reached a point of control, at which it accepts calls from other tasks. This point of control is indicated by the keyword **accept**:

```
task body screen_out is
  ...
begin
    ...
    accept call (val : character; x, y : integer) do
    ...
    end call;
  ...
end screen_out;
```

Obviously, task `screen_out` may be waiting for several calls at the same time. The Ada **select** statement provides this capability:

```
task screen_output is
  entry call_ch(val:character; x, y: integer);
  entry call_int(z, x, y: integer);
end screen_out;
task body screen_output is
  ...
  select
    accept call_ch ...  do...
    end call_ch;
  or
    accept call_int ... do ..
    end call_int;
  end select;
  ...
```

In this case, task `screen_out` will be waiting until either `call_ch` or `call_int` is called.   ∇

Due to the presence of the **select** statement, Ada is not determinate. Ada has been the preferred language for military equipment produced in the Western hemisphere for some time. Information about Ada is available from a number of Web sites (see, for example, [275]).

2.8.3 Java

For Java, communication can be selected by choosing between different libraries. Computation is strictly sequential.

Java was designed as a platform-independent language. It can be executed on any machine for which an interpreter of the internal byte-code representation of Java programs is available. This byte-code representation is a very compact representation, which requires less memory space than a standard binary machine code representation. Obviously, this is a potential advantage in system-on-a-chip applications, where memory space is limited.

Also, Java was designed to be a safe language. Many potentially dangerous features of C or C++ (like pointer arithmetic) are not available in Java. Java supports exception handling, simplifying recovery in case of run-time errors. There is no danger of memory leakages due to missing memory deallocation, since Java provides automatic garbage collection. This feature avoids potential problems in applications that must run for months or even years without ever being restarted. Java also meets the requirement to support concurrency since it includes threads (lightweight processes).

In addition, Java applications can be implemented quite fast, since Java supports object orientation and since Java development systems come with powerful libraries.

However, standard Java is not really designed for real-time and embedded systems. A number of characteristics which would make it a real-time and embedded programming language are missing:

- The size of Java run-time libraries has to be added to the size of the application itself. These run-time libraries can be quite large. Consequently, only really large applications benefit from the compact representation of the application itself.
- For many embedded applications, direct control over I/O devices is necessary (see p. 30). For safety reasons, no direct control over I/O devices is available in standard Java.
- Automatic garbage collection requires some computing time. In standard Java, the instance in time at which automatic garbage collection is started cannot be predicted. Hence, the worst-case execution time is very difficult to predict. Only extremely conservative estimates can be made.
- Java does not specify the order in which threads are executed if several threads are ready to run. As a result, worst-case execution time estimates must be even more conservative.
- Java programs are typically less efficient than C programs. Hence, Java is less recommended for resource constrained systems.

Proposals for solving the problems were made by Nilsen [402]. Proposals include hardware-supported garbage collection, replacement of the run-time scheduler, and tagging of some of the memory segments.

Currently (2017), relevant Java programming environments include the Java Enterprise Edition (J2EE), the Java Standard Edition (J2SE), the Java Micro

Edition (J2ME), and CardJava [492]. CardJava is a stripped-down version of Java with emphasis on security for SmartCard applications. J2ME is the relevant Java environment for all other types of embedded systems. Two library profiles have been defined for J2ME: CDC and CLDC. CLDC is used for mobile phones, using the so-called MIDP 1.0/2.0 as its standard for the application programming interface (API). CDC is used, for example, for TV sets and powerful mobile phones. Currently, relevant sources for Java real-time programming include book by Wellings [549], Dibble [126] and Bruno [73] as well as Web sites [16, 258] and the annual JTRES on "Java Technologies for Real-time and Embedded Systems" (see http://jtres2016.compute.dtu.dk/ for the latest edition).

2.8.4 Communication Libraries

Standard von-Neumann languages do not come with built-in communication primitives. However, communication can be provided by libraries. There is a trend toward supporting communication within some local system as well as communication over longer distances. The use of Internet Protocols is becoming more popular.

2.8.4.1 MPI

Multi-core programming with imperative programs is possible with the message passing interface MPI. MPI is a very frequently used library, initially designed for high-performance computing. It allows a choice between synchronous and asynchronous message passing. For example, synchronous message passing is possible with the MPI_Send library function [376]:

MPI_Send(buffer,count,type,dest,tag,comm) where

- buffer is the address of data to be sent,
- count is the number of data elements to be sent,
- type is the data type of data to be sent (e.g., MPI_CHAR, MPI_SHORT, MPI_INT),
- dest is the process id of the target process,
- tag is a message id (for sorting incoming messages),
- comm is the communication context (set of processes for which destination field is valid) and
- function result indicates success.

The following is an asynchronous library function:

MPI_Isend(buffer,count,type,dest,tag,comm,request) where

- buffer, count, type, dest, tag, comm are same as above, and
- the system issues a unique "request number." The programmer uses this system assigned "handle" later (in a WAIT type routine) to determine completion of the non-blocking operation.

For MPI, the partitioning of computations among various processors must be done explicitly and the same is true for the communication and the distribution of data. Synchronization is implied by communication, but explicit synchronization is also possible. As a result, much of the management code is explicit and causes a major amount of work for the programmer. Also, it does not scale well when the number of processors is significantly changed [530].

In order to apply the MPI style of communication to real-time systems, a real-time version of MPI, called MPI/RT, has been defined [476]. MPI/RT does not cover issues such as thread creation and termination. MPI/RT is conceived as a potential layer between the operating system and standard (non-real-time) MPI.

MPI is available on a variety of platforms and also considered for multiple processors on a chip. However, it is based on the assumption that memory accesses are faster than communication operations. Also, MPI is mainly targeting at homogeneous multiprocessors. These assumptions are not true for multiple processors on a chip.

MPI has recently been extended to cover shared memory-based communication as well.

2.8.4.2   OpenMP

OpenMP is a compiler-based solution for shared memory-based communication. For OpenMP, parallelism is mostly explicit, whereas computation partitioning, communication, synchronization, etc. are implicit. Parallelism is expressed with pragmas: For example, loops can be preceded by pragmas indicating that they should be parallelized.

Example 2.40: The following program demonstrates a small parallel loop [417]:

```
void a1(int n, float *a, float *b)
{int i;
#pragma omp parallel for
  for (i=1; i<n; i++) /* i is private by default */
    b[i] = (a[i] + a[i-1]) / 2.0;
}
```

Note that a simple pragma is sufficient to indicate parallel programming.   ∇

This means that OpenMP requires a relatively small amount of effort for parallelization for the user. However, this also means that the user cannot control partitioning [530]. There are some applications for MPSoCs (see, for example, Marian et al. [350]).

More techniques for multi-core programming will be described in the section on system software (see p. 225).

2.8.5   Additional Languages

Pearl [122] was designed for industrial control applications. It does include a large repertoire of language elements for controlling processes and referring to time. It requires an underlying real-time operating system. Pearl has been very popular in Europe, and a large number of industrial control projects have been implemented in Pearl. Pearl supports semaphores which can be used to protect communication based on shared buffers.

Chill [564] was designed for telephone exchange stations. It was standardized by the CCITT and used in telecommunication equipment. Chill is a kind of extended PASCAL.

IEC 60848 [223] and STEP 7 [49] are specialized languages that are used in control applications. Both provide graphical elements for describing the system functionality.

2.9   Levels of Hardware Modeling

In practice, designers start design cycles at various levels of abstraction. In some cases, these are high levels describing the overall behavior of the system to be designed. In other cases, the design process starts with the specification of electrical circuits at lower levels of abstraction. For each of the levels, a variety of languages exists, and some languages cover various levels. In the following, we will describe a set of possible levels. Some lower end levels are presented here for context reasons. Specifications should not start at those levels. The following is a list of frequently used names and attributes of levels:

- **System-level models**: The term system level is not clearly defined. It is used here to denote the entire embedded system and the system into which information processing is embedded ("the product"), and possibly also the environment (the physical input to the system, reflecting, e.g., the roads and weather conditions). Obviously, such models include mechanical as well as information processing aspects and it may be difficult to find appropriate simulators. Possible solutions include VHDL-AMS (the analog extension to VHDL), Verilog-AMS, SystemC, Modelica, COM-SOL (see https://www.comsol.com/), or MATLAB/Simulink. MATLAB/Simulink and VHDL-AMS support modeling partial differential equations, which is a key requirement for modeling mechanical systems. It is a challenge to model information processing parts of the system in such a way that the simulation model can also be used for the synthesis of the embedded system. If this is not possible, error-prone manual translations between different models may be needed.
- **Algorithmic level**: At this level, we are simulating the algorithms that we intend to use within the embedded system. For example, we might be simulating MPEG video encoding algorithms in order to evaluate the resulting video quality. For such simulations, no reference is made to processors or instruction sets.

Data types may still allow a higher precision than the final implementation. For example, MPEG standards use double precision floating point numbers. The final embedded system will hardly include such data types. If data types have been selected such that every bit corresponds to exactly one bit in the final implementation, the model is said to be **bit-true**. Translating non-bit-true into bit-true models should be done with tool support (see p. 344).

Models at this level may consist of single processes or of sets of cooperating processes.

- **Instruction set level**: In this case, algorithms have already been compiled for the instruction set of the processor(s) to be used. Simulations at this level allow counting the executed number of instructions. There are several variations of the instruction set level:

  - In a coarse-grained model, only the effect of the instructions is simulated and their timing is not considered. The information available in assembly reference manuals (instruction set architecture (ISA)) is sufficient for defining such models.
  - **Transaction-level modeling**: In transaction-level modeling (see also p. 89), transactions, such as bus reads and writes, and communication between different components are modeled. Transaction-level modeling includes fewer details than cycle-true modeling (see below), enabling significantly superior simulation speeds [105].
  - In a more fine-grained model, we might have **cycle-true instruction set simulation**. In this case, the exact number of clock cycles required to run an application can be computed. Defining cycle-true models requires a detailed knowledge about processor hardware in order to correctly model, for example, pipeline stalls, resource hazards, and memory wait cycles.

- **Register-transfer level (RTL)**: At this level, we model all the components at the register-transfer level, including arithmetic/logic units (ALUs), registers, memories, multiplexers, and decoders. Models at this level are always cycle-true. Automatic synthesis from such models is not a major challenge.
- **Gate-level models**: In this case, models contain gates as the basic components. Gate-level models provide accurate information about signal transition probabilities and can therefore also be used for power estimations. Also delay calculations can be more precise than for the RTL. However, typically no information about the length of wires and hence no information about capacitances is available. Hence, delay and power consumption calculations are still estimates.

The term "gate-level model" is sometimes also employed in situations in which gates are only used to denote Boolean functions. Gates in such a model do not necessarily represent physical gates; we are only considering the behavior of the gates, not the fact that they also represent physical components. More precisely, such models should be called "Boolean function models[23]," but this term is not frequently used.

---

[23] These models could be represented with binary decision diagrams (BDDs) [546].

- **Switch-level models**: Switch-level models use switches (transistors) as their basic components. Switch-level models use digital values models (refer to p. 84 for a description of possible value sets). In contrast to gate-level models, switch-level models are capable of reflecting bidirectional transfer of information. Switch-level models can be simulated with ternary simulation [74].
- **Circuit-level models**: Circuit theory and its components (current and voltage sources, resistors, capacitances, inductances, and frequently possible macro-models of semiconductors) form the basis of simulations at this level. Simulations involve partial differential equations. These equations are linear if and only if the behavior of semiconductors is linearized (approximated). The most frequently used simulator at this level is SPICE [533] and its variants.
- **Layout models**: Layout models reflect the actual circuit layout. Such models include **geometric** information. Layout models cannot be simulated directly, since the geometric information does not directly provide information about the behavior. Behavior can be deduced by correlating the layout model with a behavioral description at a higher level or by extracting circuits from the layout, using knowledge about the representation of circuit components at the layout level.

  In a typical design flow, the length of wires and the corresponding capacitances are extracted from the layout and **back-annotated** to descriptions at higher levels. This way, more precision can be gained for delay and power estimations. Also, layout information may be essential for thermal modeling.
- **Process and device models**: At even lower levels, we can model fabrication processes. Using information from such models, we can compute parameters (gains, capacitances, etc.) for devices (transistors). Due to a growing complexity of the fabrication process, these models are also becoming more complex.

2.10   Comparison of Models of Computation

2.10.1   Criteria

Models of computation can be compared according to several criteria. For example, Stuijk [491] compares MoCs according to the following criteria:

- **Expressiveness** and **succinctness** indicate which systems can be modeled and how compact they are.
- **Analyzability** relates to the availability of schedulability tests and scheduling algorithms. Also, analyzability is affected by the need for run-time support.
- The **implementation efficiency** is influenced by the required scheduling policy and the code size.

Figure 2.74 classifies data flow models according to these criteria.

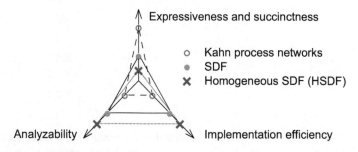

Fig. 2.74 Comparison between data flow models

This figure reflects the fact that Kahn process networks are expressive: They are Turing complete, meaning that any problem which can be computed on a Turing machine can also be computed in a KPN. Turing machines are used as the standard model of universal computers [206]. However, termination properties and upper bounds on buffer sizes of KPNs are difficult to analyze. While Kahn process networks are Turing complete, cyclo-static data flow (CSDF, see p. 70) is not Turing complete. Also, SDF graphs are not Turing complete. The underlying reason is that they cannot model control flow. However, deadlock properties and upper bounds on buffer sizes of SDF graphs are easier to analyze. Homogeneous SDF (HSDF) graphs (graphs for which all rates are equal to one) are even less expressive, but also easier to analyze.

We could compare MoCs also with respect to the type of processes supported:

- The **number of processes** can be either **static** or **dynamic**. A static number of processes simplifies the implementation and is sufficient if each process models a piece of hardware and if we do not consider "hot-plugging" (dynamically changing the hardware architecture). Otherwise, dynamic process creation (and termination) should be supported.
- Processes can either be statically **nested** or all declared at the same level. For example, StateCharts allows nested process declarations while SDL (see p. 58) does not. Nesting provides encapsulation of concerns.
- Different techniques for **process creation** exist. Process creation can result from an elaboration of the process declaration in the source code, through the fork and join mechanism (supported for example in Unix), and also through explicit process creation calls.

The expressiveness of different data flow-oriented models of computation is also shown in Fig. 2.75 [43]. MoCs not discussed in this book are indicated by dashed lines.

Fig. 2.75 Expressiveness of
data flow models

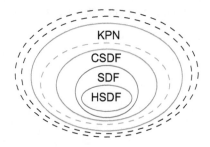

None of the MoCs and languages presented so far meets all the requirements for specification languages for embedded systems. Table 2.6 presents an overview over some of the key properties of some of the languages.

Table 2.6 Language comparison

Language	Behavioral hierarchy	Structural hierarchy	Programming language elements	Exceptions supported	Dynamic process creation
StateCharts	+	−	−	+	−
VHDL	+	+	+	−	−
SDL	+−	+−	+−	−	+
Petri nets	−	−	−	−	+
Java	+	−	+	+	+
SpecC	+	+	+	+	+
SystemC	+	+	+	+	+
Ada	+	−	+	+	+

Interestingly, SpecC and SystemC meet all listed requirements. However, some other requirements (like a precise specification of deadlines) are not included. It is not very likely that a single MoC or language will ever meet all requirements, since some of the requirements are essentially conflicting. A language supporting hard real-time requirements may well be inconvenient to use for less strict real-time requirements. A language appropriate for distributed control-dominated applications may be poor for local data flow dominated applications. Hence, we can expect that we will have to live with compromises and possibly with mixed models.

Which compromises are actually used in practice? In practice, assembly language programming was very common in the early years of embedded systems programming. Programs were small enough to handle the complexity of problems in assembly languages. The next step was the use of C or derivatives of C. Due to the increasing complexity of embedded system software, higher level languages are to follow the introduction of C. Object-oriented languages and SDL are languages which provide the next level of abstraction. Also, languages like UML are required to capture specifications at an early design stage. The trend is to move toward model-based designs [453]. In practice, languages can be used like shown in Fig. 2.76.

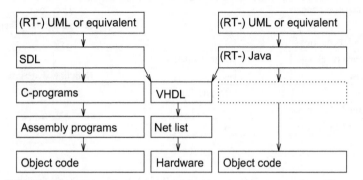

Fig. 2.76 Using various languages in combination

According to Fig. 2.76, languages like SDL or StateCharts can be translated into C. These C descriptions are then compiled. Starting with SDL or StateCharts also opens the way to implementing the functionality in hardware, if translators from these languages to VHDL are provided. Both C and VHDL will certainly survive as intermediate languages for many years. Java does not need intermediate steps but does also benefit from good translation concepts to assembly languages. In a similar way, translations between various graphs are feasible. For example, SDF graphs can be translated into a subclass of Petri nets [491]. Also, they correspond to a subclass of the **computation graph model** proposed by Karp and Miller [270]. Linking the various models of computation is facilitated by formal techniques [96].

Several languages for embedded system design are covered in a book edited by M. Radetzki [439]. Popovici et al. [432] use a combination of Simulink and SystemC.

We have skipped the discussion of algebraic languages such as LOTOS [246] and Z [480]. These languages enable precise specifications and formal proofs, but they are not executable.

2.10.2 UML™

UML™ is a language including diagrams reflecting several MoCs. Table 2.7 classifies the UML diagrams mentioned so far with respect to our table of MoCs.

Table 2.7 Models of computation available in UML™

Communication/ organization of components	Shared memory	Message passing	
		Synchronous	Asynchronous
Undefined components		Use cases	
		Sequence charts, timing diagrams	
Differential equations		–	
Finite state machines	State diagrams	–	–
Data flow	–	Data flow diagrams	
Petri nets	(Not useful)	Activity charts	
Distributed event model	–	–	
Von-Neumann model	–	–	

This figure shows how UML covers several models of computation, with a focus on early design phases. Semantics of communication is typically imprecisely defined. Therefore, our classification cannot be precise in this respect. In addition to the diagrams already mentioned, the following diagrams can be modeled:

- **Deployment diagrams**: These diagrams are important for embedded systems, and they describe the "execution architecture" of systems (hardware or software nodes).
- **Package diagrams**: Package diagrams represent the partitioning of software into software packages. They are similar to module charts in StateMate.
- **Class diagrams**: These diagrams describe inheritance relations of object classes.
- **Communication diagrams** (called **Collaboration diagrams** in UML™ 1.x): These graphs represent classes, relations between classes, and messages that are exchanged between them.
- **Component diagrams**: They represent the components used in applications or systems.
- **Object diagrams**, **interaction overview diagrams**, **composite structure diagrams**: This list consists of three types of diagrams which are less frequently used. Some of them may actually be special cases of other types of diagrams.

Available tools provide some consistency checking between the different diagram types. Complete checking, however, seems to be impossible. One reason for this is that the semantics of UML initially was left undefined. It has been argued that this was done intentionally, since one does not like to bother about the precise semantics during the early phases of the design. As a consequence, precise, executable specifications can only be obtained if UML is combined with some other, executable languages. Available design tools have combined UML with SDL [219] and C++. There are, however, also some first attempts to define the semantics of UML.

Version 1.4 of UML was not designed for embedded systems. Therefore, it lacks a number of features required for modeling embedded systems (see p. 27). In particular, the following features are missing [368]:

- the partitioning of software into tasks and processes cannot be modeled,
- timing behavior cannot be described at all,
- the presence of essential hardware components cannot be described.

Due to the increasing amount of software in embedded systems, UML is gaining importance for embedded systems as well. Hence, several proposals for UML extensions to support real-time applications have been made [132, 368]. These extensions have been considered during the design of UML 2.0. UML 2.0 includes 13 diagram types (up from nine in UML 1.4) [12]. Special profiles are taking the requirements of real-time systems into account [352]. Profiles include class diagrams with constraints, icons, diagram symbols, and some (partial) semantics. There are UML profiles for the following [352]:

- Schedulability, Performance, and Time Specification (SPT) [409],
- Testing [412],
- Quality of Service (QoS) and Fault Tolerance [412],
- a Systems Modeling Language called SysML [410],
- Modeling and Analysis of Real-Time Embedded Systems (MARTE), [411]
- UML and SystemC interoperability [446],
- the SPRINT profile for reuse of intellectual property (IP) [481].

Using such profiles, we can—for example—attach timing information to sequence charts. However, profiles may be incompatible. Also, UML has been designed for modeling and frequently leaves too many semantic issues open to allow automatic synthesis of implementations [352].

2.10.3 Ptolemy II

The Ptolemy project [435] focuses on modeling, simulation, and design of heterogeneous systems. Emphasis is on embedded systems that mix different technologies and, accordingly, also MoCs. For example, analog and digital electronics, hardware and software, and electrical and mechanical devices can be described. Ptolemy supports different types of applications, including signal processing, control applications, sequential decision making, and user interfaces. Special attention is paid to the generation of embedded software. The idea is to generate this software from the MoC which is most appropriate for a certain application. Version 2 of Ptolemy (Ptolemy II) supports the following MoCs and corresponding domains (see also p. 37):

1. Communicating sequential processes (CSP)
2. Continuous time (CT): This model is appropriate for mechanical systems and analog circuits. Hence, this model supports differential equations. Tools include extensible differential equation solvers.

3. Discrete event model (DE): This is the model used by many simulators, e.g., VHDL simulators.
4. Distributed discrete events (DDE). Discrete event systems are difficult to simulate in parallel, due to the inherent centralized queue of future events. Attempts to distribute this data structure have not been very successful so far. Therefore, this special (experimental) domain is introduced. Semantics can be defined such that distributed simulation becomes more efficient than in the DE model.
5. Finite state machines (FSM)
6. Process networks (PN), using Kahn process networks (see p. 65).
7. Synchronous dataflow (SDF)
8. Synchronous/reactive (SR) MoC: This model uses discrete time, but signals do not need to have a value at every clock tick. Esterel (see p. 57) is a language following this style of modeling.

This list clearly shows the focus on different models of computation in the Ptolemy project.

2.11 Problems

We suggest solving the following problems either at home or during a flipped classroom session:

2.1: What is a (design) model?

2.2: Prepare a list of up to six requirements for specification/modeling languages for embedded systems!

2.3: Why could our specification lead to deadlocks?

2.4: What is a "model of computation (MoC)"?

2.5: What is a "job" and how is it different from "tasks"?

2.6: Which are the two key techniques for communication in computers?

2.7: Which description techniques can be used for capturing initial ideas about the system to be designed?

2.8: Simulate trains between Paris, Brussels, Amsterdam, and Cologne, using the levi simulation software [473]! Modify the examples included with the software such that two independent tracks exist between any two stations and demonstrate an (arbitrary) schedule involving 10 trains!

2.9: Download the OpenModelica simulation software. Develop a simulation model for Newton's cradle (see, for example, https://en.wikipedia.org/wiki/Newton%27s_cradle).

2.10: Modify the answering machine of Example 2.8 such that the owner can intervene at any time during the playing of precorded text or the recording of the message.

2.11: Model your daily schedule with a timed automaton. Hours are reflected by a variable h, days by a variable d. $d = 1$ means Monday, $d = 7$ means Sunday. On a weekend ($d = 6$ or $d = 7$), you leave the sleeping state between $h = 10$ and $h = 11$, spend 1–2h getting yourself ready for the day, stay with your friend until some time in the range $h = 20$ to $h = 21$, walk back home and enter the sleeping state between $h = 22$ and $h = 23$. During the week ($d = 1$ or ... or $d = 5$), you leave the sleeping state between $h = 7$ and $h = 8$, spend 1–2h getting yourself ready for the day, study until some time in the range $h = 20$ to $h = 21$, walk back home and enter the sleeping state between $h = 22$ and $h = 23$. Model your schedule! Do not forget to increase the day d at the end of each day.

2.12: Suppose the StateCharts model in Fig. 2.77 (**left**) model is given.

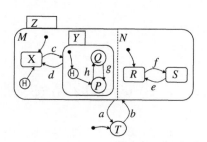

	M	N	P	Q	R	S	T	X	Y	Z
(Reset)							v			
b										
c										
f										
h										
g										
h										
e										
a										
b										
c										

Fig. 2.77 StateCharts example: **left**: graphical model; **right**: table of states

Also, suppose that we have the following sequence of input events: $b\ c\ f\ h\ g\ h\ e$ $a\ b\ c$. In the diagram in Fig. 2.77 (**right**), mark all the states the StateCharts model will be in after a particular input has been applied! Note that H denotes the history mechanism.

2.13: Are StateCharts determinate models if we follow the StateMate semantics? Please explain your answer!

2.14: Is SDL a determinate language? Please explain your answer!

2.15: Let us assume that you have been asked to help modeling the flow of visitors in the hypothetical Museum of Fine Future Information Nuggets (MUFFIN). We consider a steady state with no visitors entering or exiting the museum. The museum will have three exhibition halls. In front of each hall, there is space for a waiting line. The exit of this space is connected to the entry of the hall. Each of the hall exits is connected to each entry of the waiting spaces. Visitors leaving one of the halls are free to chose any of the other halls as their next one. We assume that each hall can be described as a process in a meaningful way, with some randomness of the time that a visitor stays in a hall. Assume that you would like to model this situation is SDL. Show a diagram with explicit processes and FIFO queues!

2.16: Download the levi simulation software for KPNs [471] and develop a KPN model computing Fibonacci numbers in a distributed fashion (i.e., just using a single KPN node is illegal).

2.17: Which three types of Petri nets did we discuss in this book?

2.18: One of the types of Petri nets allows several non-distinguishable tokens per place. Which components are used in a mathematical model of such nets? Hint: $N = (P, \ldots\ldots)$

2.19: Draw the following condition/event system: $N = (C, E, F)$, given

- *Conditions:* $C = \{c_1, c_2, c_3, c_4\}$,
- *Events:* $E = \{e_1, e_2, e_3\}$,
- *Relation:*

$$F = \{(c_1, e_1), (c_1, e_2), (e_1, c_2), (e_1, c_3), (e_2, c_2), (e_2, c_3), (e_2, c_4), (c_2, e_3), \\ (c_3, e_3), (c_4, e_3), (e_3, c_1), (e_3, c_4)\}$$

Specify the precondition of e_3 as well as the postcondition of e_1. Is N *simple* or/and *pure*? Given it is not, which edge(s) need(s) to be removed in order to turn N into a pure net? Substantiate or prove your answers **concisely**.

2.20: What does a compact model of the dining philosophers problem look like?

2.21: CSA theory leads to 2, 3, and 4 logic strengths, corresponding to 4, 7, and 10 logic values. How many strengths and values are we using in IEEE 1164? Please show the partial order among the values of IEEE 1164 in a diagram! Which of the values of IEEE 1164 are not included in the partial order and what is the meaning of these values?

2.22: Suppose that a bus as shown in Fig. 2.78 is given. Rectangles containing an &-sign denote AND gates.

Fig. 2.78 Bus driven by
tristate outputs

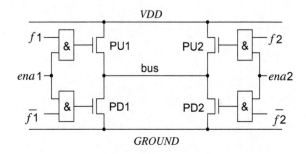

Which of the IEEE 1164 values will be on the bus if both enable inputs are set to
'0' ($ena1 = ena2 = $ '0')?

Which of the IEEE 1164 values will be on the bus if $ena1 = $ '0', $ena2 = $ '1' and
$f2 = $ '1'?

2.23: Which of the following circuits can be modeled with IEEE 1164: complementary CMOS outputs, outputs with a depletion transistor, open collector outputs,
tristate outputs, precharging on buses (if depletion transistors are used as well)?

2.24: Which of the following languages use asynchronous message passing: StateCharts, SDL, VHDL, CSP, Petri nets, MPI?

2.25: Which of the following languages use a broadcast mechanism for updating
variables: StateCharts, SDL, Petri nets?

2.26: Which of the following diagram types are supported by UML: sequence
charts, record charts, Y-charts, use cases, activity diagrams, circuit diagrams?

2.27: UML™ is a frequently used modeling technique. In the table below, enter
models of computation for the components in the left column and for communication
in the top row. Then, enter as many UML diagram types as feasible into the remaining
table cells.

Communication/organiza-tion of components			

Chapter 3
Embedded System Hardware

3.1 Introduction

It is one of the characteristics of embedded and cyber-physical systems that both hardware and software must be taken into account. The achievable performance (and hence, the user perception) depends crucially on the available hardware platform. Moreover, motivation for this was already included in the Preface: "*The development of embedded systems cannot ignore the underlying hardware characteristics. Timing, memory usage, power consumption, and physical failures are important*" [86].

The reuse of available hard- and software components is at the heart of the **platform-based design methodology** (see also p. 283). Consistent with the need to consider available hardware components and with the design information flow shown in Fig. 3.1, we are now going to describe some of the essentials of embedded system hardware.

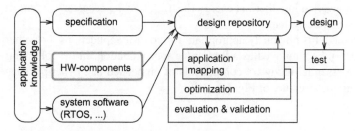

Fig. 3.1 Simplified design information flow

Hardware for embedded systems is much less standardized than hardware for personal computers. Due to the huge variety of embedded system hardware, it is impossible to provide a comprehensive overview of all types of hardware components. Nevertheless, we will try to provide a survey of some of the essential components which can be found in most systems.

© Springer International Publishing AG 2018
P. Marwedel, *Embedded System Design*, Embedded Systems,
DOI 10.1007/978-3-319-56045-8_3

In many cyber-physical systems, especially in control systems, hardware is used in a loop (see Fig. 3.2). We will use this loop to structure the presentation of components in this chapter.

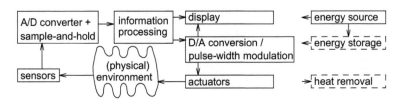

Fig. 3.2 Hardware in the loop

In this (control) loop, information about the physical environment is made available through **sensors**. Typically, sensors generate continuous sequences of analog values. In this book, we will restrict ourselves to information processing where digital computers process discrete sequences of values. Appropriate conversions are performed by two kinds of circuits: sample-and-hold circuits and analog-to-digital converters (ADCs). After such conversion, information can be processed digitally. Generated results can be displayed and also be used to control the physical environment through actuators. Since many actuators are analog actuators, conversion from digital to analog signals may also be needed. We will see how this conversion can be achieved either by digital-to-analog converters (DACs) or indirectly by pulse-width modulation (PWM).

Due to the prevailing *electronic* information processing, we assume that we require electrical energy. Some source of this energy must be available. If our energy source does not provide energy permanently, we may need to store energy, e.g., in rechargeable batteries or capacitors. During system operation, much of the electrical energy will be converted into thermal energy (heat). It may be necessary to remove thermal energy from the system.

This model is obviously appropriate for control applications. For other applications, it can be employed as a first-order approximation. In the following, we will describe essential hardware components of embedded and cyber-physical systems following the structure of Fig. 3.2.

3.2 Input

3.2.1 Sensors

We start with a brief discussion of sensors. Sensors can be designed for virtually every physical quantity. There are sensors for weight, velocity, acceleration, electrical current, voltage, temperature, etc. A wide variety of physical effects can be exploited in the construction of sensors [145]. Examples include the law of induction (generation of voltages in an electric field), and photoelectric effects. There are also sensors for chemical substances [146].

Recent years have seen the design of a huge range of sensors, and much of the progress in designing smart systems can be attributed to modern sensor technology. The availability of sensors has enabled the design of sensor networks (see, for example, Tiwari et al. [518]), a key element of the Internet of Things.

It is impossible to cover this subset of cyber-physical hardware technology comprehensively, and we can only give characteristic examples:

- **Acceleration sensors**: Fig. 3.3 shows a small sensor manufactured using microsystem technology. The sensor contains a small mass in its center. When accelerated, the mass will be displaced from its standard position, thereby changing the resistance of the tiny wires connected to the mass.

Fig. 3.3 Acceleration sensor (courtesy S. Bütgenbach, IMT, TU Braunschweig), ©TU Braunschweig, Germany

Acceleration sensors are included in the powerful inertial measurement units (IMUs) (see, for example, Siciliano et al. [462], Sect. 20.4). They contain gyros and accelerometers, and they capture up to six degrees of freedom, comprising position (x, y, and z) and orientation (roll, pitch, and yaw).
- **Image sensors**: There are essentially two kinds of image sensors: charge-coupled devices (CCDs) and CMOS sensors. In both cases, arrays of light sensors are used. The architecture of CMOS sensor arrays is similar to that of standard memories: Individual pixels can be randomly addressed and read out. CMOS sensors use standard CMOS technology for integrated circuits. Due to this, sensors and logic circuits can be integrated on the same chip. This allows some preprocessing to be done already on the sensor chip, leading to so-called smart sensors. CMOS sensors require only a single standard supply voltage, and interfacing in general is easy. Therefore, CMOS-based sensors can be cheap.

In contrast, CCD technology is optimized for optical applications. In CCD technology, charges must be transferred from one pixel to the next until they can finally be read out at an array boundary. This sequential charge transfer also gave CCDs their name. For CCD sensors, interfacing is more complex.

Selecting the most appropriate image sensor depends on several constraints, which change as technology evolves. The image quality of CMOS sensors has been improved over the recent years, and the initial image superiority of CCDs became questionable. Therefore, achieving a good image quality is feasible with CCD and with CMOS sensors. Due to their faster readout speed, CMOS sensors are preferred for cameras with live view modes or video recording functionality [387]. Also, CMOS sensors are preferred for low-cost devices if smart sensors are to be designed. Several application areas for CCDs have disappeared, but they are still used in areas such as scientific image acquisition.

- **Biometric sensors**: Demands for higher security standards as well as the need to protect mobile and removable equipment have led to an increased interest in authentication. Due to the limitations of password-based security (e.g., stolen and lost passwords), biometric sensors and biomedical authentication receive attention. Biometric authentication tries to identify whether or not a certain person is actually the person she or he claims to be. Methods for biometric authentication include iris scans, finger print sensors, and face recognition. Finger print sensors are typically fabricated using the same CMOS technology [550] which is used for manufacturing integrated circuits. CCD and CMOS image sensors can be used for face recognition. False accepts as well as false rejects are an inherent problem of biometric authentication (see definitions on p. 248). In contrast to password-based authentication, exact matches are not possible.
- **Artificial eyes**: Artificial eye projects have received significant attention. Some projects have an impact on the eye, but others provide vision in an indirect way. For example, the Dobelle Institute experimented with a camera attached to a computer sending electrical pulses to a direct brain contact [509]. More recently, the less invasive translation of images into audio has been preferred.
- **Radio frequency identification (RFID)**: RFID technology is based on the response of a **tag** to radio frequency signals [217]. The tag consists of an integrated circuit and an antenna, and it provides its identification to **RFID readers**. The maximum distance between tags and readers depends on the type of the tag. The technology is used to identify objects, animals, or people and is a key enabler for the Internet of Things.
- **Rain sensors**: In order to remove distraction from drivers, some cars contain rain sensors. Using these, the speed of the wipers is adjusted to the amount of rain.
- **Other sensors:** Other common sensors include pressure sensors, proximity sensors, engine control sensors, and Hall effect sensors.

Sensors are generating **signals**. Mathematically, the following definition applies:

Definition 3.1: A **signal** σ is a mapping from a time domain D_T to a value domain D_V:

$$\sigma : D_T \to D_V$$

Signals may be defined over a continuous or a discrete time domain as well as over a continuous or a discrete value domain.

3.2.2 Discretization of Time: Sample-and-Hold Circuits

All known digital computers work in a **discrete** time domain D_T. This means that they can process discrete sequences or **streams** of values. Hence, incoming signals over the continuous time domain must be converted to signals over the discrete time domain. This is the purpose of **sample-and-hold circuits**. Figure 3.4 (**left**) shows a simple sample-and-hold circuit. In essence, the circuit consists of a clocked transistor

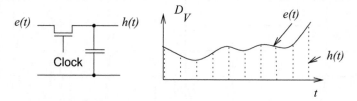

Fig. 3.4 Sample-and-hold phase: **left**: circuit; **right**: signals

and a capacitor. The transistor operates like a switch. Each time the switch is closed by the clock signal, and the capacitor is charged so that its voltage $h(t)$ is practically the same as the incoming voltage $e(t)$. After opening the switch again, this voltage will remain essentially unchanged until the switch is closed again. Each of the values stored on the capacitor can be considered as an element of a discrete sequence of values $h(t)$, generated from a continuous function $e(t)$ (see Fig. 3.4 (**right**)). If we sample $e(t)$ at times $\{t_s\}$, then $h(t)$ will be defined only at those times.

An ideal sample-and-hold circuit would be able to change the voltage at the capacitor in an arbitrarily short amount of time. This way, the input voltage at a particular instance in time could be transferred to the capacitor, and each element in the discrete sequence would correspond to the input voltage at a particular point in time. In practice, however, the transistor has to be kept closed for a short time window in order to really charge or discharge the capacitor. The voltage stored on the capacitor will then correspond to a voltage reflecting that short time window.

3.2.3 Fourier Approximation of Signals

Would we be able to reconstruct the original signal $e(t)$ from the sampled signal $h(t)$? In order to answer this question, we revert to the fact that arbitrary signals can be approximated by summing (possibly phase-shifted) sine functions of different frequencies (Fourier approximation)[1].

---

[1]This presentation is based on the assumption that a comprehensive coverage of Fourier approximations cannot be included in our course. Therefore, only the impact of these approximations is demonstrated by examples. Knowing the theory behind these examples would be beneficial.

Example 3.1: A square wave be approximated by Eq. (3.1) [418].

$$e'_K(t) = \sum_{k=1,3,5,7,9,...}^{K} \left(\frac{4}{\pi k}\sin(\frac{2\pi kt}{T})\right) \tag{3.1}$$

In this equation, T is the period and approximation is improved for increasing K. Figures 3.5 and 3.6 visualize Eq. (3.1).

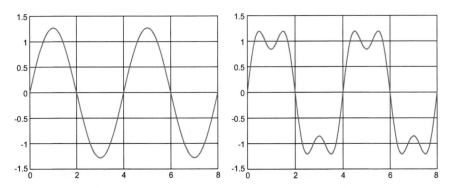

Fig. 3.5 Approximation of a square wave by sine waves for $K = 1$ (left) and $K = 3$ (right)

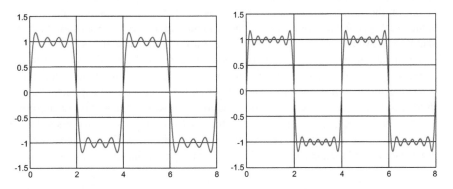

Fig. 3.6 Approximation of a square wave by sine waves for $K = 7$ (left) and $K = 11$ (right)

The larger difference between the square wave and its approximation at the jump discontinuities of the square wave is called **Gibbs Phenomenon** [418].   ∇

A signal transformation Tr is **linear** if for all signals $e_1(t)$ and $e_2(t)$ we have:

$$Tr(e_1 + e_2) = Tr(e_1) + Tr(e_2) \tag{3.2}$$

Next, we restrict ourselves to linear systems. Then, in order to answer the question raised above, we study sampling each of the sine waves independently.

Example 3.2: Consider signals described by either of the two functions e_3 or e_4:

$$e_3(t) = \sin\left(\frac{2\pi t}{8}\right) + 0.5\sin\left(\frac{2\pi t}{4}\right) \tag{3.3}$$

$$e_4(t) = \sin\left(\frac{2\pi t}{8}\right) + 0.5\sin\left(\frac{2\pi t}{4}\right) + 0.5\sin\left(\frac{2\pi t}{1}\right) \tag{3.4}$$

The sine waves used in these functions have periods of $T = 8, 4$, and 1, respectively (this can be seen by comparing these sine waves with those of Eq. (3.1)). A graphical representation of these functions is shown in Fig. 3.7. Suppose that we will be

Fig. 3.7 Visualization of functions $e_3(t)$ (blue) and $e_4(t)$ (red)

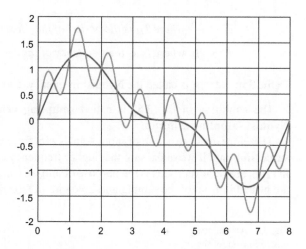

sampling these signals at integer times. It then so happens that both signals have the same value whenever they are sampled. Obviously, it is not possible to distinguish between $e_3(t)$ and $e_4(t)$ if we sample at these instances in time and if only the sampled signal is available.   ∇

In general, sampled signals will not allow us to distinguish between some slow signal $e_3(t)$ and some other faster varying signal $e_4(t)$ if $e_3(t)$ and $e_4(t)$ are identical each time we are sampling the signals. The fact that two or more unsampled signals can have the same sampled representation is called **aliasing**. We are not sampling $e_4(t)$ frequently enough to notice, for example, that it has slope changes between integer times. So, from this counterexample, we can conclude that **reconstruction of the original unsampled signal is not feasible unless we have additional knowledge about the frequencies or the waveforms present in the input signal**.

How frequently do we have to sample signals to be able to distinguish between different sine waves?

Let us assume that we are sampling the input signal at constant time intervals, such that T_s is the **sampling period**:

$$\forall s : T_s = t_{s+1} - t_s \tag{3.5}$$

Let

$$f_s = \frac{1}{T_s} \tag{3.6}$$

be the **sampling rate** or **sampling frequency**.

Then, sampling theory provides us with the following theorem (see, e.g., [418]):

Theorem 3.1 (Sampling theorem): *Given the above definitions of variables, **aliasing is avoided if we restrict the frequencies of the incoming signal to less than half of the sampling frequency** f_s:*

$$T_s < \frac{T_N}{2} \text{ where } T_N \text{ is the period of the "fastest" sine wave, or} \tag{3.7}$$

$$f_s > 2f_N \text{ where } f_N \text{ is the frequency of the "fastest" sine wave} \tag{3.8}$$

Definition 3.2: f_N is called the Nyquist frequency, and f_s is the sampling rate.

The condition in Eq. (3.8) is called **sampling criterion**, and sometimes the **Nyquist sampling criterion**.

Therefore, reconstruction of input signals $e(t)$ from discrete samples $h(t)$ can be successful only if we make sure that higher frequency components, such as the one in $e_4(t)$, are removed. This is the purpose of anti-aliasing filters. Anti-aliasing filters are placed in front of the sample-and-hold circuit (see Fig. 3.8).

Fig. 3.8 Anti-aliasing placed in front of the sample-and-hold circuit

Figure 3.9 demonstrates the ratio between the amplitudes of the output and the input waves as a function of the frequency for this filter.

Fig. 3.9 Ideal and realizable anti-aliasing filters (low-pass filters)

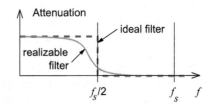

Ideally, such a filter would remove all frequencies at and above half the sampling frequency and keep all other components unchanged. This way, it would convert signal $e_4(t)$ into signal $e_3(t)$.

In practice, such ideal filters (so-called **brick-wall filters**) do not exist[2]. Realizable filters will already start attenuating frequencies smaller than $f_s/2$ and will still not eliminate all frequencies larger than $f_s/2$ (see Fig. 3.9). Attenuated high-frequency components will exist even after filtering. For frequencies smaller than $f_s/2$, there may also be some "overshooting," i.e., frequencies for which there is some amplification of the input signal.

The design of good anti-aliasing filters is an art by itself. This art has been studied, for example, in great detail for high-quality audio equipment, involving detailed hearing tests. Many of the perceived differences between high-quality equipments have been attributed to the design of such filters.

3.2.4 Discretization of Values: Analog-to-Digital Converters

Since we are restricting ourselves to digital computers, we must also replace signals that map time to a continuous value domain D_V by signals that map time to a discrete value domain D_V'. This conversion from analog to digital values is done by analog-to-digital converters (ADCs). There is a large range of ADCs with varying speed/precision characteristics. Typically, fast ADCs have a low precision and high precision converters are slow.

We will present several converters in the next subsections.

3.2.4.1   Flash ADC

This type of ADCs uses a large number of comparators. Each comparator has two inputs, denoted as + and -. If the voltage at input + exceeds that at input -, the output corresponds to a logical '1' and it corresponds to a logical '0' otherwise[3].

In the ADC, all - inputs are connected to a voltage divider. If input voltage $h(t)$ exceeds $\frac{3}{4}V_{ref}$, the comparator at the top of Fig. 3.10 (**a**) will generate a '1'. The encoder at the output of the comparators will try to identify the most significant '1' and will encode this case as the largest output value. The case $h(t) > V_{ref}$ should normally be avoided since V_{ref} is typically close to the supply voltage of the circuit and input voltages exceeding the supply voltage can lead to electrical problems. In our case, input voltages larger than V_{ref} generate the largest digital value as long as the converter does not fail due to the high input voltage.

Now, if input voltage $h(t)$ is less than $\frac{3}{4}V_{ref}$, but still larger than $\frac{2}{4}V_{ref}$, the comparator at the top of Fig. 3.10 will generate a '0', while the next comparator will still signal a '1'. The encoder will encode this as the second-largest value.

---

[2]This would require knowing the signal to be filtered for an infinite amount of time.

[3]In practice, the case of equal voltages is not relevant, as the actual behavior for very small differences between the voltages at the two inputs depends on many factors (such as temperatures and manufacturing processes.) anyway.

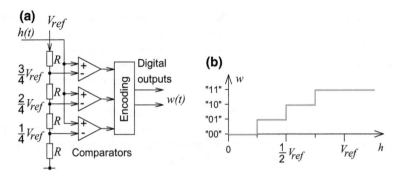

Fig. 3.10 Flash ADC: **a**: schematic; **b**: w as a function of h

Similar arguments hold for cases $\frac{1}{4}V_{ref} < h(t) < \frac{2}{4}V_{ref}$, and $0 < h(t) < \frac{1}{4}V_{ref}$, which will be encoded as the third-largest and the smallest value, respectively. Figure 3.10 (**b**) shows the relation between input voltages and generated digital values.

The outputs of the comparators encode numbers in a special way: if a certain comparator output is equal to '1', then all the less significant outputs are all equal to '1'. The encoder transforms this representation of numbers into the usual representation of natural numbers. The encoder is actually a so-called priority encoder, encoding the most significant input number carrying a '1' in binary[4].

The circuit can convert positive analog input voltages into digital values. Converting both positive and negative voltages and generating two's complement numbers requires some extensions.

One nice property of the flash ADC is the fact that it is automatically **monotonic**: For any increase in the analog voltage from 0 to the maximum, the corresponding digital value increases as well. This property is maintained even if the actual value of the resistors would deviate from the nominal value. However, such a deviation would have an impact on the precision of the linear relation expected between analog and digital values.

Unfortunately, the chain of resistors forms a conducting path, which exists even if the converter is not used. This could make it impossible to use this converter for low-power equipment.

In general, ADCs are also characterized by their **resolution**.

This term has several different but related meanings [13]. The resolution (measured in bits) is the number of bits produced by an ADC. For example, ADCs with a resolution of 16 bits are needed for many audio applications. However, the resolution is also measured in Volts, and in this case it denotes the difference between two input voltages causing the output to be incremented by 1:

---

[4]Such encoders are also useful for finding the most significant '1' in the mantissa of floating-point numbers.

$$Q = \frac{V_{FSR}}{n} \tag{3.9}$$

Where:

V_{FSR} : is the difference between the largest and the smallest voltage,

Q : is the resolution in Volts per step, and

n : is the number of voltage intervals (**not** the number of bits).

Example 3.3: For the ADC of Fig. 3.10, the resolution is two bits or $\frac{1}{4}V_{ref}$ Volts, if we assume V_{ref} as the largest voltage. ▽

The key advantage of the flash ADC is its speed. It does not need any clock. The delay between the input and the output is very small and the circuit can be used easily, for example, for high-speed video applications. The disadvantage is its hardware complexity: We need $n - 1$ comparators in order to distinguish between n values. Imagine using this circuit in generating digital audio signals for CD recorders. We would need $2^{16} - 1$ comparators! High-resolution ADCs must be built in a different way.

3.2.4.2 Successive Approximation

Distinguishing between a large number of digital values is possible with ADCs using successive approximation. The circuit is shown in Fig. 3.11.

Fig. 3.11 Circuit using successive approximation

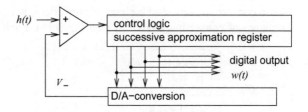

The key idea of this circuit is to use binary search. Initially, the most significant output bit of the successive approximation register is set to '1', and all other bits are set to '0'. This digital value is then converted to an analog value, corresponding to $0.5*$ the maximum input voltage[5]. If $h(t)$ exceeds the generated analog value, the most significant bit is kept at '1', otherwise it is reset to '0'.

This process is repeated with the next bit. It will remain set to '1' if the input value is either within the second or the fourth quarter of the input value range. The same procedure is repeated for all the other bits.

Figure 3.12 shows an example. Initially the most significant bit is set to '1'. This value is kept, since the resulting V_- is less than $h(t)$. Then, the second-most significant bit is set to '1'. It is reset to '0', since the resulting V_- is exceeding $h(t)$. Next, the

---

[5]Fortunately, the conversion from digital to analog values (D/A-conversion) can be implemented very efficiently and can be very fast (see p. 176).

Fig. 3.12 Successive
approximation

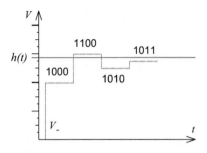

third-most significant bit is tried. It is set to '1', and this value is kept. Finally, the least significant bit is also set, and it remains set after the comparison has been completed. Obviously, $h(t)$ must be constant during the conversion, otherwise the whole procedure would be jeopardized. This requirement is met if we employ a sample-and-hold circuit as shown above. The resulting digital signal is called $w(t)$.

The key advantage of the successive approximation technique is its hardware efficiency. In order to distinguish between n digital values, we need $\lceil log_2(n) \rceil$ bits in the successive approximation register and the D/A converter. The disadvantage is its speed, since it needs $O(log_2(n))$ steps. These converters can, therefore, be used for high-resolution applications, where moderate speeds are required. Examples include audio applications.

3.2.4.3   Pipelined Converters

These converters consist of a chain of converters, where each stage in the chain is in charge of converting a few bits (see Fig. 3.13). Each stage passes the remaining

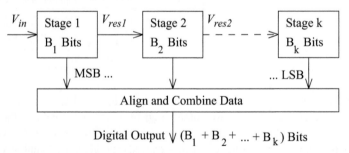

Fig. 3.13 Pipelined ADC [279]

residue of the voltage to the next stage (if any). For example, each stage could convert a single bit and subtract the corresponding voltage. The resulting residue would typically be scaled up by a factor of two (in order to avoid too small voltages) and be passed on to the next stage. Typically, each stage would include a flash ADC of a few bits and a D/A converter to compute the voltage to be subtracted. Resulting digital values must be aligned in time. Required hardware resources increase linearly with the number of bits. With this structure, a good throughput can be achieved, but the latency is larger than for flash converters.

3.2.4.4 Other Converters

Integrating converters use (at least) two phases for the measurement. During the first phase of length t_1, the integral of the input voltage over time is computed[6]. For constant inputs, the resulting value V_{out} is proportional to the input voltage ($V_{out} \sim V_{in} * t_1$). During the second phase, this value is decreased at a constant rate and the time to reach a value of zero is counted. The final count is proportional to the input voltage. Hence, using proper scaling, the final count represents the input voltage. If the input voltage contains some noise, its impact is likely to be averaged out during the first integration phase. Hence, these converters are capable of compensating noise. They are typically found in slow, high-resolution multi-meters.

For **folding ADCs**, the input voltage range is divided into 2^m segments [101]. A coarse-grained converter detects the segment of the current input voltage, yielding the m most significant output bits. A fine-grained converter computes the value within a segment, yielding the less significant output bits.

For **Delta-Sigma ADCs** ($\Delta\Sigma$ ADCs), the name indicates that signal differences (Δs) are encoded and that they are summed up (Σ). A description of these converters is beyond the scope of this book. For details refer to Khorramabadi [280].

3.2.4.5 Comparison of ADCs

Figure 3.14 provides an overview of the speed/resolution trade-offs of ADCs, using a trade-off analysis of Vogels et al. [534]. Flash ADCs are clearly the fastest, but provide only a small resolution. Pipelining is frequently superior to successive approximation. Another overview of ADCs is provided by IEEE TV [414].

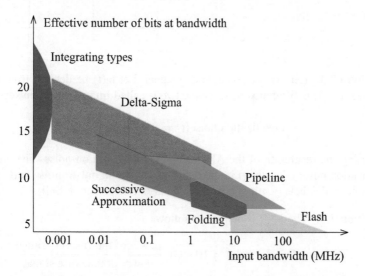

Fig. 3.14 Comparison of the speed/resolution characteristics of various ADCs [534]

---

[6]This can be done with a capacitor in the feedback loop of an operational amplifier (see p. 383).

3.2.4.6 Quantization Noise

Figure 3.15 shows the behavior of a flash ADC when the input signal is that of Eq. (3.3). Only the behavior for a positive input signal is shown. The figure includes the voltage corresponding to the digital value, the original voltage, and the difference between the two. Obviously, the converter is "truncating" the digital representation of the analog signal to the number of available bits (i.e., the digital value is always less than or equal to the analog value). This is a consequence of the way in which the flash converter is doing comparisons. "Rounding" converters would need an internal correction by "half a bit." Effectively, the digital signal encodes values corresponding to the sum of the original analog values and the difference $w(t) - h(t)$. This means, it appears **as if the difference between the two signals had been added to the original signal**. This difference is a signal called **quantization noise**:

Fig. 3.15 $h(t)$ (blue), $w(t)$ (red), $w(t) - h(t)$ (black)

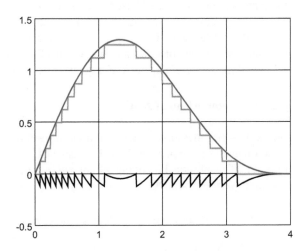

Definition 3.3: Let $h(t)$ be some analog signal. Let $w(t)$ be derived from $h(t)$ by quantization. The difference between the two is called **quantization noise**:

$$\text{quantization noise}(t) = w(t) - h(t) < Q \tag{3.10}$$

Increasing the resolution of the ADC decreases quantization noise. The impact of quantization noise is captured in the definition of the **signal-to-noise ratio** (SNR), measured in decibels (tenth of a Bel, named after Alexander G. Bell).

Definition 3.4: The SNR is defined as follows:

$$\text{SNR (in dB = decibels)} = 10 * log \frac{\text{power of the "useful" signal}}{\text{power of the noise signal}} \tag{3.11}$$

$$= 20 * log \frac{\text{voltage of the "useful" signal}}{\text{voltage of the noise signal}} \tag{3.12}$$

We have used that, for any given impedance R, the power of a signal is proportional to the square of the voltage. Decibels are no physical units, since the SNR is dimensionless.

For any signal $h(t)$, the power of the quantization noise is equal to $\alpha * Q$, where $\alpha \leq 1$ depends on the waveform of $h(t)$. If $h(t)$ can always be represented exactly by a digital value, then $\alpha = 0$. If $h(t)$ is always "just a little" below the next value that can be represented, α may be close to 1.

Example 3.4: The SNR of 16-bit CD audio is (for $\alpha \sim 1$) about $20 * log(2^{16}) = 96\,dB$. Values of $\alpha < 1$ and imperfect ADCs change this number. $\qquad \nabla$

3.3 Processing Units

Let us now discuss the next element in the control loop of Fig. 3.2, processing units. For information processing in embedded systems, we will consider ASICs (application-specific integrated circuits) using hardwired multiplexed designs, reconfigurable logic, and several types of programmable processors. We will consider ASICs first.

3.3.1 Application-Specific Circuits (ASICs)

For high-performance applications and for large markets, application-specific integrated circuits (ASICs) can be designed. In general, ASICs are very energy-efficient (see Sect. 3.7.1 on p. 186). However, the cost of designing and manufacturing such chips is quite high. The cost of the mask set (which is used for transferring geometrical patterns onto the chip) has grown[7].

However, it is feasible to decrease this cost by using less advanced semiconductor fabrication technologies and by using multi-project wafers (MPW) containing several designs. But there is a lack of flexibility: Correcting design errors typically requires a new mask set and a new fabrication run (unless the ASIC contains processors with writable memories). This approach also has to cope with potentially large design efforts requiring dedicated skills and expensive tools. Therefore, ASICs are appropriate only under special circumstances, such as large market volumes, ultimate energy efficiency demands, special voltage or temperature ranges, mixed analog/digital signals, or security-driven designs. Consequently, the design of ASICs is not covered in this book.

---

[7]According to http://anysilicon.com/semiconductor-wafer-mask-costs/, the average cost is currently about $ 1.5 M for a leading-edge 28-nm technology.

3.3.2 Processors

The key advantage of processors is their flexibility. With processors, the overall behavior of embedded systems can be changed by just changing the software running on those processors. Changes of the behavior may be required in order to correct design errors, to update the system to a new or changed standard, or to add features to the previous system. Because of this, processors have found widespread use in embedded systems. In particular, processors which are available commercially "off-the-shelf" (COTS) have become very popular.

Embedded processors must be used in a resource-aware manner, i.e., we need to care about resources required for running applications on them. Furthermore, they do not need to be instruction set compatible with commonly used personal computers (PCs) or servers. Therefore, their architectures may be different from those processors. Efficiency has a number of different aspects (see p. 11), which are discussed next.

3.3.2.1 Energy Efficiency

The energy E for a certain application is closely related to the power P required per operation, since

$$E = \int P dt \tag{3.13}$$

Let us assume that we start with some design having a power consumption of $P_0(t)$, leading to an energy consumption of

$$E_0 = \int_0^{t_0} P_0(t) dt$$

after t_0 units of execution time. Suppose that a modified design finishing computations already at time t_1 comes with a power consumption of $P_1(t)$ and an energy consumption of

$$E_1 = \int_0^{t_1} P_1(t) dt$$

If $P_1(t)$ is not too much larger than $P_0(t)$, then a reduction of the execution time also reduces the energy consumption. However, in general this is not necessarily always true. The situation is also shown in Fig. 3.16: E_1 may be smaller than E_0, but E_1 can also be larger than E_0.

Minimization of power and energy consumption are both important. Power consumption has an effect on the size of the power supply, the design of the voltage regulators, the dimensioning of the interconnect, and short-term cooling. Minimizing the energy consumption is required especially for mobile applications, since

Fig. 3.16 Comparison of
energies E_0 and E_1

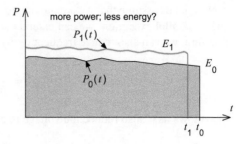

battery technology is only slowly improving, and since the cost of energy may be quite high. Also, a reduced energy consumption decreases cooling requirements and improves the reliability (since the lifetime of electronic circuits decreases for high temperatures).

Next, we would like to demonstrate that for CMOS technology, it is preferable to replace high-speed sequential computations by reduced speed parallel computations. This is shown by—first of all—considering the power consumption of CMOS devices. The **dynamic power consumption** is the power consumption caused by switching (in contrast to the **static power consumption** which exists even if no switching takes place). The dynamic power consumption P_{dyn} of CMOS circuits is given by [91]:

$$P_{dyn} = \alpha \ C_L \ V_{dd}^2 f \tag{3.14}$$

where α is the switching activity, C_L is the load capacitance, V_{dd} is the supply voltage, and f is the clock frequency. This means that the power consumption of CMOS processors increases (at least)[8] quadratically with the supply voltage V_{dd}.

The delay of CMOS circuits can be approximated as [91, 92]:

$$\tau = kC_L \frac{V_{dd}}{(V_{dd} - V_t)^2} \tag{3.15}$$

where k is a constant, and V_t is the threshold voltage. V_t has an impact on the transistor input voltage required to switch the transistor on. For example, for a maximum supply voltage of $V_{dd,max}$=3.3 Volts, V_t may be in the order of 0.8 Volts. Consequently, the maximum clock frequency is a function of the supply voltage. However, decreasing the supply voltage reduces the power quadratically, while the run-time of algorithms is only linearly increased (ignoring the effects of the memory system).

We can use this to reduce the amount of energy required for a certain amount of computations. Let us assume that we are initially performing computations sequentially at voltage V_{dd}, constant power P, clock frequency f, run-time of t, and energy $E = P * t$.

---

[8]In practice, the increase may actually come with a larger exponential.

Now, let us assume that we are moving toward executing β operations in parallel. Due to parallel execution, we can extend the time for each operation by a factor of β. In turn, we can also reduce frequency f by a factor of β and use a new frequency

$$f' = \frac{f}{\beta} \tag{3.16}$$

This allows us to also reduce the voltage to a new voltage

$$V'_{dd} = \frac{V_{dd}}{\beta} \tag{3.17}$$

This reduces the power P^0 per operation quadratically:

$$P^0 = \frac{P}{\beta^2} \tag{3.18}$$

Due to executing β operations in parallel, the overall power P' can be computed as:

$$P' = \beta * P^0 = \frac{P}{\beta} \tag{3.19}$$

The time t' to execute operations in parallel is the same as the time to compute them sequentially ($t' = t$). Hence, the energy to execute the operations in parallel is

$$E' = P' * t = \frac{E}{\beta} \tag{3.20}$$

We conclude that it is more energy-efficient to execute β operations in parallel instead of computing them sequentially. However, our derivation contains a number of approximations. On the one hand, power may be depending even cubically on the voltage and we have ignored the fact that memory speed is frequently a limiting constraint. Faster processor clock speeds might just lead to more waiting for memory accesses (but there may be also conflicts for memory access from multiple cores). On the other hand, we need to be able to find β operations which can be executed in parallel. Overall, we keep in mind that parallel execution is a means for deriving energy-efficient implementations, regardless of which hardware technology we are using.

Architectures must be optimized for their energy efficiency, and we must make sure that we are not losing efficiency in the software generation process. For example, compilers generating 50% overhead in terms of the number of cycles will take us further away from the efficiency of ASICs, possibly by even more than 50%, if the supply voltage and the clock frequency must be increased in order to meet timing deadlines.

There is a large amount of techniques available that can make processors energy-efficient and energy efficiency should be considered at various levels of abstraction, from the design of the instruction set down to the design of the chip manufacturing process [78]. Gated clocking and power gating are examples of such techniques. With gated clocking, parts of the processor are disconnected from the clock during idle periods. In a similar way, the power can be disconnected for some components. For example, direct memory access (DMA) hardware or bus bridges can be disconnected if they are not needed. Also, there are attempts, to get rid of the clock for major parts of the processor altogether. There are two contrasting approaches: globally synchronous, locally asynchronous processors and globally asynchronous, locally synchronous processors (GALS) [250]. Further information about low-power design techniques is available in a book by E. Macii [343] and in the PATMOS proceedings (see http://www.patmos-conf.org/).

At least three techniques can be applied at a rather high level of abstraction:

- **Parallel execution**: According to Eq. (3.20), parallel execution is an effective means of improving the overall energy efficiency.
- **Dynamic power management (DPM)**: With this approach, processors have several power-saving states in addition to the standard operating state. Each power-saving state has a different power consumption and a different time for transitions into the operating state. Figure 3.17 shows the three states for the StrongArm SA 1100 processor.

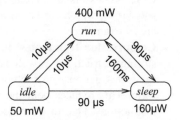

Fig. 3.17 Dynamic power management states of the StrongArm processor SA 1100 [47]

The processor is fully operational in the *run* state. In the *idle* state, it is just monitoring the interrupt inputs. In the *sleep* state, on-chip activity is shutdown, the processor is reset, and the chip's power supply is shut off [565]. A separate I/O-power supply provides power to power manager hardware. The processor can be restarted by the power manager hardware by a preprogrammed wake-up event. Note the large difference in the power consumption between the *sleep* state and the other states, and note also the large delay for transitions from the *sleep* to the *run* state.

- **Dynamic voltage and frequency scaling (DVFS)**: Eq. (3.14) can be exploited in a technique called **dynamic voltage and frequency scaling (DVFS)**. For example, the Crusoe™ processor by Transmeta [283] provided 32 voltage levels between 1.1 and 1.6 Volts, and the clock could be varied between 200 MHz and 700 MHz in increments of 33 MHz. Transitions from one voltage/frequency pair to the next took about 20 ms. Design issues for DVFS-capable processors are described in a paper by Burd and Brodersen [77]. According to the same paper, potential power savings will exist even for future technologies with a decreased maximum V_{dd}, since the threshold voltages will also be decreased (unfortunately, this will lead to increased leakage currents, increasing the static power consumption).

3.3.2.2   Code Size Efficiency

Minimizing the code size is very important for embedded systems, since large hard disk drives (HDDs) or solid-state disks (SSDs) are typically not available and since the capacity of memory is typically also very limited[9]. This is even more pronounced for **systems on a chip** (SoCs). For SoCs, the memory and processors are implemented on the same chip. In this particular case, memory is called **embedded memory**. Embedded memory may be more expensive to fabricate than separate memory chips, since the fabrication processes for memories and processors must be compatible. Nevertheless, a large percentage of the total chip area may be consumed by the memory. There are several techniques for improving the code size efficiency:

- **CISC machines**: Standard RISC processors have been designed for speed, not for code size efficiency. Earlier complex instruction set processors (CISC machines) were actually designed for code size efficiency, since they had to be connected to slow memories. Caches were not frequently used. Therefore, "old-fashioned" CISC processors are finding applications in embedded systems. ColdFire processors [164], which are based on the Motorola 68000 family of CISC processors, are an example.
- **Compression techniques**: In order to reduce the amount of silicon needed for storing instructions as well as in order to reduce the energy needed for fetching these instructions, instructions are frequently stored in memory in compressed form. This reduces both the area as well as the energy necessary for fetching instructions. Due to the reduced bandwidth requirements, fetching can also be faster. A (hopefully small and fast) decoder is placed between the processor and the (instruction) memory in order to generate the original instructions on the fly (see Fig. 3.18 (**right**))[10]. Instead of using a potentially large memory of uncompressed instructions, we are storing the instructions in a compressed format.

---

[9]The availability of large flash memories makes memory size constraints less tight.

[10]We continue denoting multiplexers, arithmetic units, and memories by shape symbols, due to their widespread use in technical documentation. For memories, we adopt shape symbols including an explicit address decoder (included in the shape symbols for the ROMs on the right). These decoders identify the address input.

Fig. 3.18 Schemes for
instruction fetch: **left**:
uncompressed; **right**:
compressed

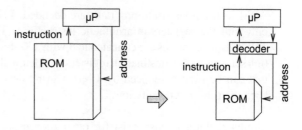

The goals of compression can be summarized as follows:

- We would like to save ROM and RAM areas, since these may be more expensive than the processors themselves.
- We would like to use some encoding technique for instructions and possibly also for data with the following properties:
  - There should be little or no run-time penalty for these techniques.
  - Decoding should work from a limited context (it is, for example, impossible to read the entire program to find the destination of a branch instruction).
  - Word sizes of the memory, of instructions, and of addresses must be taken into account.
  - Branch instructions branching to arbitrary addresses must be supported.
  - Fast encoding is only required if writable data is encoded. Otherwise, fast decoding is sufficient.

There are several variations of this scheme:

- For some processors, there is a **second instruction set**. This second instruction set has a narrower instruction format. An example of this is the ARM® processor family. The original ARM instruction set is a 32-bit instruction set and includes **predicated execution**[11]. This means an instruction is executed if and only if a certain condition concerning values in the condition code registers is met. This condition is encoded in the first four bits of the instruction format. Most ARM processors also provide a second instruction set, with 16-bit wide instructions, called THUMB instructions. THUMB instructions are shorter, since they do not support predication, use shorter and less register fields, and use shorter immediate fields (see Fig. 3.19). THUMB instructions are dynamically converted into

Fig. 3.19 Re-encoding
THUMB into ARM
instructions

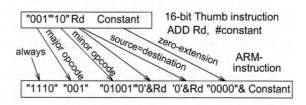

---

[11]The more recently introduced 64 bit instruction set places less emphasis on predicated execution.

ARM instructions while programs are decoded. THUMB instructions can use only half the registers in arithmetic instructions. Therefore, register fields of THUMB instructions are concatenated with a '0'-bit[12]. In the THUMB instruction set, source and destination registers are identical and the length of constants that can be used is reduced by 4 bits. During decoding, pipelining is used to keep the run-time penalty low.

Similar techniques also exist for other processors. The disadvantage of this approach is that the tools (compilers, assemblers, debuggers, etc.) must be extended to support a second instruction set. Therefore, this approach can be quite expensive in terms of software development cost.

- A second approach is the use of **dictionaries**. With this approach, each instruction pattern is stored only once. For each value of the program counter, a look-up table provides a pointer to the corresponding instruction in the instruction table, the dictionary (see Fig. 3.20). This approach relies on using only very few

Fig. 3.20 Dictionary approach for instruction compression

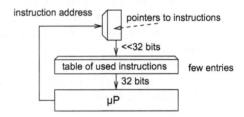

different instruction patterns. Therefore, only few entries are required for the instruction table. Hence, the bit width of the pointers can be quite small. Many variations of this scheme exist. Some are called *two-level control store* [115], *nanoprogramming* [490], or *procedure ex-lining* [528].

Beszedes [55] and Latendresse [310] provide overviews of compression techniques. In addition, Bonny et al. [61] published a Huffman-based technique.

3.3.2.3   Execution-Time Efficiency

In order to meet time constraints without having to use high clock frequencies, architectures can be customized to certain application domains, such as digital signal processing (DSP). One can even go one step further and design application-specific instruction set processors (ASIPs). As an example of domain-specific processors, we will consider processors for DSP. In digital signal processing, digital filtering is a very frequent operation. Let us assume that we are extending the processing pipeline of Fig. 3.8 on p. 132 by such filtering, as shown in Fig. 3.21.

---

[12]Using VHDL-notation (see p. 84), concatenation is denoted by an &-sign and constants are enclosed in quotes in Fig. 3.19.

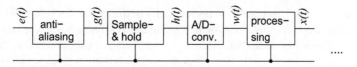

Fig. 3.21 Naming conventions for signals

Equation (3.21) describes a digital filter generating an output signal $x(t)$ from an input signal $w(t)$. Both signals are defined over the (usually unbounded) domain $\{t_s\}$ of sampling instances. We write x_s instead of $x(t_s)$ and w_{s-k} instead of $w(t_{s-k})$:

$$x_s = \sum_{k=0}^{n-1} w_{s-k} * a_k \qquad (3.21)$$

Output element x_s corresponds to a weighted average over the last n signal elements of w and can be computed iteratively, adding one product at a time. Processors for DSP are designed such that each iteration can be encoded as a single instruction.

Example 3.5: Figure 3.22 shows the architecture of an ADSP 2100 DSP processor.

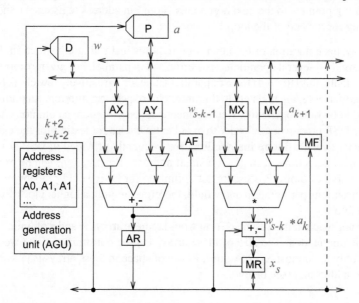

Fig. 3.22 Internal architecture of the ADSP 2100 processor

The processor has two memories, called D and P. A special address generating unit (AGU) can be used to provide the pointers for accessing these memories. There are separate units for additions and multiplications, each with their own argument registers AX, AY, AF, MX, MY, and MF. The multiplier is connected to a second adder in order to compute the series of multiplications and additions quickly.

For this processor, one iteration is essentially performed in a single cycle. For this purpose, the two memories are allocated to hold the two arrays w and a and address registers are allocated such that relevant pointers can be easily updated in the AGU. Partial sums are stored in MR. The pipelined computation involves registers A1, A2, MX, and MY. This allocation of resources enables the execution of the following code:

```
/* outer loop over sampling times tₛ */
{ MR:=0;  A1:=1;  A2:=s-1;  MX:=w[s];  MY:=a[0];
    for (k=0;  k <= (n − 1); k++)
      {MR:=MR + MX * MY;   MX:=w[A2];   MY:=a[A1];
        A1++; A2--; }
  x[s]:=MR;}
```

The outer loop corresponds to the progressing time. A single instruction encodes the inner loop body, comprising the following operations:

- reading of two arguments from argument registers MX and MY, multiplying them, and adding the product to register MR storing partial sums,
- fetching the next elements of arrays a and w from memories P and D and storing them in argument registers MX and MY,
- updating pointers to the next arguments, stored in address registers A1 and A2,
- testing for the end of the loop.

This way, each iteration of the inner loop requires just a single instruction. In order to achieve this, several operations are performed in parallel. For given computational requirements, this (limited) form of parallelism leads to relatively low clock frequencies. Furthermore, the registers in this architecture perform different functions. They are said to be **heterogeneous**. Heterogeneous register files are a common characteristic for DSP processors. In order to avoid extra cycles for testing for the end of the loop, **zero-overhead loop instructions** are frequently provided in DSP processors. With such instructions, a single or a small number of instructions can be executed a fixed number of times. Processors not optimized for DSP would probably need several instructions per iteration and would, therefore, require a higher clock frequency if available.

The approach in its presented form would require arrays w and x of unlimited size if $\{t_s\}$ is unbounded. The size of these arrays can be constrained since we need to access only the n most recent values. Reuse of space in these arrays is possible with modulo addressing (see below).   ∇

3.3.2.4 Digital Signal Processing (DSP)

In addition to allowing single instruction realizations of loop bodies for filtering, DSP processors provide a number of other application domain-oriented features:

- **Specialized addressing modes**: In the filter application described above, only the last n elements of w need to be available. Ring buffers can be used for that.

These can be implemented easily with modulo addressing. In modulo addressing, addresses can be incremented and decremented until the first or last element of the buffer is reached. Additional increments or decrements will result in addresses pointing to the other end of the buffer.

- **Separate address generation units**: Address generation units (AGUs) are typically directly connected to the address input of the data memory (see Fig. 3.23).

Fig. 3.23 AGU using special address registers

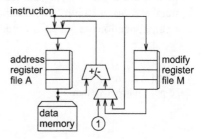

Addresses which are available in address registers can be used in register-indirect addressing modes. This saves machine instructions, cycles, and energy. In order to increase the usefulness of address registers, instruction sets typically contain auto-increment and auto-decrement options for most instructions using address registers.

- **Saturating arithmetic**: Saturating arithmetic changes the way overflows and underflows are handled. In standard binary arithmetic, wraparound is used for the values returned after an overflow or underflow. Table 3.1 shows an example in which two unsigned four-bit numbers are added. A carry is generated which cannot be returned in any of the standard registers. The result register will contain a pattern of all zeros. No result could be further away from the true result than this one.

Table 3.1 Wraparound versus saturating arithmetic for unsigned integers

	0 1 1 1
+	1 0 0 1
Standard *wrap-around* arithmetic	1 0 0 0 0
saturating arithmetic	1 1 1 1

In saturating arithmetic, we try to return a result which is as close as possible to the true result. For saturating arithmetic, the largest value is returned in the case of an overflow and the smallest value is returned in the case of an underflow. This approach makes sense especially for video and audio applications: The user will hardly recognize the difference between the true result value and the largest value that can be represented. Also, it would be useless to raise exceptions if overflows occur, since it is difficult to handle exceptions in real time. Note that we need to know whether we are dealing with signed or unsigned add instructions in order to return the right value.

- **Fixed-point arithmetic**: Floating-point hardware increases the cost and power consumption of processors. Consequently, it has been estimated that 80% of the DSP processors do not include floating-point hardware [1]. However, in addition to supporting integers, many such processors do support fixed-point numbers. Fixed-point data types can be specified by a 3-tuple $(wl, iwl, sign)$, where wl is the total word-length, iwl is the integer word-length (the number of bits left of the binary point), and sign $s \in \{s, u\}$ denotes whether we are dealing with unsigned or signed numbers. See also Fig. 3.24. Furthermore, there may be different rounding modes (e.g., truncation) and overflow modes (e.g., saturating and wraparound arithmetic).

Fig. 3.24 Parameters of a fixed-point number system

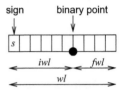

For fixed-point numbers, the position of the binary point is maintained after multiplications (some low order bits are truncated or rounded). For fixed-point processors, this operation is supported by hardware.

- **Real-time capability**: Some of the features of modern processors used in PCs are designed to improve the average execution time of programs. In many cases, it is difficult if not impossible to formally verify that they improve the worst-case execution time. In such cases, it may be better not to implement these features. For example, it is difficult (though not impossible [4]) to guarantee a certain speedup resulting from the use of caches. Therefore, caches are sometimes not used for embedded applications. Also, virtual addressing and demand paging[13] are frequently not found in embedded systems. Techniques for computing worst-case execution times will be presented in Sect. 5.2.2.
- **Multiple memory banks or memories**: The usefulness of multiple memory banks was demonstrated in the ADSP 2100 example: The two memories D and P allow fetching both arguments at the same time. Several DSP processors come with two memory banks.
- **Heterogeneous register files**: Heterogeneous register files were already mentioned for the filter application.
- **Multiply/accumulate instructions**: These instructions perform multiplications followed by additions. They were also already used in the filter application.

Due to the importance of signal processing, instructions for DSP have been added to many instruction sets.

---

[13] See Appendix C on page 385 for an introduction to paging.

3.3.2.5 Microcontrollers

A large number of the processors in embedded systems are in fact microcontrollers. Microcontrollers are typically not very complex and can be used easily. Due to their relevance for designing control systems, we introduce one of the most frequently used processors: the Intel 8051, available from many vendors (but no longer from Intel). This processor has the following characteristics:

- 8-bit CPU, optimized for control applications,
- large set of operations on Boolean data types,
- program address space of 64 k bytes,
- separate data address space of 64 k bytes,
- 4 k bytes of program memory on chip, 128 bytes of data memory on chip,
- 32 I/O lines, each of which can be addressed individually,
- 2 counters on the chip,
- universal asynchronous receiver/transmitter for serial lines available on the chip,
- clock generation on the chip,
- many variations commercially available.

All these characteristics are quite typical for microcontrollers.

3.3.2.6 Multimedia Processors/Instruction Sets

Registers and arithmetic units of many modern architectures are at least 64 bits wide. Therefore, two 32-bit data types ("double words"), four 16-bit data types ("words"), or eight 8-bit data types ("bytes") can be packed into a single register (see Fig. 3.25).

Fig. 3.25 Using 64-bit registers for packed words

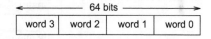

| ◄─────── 64 bits ───────► |
| word 3 | word 2 | word 1 | word 0 |

Arithmetic units can be designed such that they suppress carry bits at double word, word, or byte boundaries. Multimedia instruction sets exploit this fact by supporting operations on packed data types. Such instructions are sometimes called single instruction, multiple-data (SIMD) instructions, since a single instruction encodes operations on several data elements. With bytes packed into 64-bit registers, speedups of up to about eight over non-packed data types are possible. Data types are typically stored in packed form in memory. Unpacking and packing are avoided if arithmetic operations on packed data types are used. Furthermore, multimedia instructions can usually be combined with saturating arithmetic and, therefore, provide a more efficient form of overflow handling than standard instructions. Hence, the overall speedup achieved with multimedia instructions can be significantly larger than the factor of eight enabled by operations on packed data types. Due to the advantages of operations on packed data types, new instructions have been added to several processors. For example, so-called **streaming SIMD extensions** (SSE) have been added to Intel's family of Pentium®-compatible processors [239]. New instructions have also been called **short vector instructions** and introduced by Intel® as advanced vector extensions (AVX) [240].

3.3.2.7 Very Long Instruction Word (VLIW) Processors

Computational demands for embedded systems are increasing, especially when multimedia applications, advanced coding techniques, or cryptography are involved. Performance improvement techniques used in high-performance microprocessors are not appropriate for embedded systems: Driven by the need for instruction set compatibility, processors found, for example, in PCs spend a huge amount of resources and energy on automatically finding parallelism in application programs. Still, their performance is frequently not sufficient. For embedded systems, we can exploit the fact that instruction set compatibility with PCs is not required. Therefore, we can use instructions which explicitly identify operations to be performed in parallel. This is possible with **explicit parallelism instruction set computers** (EPICs). With EPICs, detection of parallelism is moved from the processor to the compiler. This avoids spending silicon and energy on the detection of parallelism at run-time. As a special case, we consider very long instruction word (VLIW) processors. For VLIW processors, several operations or instructions are encoded in a long instruction word (sometimes called **instruction packet**) and are assumed to be executed in parallel. Each operation/instruction is encoded in a separate field of the instruction packet. Each field controls certain hardware units. Four such fields are used in Fig. 3.26, each one controlling one of the hardware units.

Fig. 3.26 VLIW architecture (example)

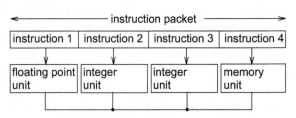

For VLIW architectures, the compiler has to generate instruction packets. This requires that the compiler is aware of the available hardware units and to schedule their use.

Instruction fields must be present, regardless of whether or not the corresponding functional unit is actually used in a certain instruction cycle. As a result, the code density of VLIW architectures may be low if insufficient parallelism is detected to keep all functional units busy. The problem can be avoided if more flexibility is added. For example, the Texas Instruments TMS 320C6xx family of processors implements a variable instruction packet size of up to 256 bits. In each instruction field, one bit is reserved to indicate whether or not the operation encoded in the next field is still assumed to be executed in parallel (see Fig. 3.27). No instruction bits are wasted for unused functional units.

Fig. 3.27 Instruction packets for TMS 320C6xx

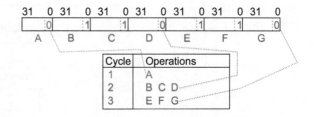

Due to its variable length instruction packets, TMS 320C6xx processors do not quite correspond to the classical model of VLIW processors. Due to their explicit description of parallelism, they are EPIC processors, though.

Partitioned Register Files

Implementing register files for VLIW and EPIC processors is far from trivial. Due to the large number of operations that can be performed in parallel, a large number of register accesses has to be provided in parallel. Therefore, a large number of ports is required. However, the delay, size, and energy consumption of register files increase with their number of ports. Hence, register files with very large numbers of ports are inefficient. As a consequence, many VLIW/EPIC architectures use partitioned register files. Functional units are then only connected to a subset of the register files. As an example, Fig. 3.28 shows the internal structure of the TMS 320C6xx processors. These processors have two register files, and each of them is connected to half of the functional units. During each clock cycle, only a single path from one register file to the functional units connected to the other register file is available.

Fig. 3.28 Partitioned register files for TMS 320C6xx

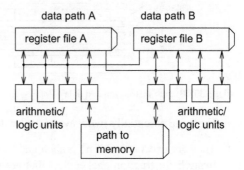

Alternative partitionings are considered by Lapinskii et al. [307].

Many DSP processors are actually VLIW processors. As an example, we are considering the experimental M3-DSP processor [156]. The M3-DSP processor is a VLIW processor containing (up to) 16 parallel data paths. These data paths are connected to a group memory, providing the necessary arguments in parallel (see Fig. 3.29).

Fig. 3.29 M3-DSP
(simplified)

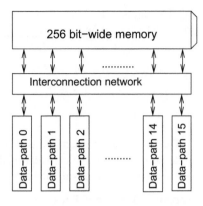

Predicated Execution

A potential problem of VLIW and EPIC architectures is their possibly large **delay penalty**: This delay penalty might originate from branch instructions found in some instruction packets. Instruction packets normally must pass through pipelines. Each stage of these pipelines implements only part of the operations to be performed by the instructions executed. The fact that branch instructions exist cannot be detected in the first stage of the pipeline. When the execution of the branch instruction is finally completed, additional instructions have already entered the pipeline (see Fig. 3.30).

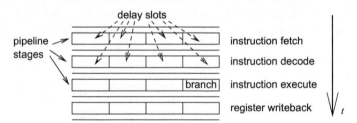

Fig. 3.30 Branch instruction and delay slots

There are essentially two ways to deal with these additional instructions:

1. They are executed as if no branch had been present. This case is called **delayed branch**. Instruction packet slots that are still executed after a branch are called **branch delay slots**. These branch delay slots can be filled with instructions which would be executed before the branch if there were no delay slots. However, it is normally difficult to fill all delay slots with useful instructions and some must be filled with no-operation instructions (NOPs). The term **branch delay penalty** denotes the loss of performance resulting from these NOPs.
2. The pipeline is stalled until instructions from the branch target address have been fetched. There are no branch delay slots in this case. In this organization, the branch delay penalty is caused by the stall.

Branch delay penalties can be significant and efficiency can be improved by avoiding branches if possible. In order to avoid branches originating from if-statements,

predicated instructions have been introduced. Predicated instructions have already been explained on p. 145. Predication can be used to implement small if-statements efficiently: The condition is stored in one of the condition registers and if-statement bodies are implemented as predicated instructions which depend on this condition. This way, if-statement bodies can be evaluated in parallel with other operations and no delay penalty is incurred.

The Crusoe™ processor is a (commercially finally unsuccessful) example of an EPIC processor designed for PCs [283]. Its instruction set includes 64-bit and 128-bit VLIW instructions. Efforts for making EPIC instruction sets available in the PC sector resulted in Intel's IA-64 instruction set [241] and its implementation in the Itanium® processor. Due to legacy problems, it has been used mainly in the server market. Many MPSoCs (see p. 159) are based on VLIW and EPIC processors.

3.3.2.8 Multi-core Processors

Processor features for single processors described above have helped to design high-performance processors in a resource-aware manner. However, it turned out that a further performance increase for single processors hits the **power wall**: a further increase in clock speeds would result in a too large power consumption and in too hot circuits. Further increase in the level of VLIW parallelism was not feasible either. Due to advances in fabrication technology, it is now feasible to manufacture multiple processors on the same semiconductor die. Multiple processors integrated on the same chip are called **multi-cores**. This is in contrast to multiprocessor systems which have been used in computing centers for decades. The integration of multiple cores on the same die enables a much faster communication, compared to multiprocessor systems. Also, this approach facilitates the sharing of resources (such as caches) among the cores. As an example, Fig. 3.31 demonstrates the architecture of the Intel® Core™ Duo [516].

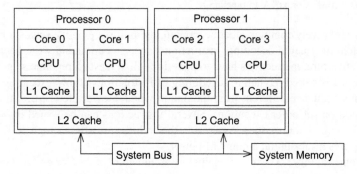

Fig. 3.31 Intel® Core™ Duo processor

In this case, L1 caches are private, whereas L2 are shared. Implementing efficient accesses to caches needs some consideration [516]. With such architectures, cache coherence is becoming an issue also within one die. This means, we have to know whether updates of data and possibly also instructions by one core are seen by the others. Protocols for automatic cache coherence, such as the MESI protocol, are known for many years in computer architecture [205]. Now, they have to be implemented on the chip. Scalability is an issue: for how many cores can we reasonably provide enough bandwidth in the communication architecture to always keep caches coherent? Also, the system memory bandwidth may be insufficient for a growing number of cores. Architectures other than the above Intel architecture exist.

In the architecture of Fig. 3.31, all processors are of the same type. Such an architecture is called a **homogeneous multi-core** architecture. Advantages of homogeneous multi-core architectures include the fact that the design effort is limited (processors will be replicated) and that software can easily be migrated from one processor to another one. This is very useful in case one of the cores fails.

In contrast to homogeneous multi-core architectures, there are also **heterogeneous multi-core architectures**. In this case, processors are of different types. In this way, we can select processors which are best suited for certain applications or run-time scenarios. Typically, heterogeneous architectures must be used to achieve the best energy efficiency that is feasible.

In order to find a good compromise between homogeneous and (totally) heterogeneous architectures, architectures with a single instruction set but different internal architectures, so-called **single-ISA heterogeneous multi-cores** [302], have been proposed. The ARM® big.LITTLE architecture is a very prominent example of this.

Figure 3.32 contains the pipeline architecture of the Cortex® A-15 processor [159].

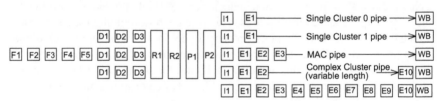

Fig. 3.32 ARM® Cortex® A-15 pipeline

It is a relatively complex pipeline, containing multiple pipeline stages for instruction fetch, instruction decoding, instruction issue, execution and write-back. Using this architecture, instructions have to pass through at least 15 pipeline stages before their result is stored. Dynamic scheduling of instructions allows executing instructions in a sequence different from the one in which they are fetched from memory (so-called out-of-order execution). Several instructions can be issued in one clock cycle (so-called multi-issue). As a result, the architecture offers a high performance, but requires a substantial amount of power.

In contrast, Fig. 3.33 shows the pipeline architecture of the Cortex® A-7 processor [159].

Fig. 3.33 ARM® Cortex® A-7 pipeline

It is a relatively simple pipeline. Overall, instructions pass through eight to eleven stages, they are always processed in the order in which they are fetched from memory and there is only a limited set of situations in which two instructions are issued at the same time. Due to this, the architecture requires very little power, but has a limited performance.

The resulting trade-offs between the two architectures shown in Fig. 3.34 [159].

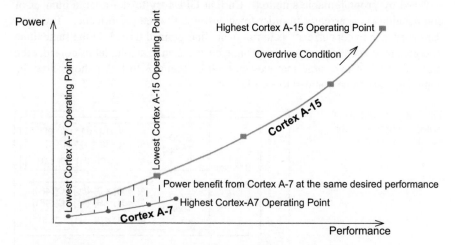

Fig. 3.34 DVFS curves for a large, representative workload on single A7 or A15

Obviously, the Cortex® A-15 is more appropriate for more demanding high-performance applications, e.g., in video processing. The Cortex® A-7 is more appropriate for "always-on applications" such as low-volume message processing. It would be a waste of energy if mobile phones would only contain Cortex® A-15 cores. Therefore, today's multi-core chips typically are heterogeneous in the sense that they contain a mixture high-performance and energy-efficient processors, as shown in Fig. 3.35.

Fig. 3.35 ARM® big.
LITTLE architecture
comprising Cortex® A-7 and
Cortex® A-15 cores

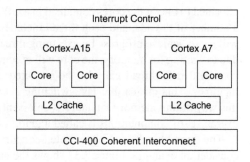

3.3.2.9   Graphics Processing Units (GPU)

In the last century, many computers used specialized graphics processing units (GPUs) in order to generate an appealing graphical representation of computer output. This hardwired solution suffered from being unable to support non-standard computer graphics algorithms. Therefore, these highly specialized GPUs have been replaced by programmable solutions. Current GPUs try to run a large number of computations concurrently in order to achieve the desired performance. The standard approach to concurrency is to run many fine-grained threads at the same time. The goal is to keep many processing units busy. As an example, let us consider the multiplication of two large matrices on a GPU. Figure 3.36 [205] shows how the computations can be mapped to a GPU.

Fig. 3.36 Partitioning of matrix multiplication for execution of a GPU

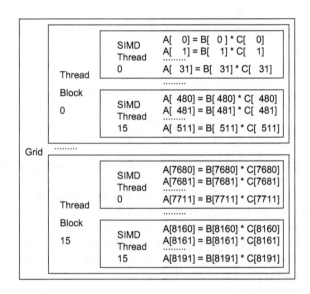

The matrix is partitioned into so-called thread blocks. Each thread block can be allocated to one of the cores contained in a GPU. Each thread block, in turn, contains a number of threads, and each thread includes a number of instructions. Each core will try to achieve progress by executing threads. If some thread gets blocked, e.g., due to waiting for memory, the core will execute some other thread. The instructions contained in a thread can also be executed concurrently, e.g., by using multiple pipelines. This means that fast switching between the execution of threads and in this way hiding memory latencies is an essential feature for GPUs. In Fig. 3.36, the overall set of computations is called a **grid**.

The thread blocks can be executed concurrently on contemporary GPUs. Let us consider an example. Figure 3.37 shows the architecture of the ARM® Mali™ T880 GPU [20].

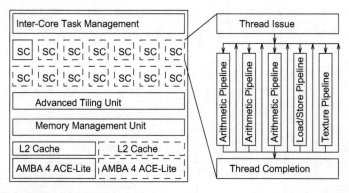

Fig. 3.37 ARM® Mali™ T880 GPU

The architecture is defined as intellectual property (IP), comprising a synthesizable model. In this model, the number of SC cores is configurable between one and sixteen. Each core includes several pipelines for the execution of arithmetic, load/store, or texture-related instructions. In the thread issue hardware, as many threads as possible are issued each clock phase. The GPU also contains additional components such as a memory management unit (see Appendix C), up to two caches and an AMBA® bus interface. Programming support includes an interface to the OpenGL library [459] and to OpenCL (see https://www.khronos.org/opencl/).

In general, GPU computing achieves high performances in an energy-efficient way (see also Sect. 3.7.1 on p. 186).

3.3.2.10 Multiprocessor Systems-on-a-Chip (MPSoCs)

Going one step further, heterogeneous multi-core systems have also been merged with GPUs. Figure 3.38 shows a contemporary heterogeneous multi-core system, also comprising a Mali GPU [19].

Mapping techniques for such processors are important, since examples demonstrate that a power efficiency close to that of ASICs can be achieved. For example, for IMEC's ADRES processor, an efficiency of $55 * 10^9$ operations per Watt (about 50% of the power efficiency of ASICs) has been predicted [238, 347]. However, the design effort for such architectures is larger than in the homogeneous case.

The architecture shown in Fig. 3.38 does not only contain processor cores. Rather, it comprises a number of additional system components, such as memory management units (see Appendix C) and interfaces for peripheral devices. Overall, the idea behind this integration is to avoid extra chips for such functionality. As a result, a whole system is integrated on one chip. Therefore, we are calling such an architecture a system-on-a-chip (SoC) or even a multiprocessor system-on-a-chip (MPSoC) architecture.

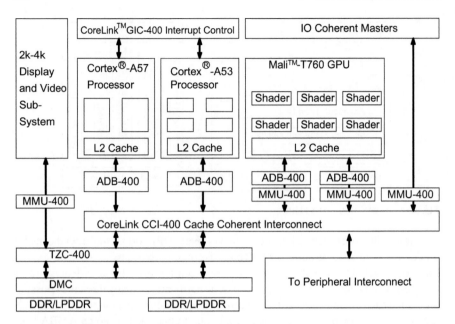

Fig. 3.38 ARM® big.LITTLE System on Chip (SoC)

The number and the diversity of components can be even larger. For example, there may be specialized processors for mobile communication or image processing. Figure 3.39 contains a simplified version of the floor-plan of the SH-MobileG1 chip [198].

Fig. 3.39 Floor-plan of the
SH-MobileG1 chip

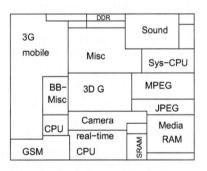

The chip demonstrates that highly specialized processors are being used: There are special processors for MPEG- and JPEG-encoding, for GSM- and 3G mobile communication, etc. In order to save energy, unused areas are typically powered down.

Interestingly, there are MPSoCs comprising a combination of processors which we introduced earlier: 66AK2x MPSoCs from Texas Instruments contain ARM® and C66xxx processors [507] (see Fig. 3.40) demonstrating the relevance of presented processors.

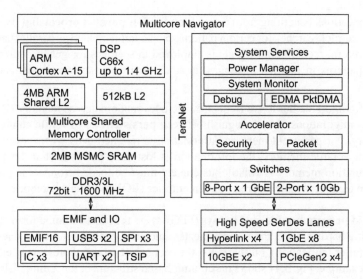

Fig. 3.40 MPSoC 66AK from Texas Instruments® containing ARM® processors and C6xx processors

3.3.3 Reconfigurable Logic

In many cases, full-custom hardware chips (ASICs) are too expensive and software-based solutions are too slow or too energy consuming. Reconfigurable logic provides a solution if algorithms can be efficiently implemented in custom hardware. It can be almost as fast as special-purpose hardware, but in contrast to special-purpose hardware, the performed function can be changed by using configuration data. Due to these properties, reconfigurable logic finds applications in the following areas:

- **Fast prototyping**: Modern ASICs can be very complex, and the design effort can be large and take a long time. It is, therefore, frequently desirable to generate a prototype, which can be used for experimenting with a system which behaves "almost" like the final system. The prototype can be more costly and larger than the final system. Also, its power consumption can be larger than the final system, some timing constraints can be relaxed, and only the essential functions need to be available. Such a system can then be used for checking the fundamental behavior of the future system.
- **Low-volume applications**: If the expected market volume is too small to justify the development of special-purpose ASICs, reconfigurable logic can be the right hardware technology for applications, for which software would be too slow or too inefficient.
- **Real-time systems**: the timing of reconfigurable logic-based designs is typically known very precisely. Therefore, they can be used to implement timing-predictable systems.

- Applications benefiting from a very **high level of parallel processing**: For example, parallel searches for certain patterns can be implemented as parallel hardware. Therefore, reconfigurable logic is employed in searches for genetic information, for patterns in Internet messages, in stock data, in seismic analysis, and many more.

Reconfigurable hardware frequently includes random access memory (RAM) to store configurations. We distinguish between **persistent** and **volatile** configuration memory. For persistent memory, information is retained even when power is shut off. For volatile memory, the information is lost once power is shut down. If the configuration memory is volatile, its content must be loaded from some persistent storage technology such as read-only memories (ROMs) or flash memories at start-up.

Field programmable gate arrays (FPGAs) are the most common form of reconfigurable hardware. As the name indicates, such devices are programmable "in the field" (after fabrication). Furthermore, they consist of arrays of processing elements. As an example, Fig. 3.41 shows the column-based structure of the Xilinx® UltraScale architecture [573][14]. Some columns contain I/O interfaces, clock devices, and/or RAM. Other columns comprise **configurable logic blocks** (CLB)s, special hardware for digital signal processing, and some RAM. CLBs are the key components. They provide configurable functions. The architecture of Xilinx® UltraScale CLBs is shown in Fig. 3.42 [574].

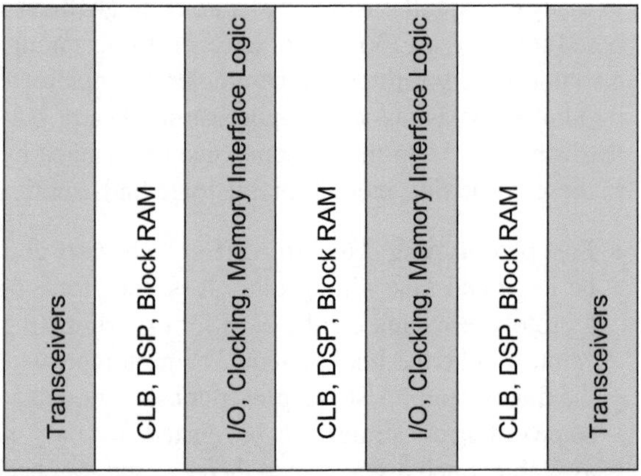

Fig. 3.41 Floor-plan of column-based Xilinx® UltraScale FPGAs

In this architecture, each CLB contains eight blocks. Each block comprises a RAM which is used to implement logic functions by a look-up table (LUT, shown in red), two registers, multiplexers, and some additional logic[15]. Each LUT has six address inputs and two outputs. It can implement any single Boolean function of six

---

[14]Rotation of this figure would improve its readability, but would contradict the official designation of this layout style.

[15]An intermediate hierarchy level called *slices* was present in earlier devices from Xilinx® but is not relevant for the discussion in this book.

Fig. 3.42 Xilinx® UltraScale CLB

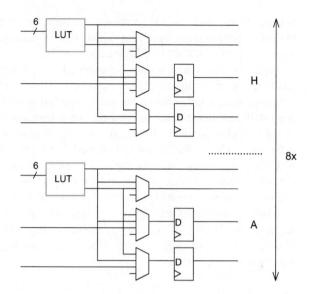

variables or two functions of five variables (provided that the two functions share input variables). This means, all 2^{64}, respectively, all 2^{32} Boolean functions of six or five inputs can be implemented! This is the key means for achieving **configurability**. In addition, the logic contained in such a block can also be configured. This includes the control of the two registers, which can be programmed to store results of the LUT or some direct input values. Blocks in a CLB can be combined to form adders, multiplexers, shift registers, or memories. Configuration data determines the setting of multiplexers in the CLBs, the clocking of registers and RAM, the content of RAM components, and the connections between CLBs. Some of the LUTs can also be used as RAM. A single CLB can store up to 512 bits.

Several CLBs can be combined to create, for example, adders having a larger bit width, memories having a larger capacity or complex logic functions.

Currently available FPGAs comprise a large number of specialized blocks, such as hardware for digital signal processing (DSP), some memory, high-speed I/O devices for various I/O standards, a decryption facility for FPGA configuration data, debugging support, ADCs, high-speed clocking, etc. For example, Virtex® UltraScale™ VU13P devices include 1728 k LUTs, 48 Mbit distributed RAM, 94.5 Mbit "Block RAM", 360 Mbit "UltraRAM", about 12 k specialized DSP devices, 4 PCIe® devices, Ethernet interfaces, and up to 832 I/O pins [572].

Integration of reconfigurable computing with processors and software is simplified if processors are available in the FPGAs. There may be either **hard cores** or **soft cores**. For hard cores, the layout contains a special area implementing a core in a dense way. This area cannot be used for anything but the hard core. Soft cores are available as synthesizable models which are mapped to standard CLBs. Soft cores are more flexible, but less efficient than hard cores. Soft cores can be implemented on any FPGA chip. The MicroBlaze processor [570] is an example of such cores. At the time of writing this book, hard cores are available, for example, on Zynq UltraScale+ MPSoCs. They contain up to four ARM® Cortex-A53 cores, two ARM Cortex-R5 cores, and a Mali-400MP2 GPU processor [573].

Typically, configuration data is generated from a high-level description of the functionality of the hardware, for example in VHDL. FPGA vendors provide the necessary design kits. Ideally, the same description could also be used for generating ASICs automatically. In practice, some interaction is required. Exploitation of the available parallelism typically requires manually parallelized applications, since automatic parallelization is frequently very limited. The parallelism offered by FPGAs is typically not fully exploited if all computations are mapped to processor cores. Overall, FPGAs allow implementing a huge variety of hardware devices without any need to create hardware other than FPGA boards.

Alternate providers of FPGAs include Altera® (see http://www.altera.com, acquired by Intel®), Lattice Semiconductor (see http://www.latticesemi.com), Microsemi (formerly Actel, see http://www.microsemi.com), and Quicklogic (see http://www.quicklogic.com).

3.4 Memories

3.4.1 Conflicting Goals

Data, programs, and FPGA configurations must be stored in some kind of memory. Memories must have a capacity as large as required by the applications, provide the expected performance and still be efficient in terms of cost, size, and energy consumption. Requirements for memories also include the expected reliability and access granularity (e.g., bytes, words, pages). Furthermore, we distinguish between **persistent** and **volatile** memory (see p. 162).

The mentioned requirements are conflicting, as has already been observed by Burks, Goldstine, and von Neumann in 1946 [79]:

"Ideally one would desire an indefinitely large memory capacity such that any particular ... word ... would be immediately available – i.e., in a time which is ... shorter than the operation time of a fast electronic multiplier. ... It does not seem possible physically to achieve such a capacity."

Access times of some currently available memories can be estimated with CACTI. These estimates are based on the tentative generation of a memory layout and the extraction of capacitances [561]. Figure 3.43 shows the results for a range of exponentially increasing sizes [36].

Obviously, the access time increases as a function of the capacity of memories: the larger the memory, the longer it takes to access information. In addition, Fig. 3.43 also includes the energy consumption. Large memories also tend to be energy-inefficient. The impact of the capacity of the memory on the energy consumption is even larger than the impact on the access time.

Fig. 3.43 Delay and access
time of random access
memory as predicted by
CACTI

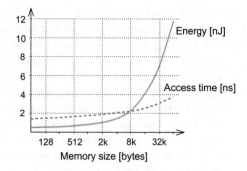

For a number of years, the difference in speeds between processors and memories
increased (see Fig. 3.44) until processor clock rates saturated (around 2003).

Fig. 3.44 Historical
speed gap increase (until
about 2003)

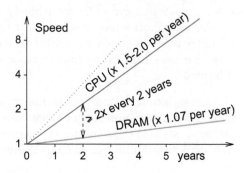

While the speed of memories increased by only a factor of about 1.07 per year,
overall processor performance increased by a factor of 1.5–2 per year [341]. Overall, the gap between processor performance and memory speeds has become large.
Accordingly, a further increase of the overall performance is made at least very difficult due to memory access times. This fact has also been called the **memory wall**
[341]. Further increase of clock rates of single processors has come to a standstill, but
the large gap remains which existed when clock speeds became essentially saturated
and multi-cores require additional memory bandwidth. As a result, we have to find
compromises between the different requirements for the memory architecture.

3.4.2 Memory Hierarchies

Due to the observed conflicts, Burks, Goldstine, and von Neumann wrote already
in 1946 [79]: *"We are therefore forced to recognize the possibility of constructing a
hierarchy of memories, each of which has greater capacity than the preceding but
which is less quickly accessible"*.

The exact structure of the hierarchy depends on technological parameters and also
on the application area. Typically, we can identify at least the following levels in the
memory hierarchy:

- **Processor registers** can be seen as the fastest level in the memory hierarchy, with only a limited capacity of at most a few hundred words.
- The **working memory** (or **main memory**) of computer systems implements the storage implied by processor memory addresses. Usually it has a capacity between a few Megabytes and some Gigabytes and is volatile.
- Typically, there is a large access speed difference between the main memory and registers. Hence, many systems include some type of buffer memory. Frequently used buffer memories include **caches**, **translation look-aside buffers** (TLBs, see Appendix C), and **scratchpad memory** (SPM). In contrast to PC-like systems and compute servers, the architecture of these small memories should guarantee a predictable real-time performance. A combination of small memories containing frequently used data and instructions and a larger memory containing the remaining data and instructions is generally also more energy efficient than a single, large memory.
- Memories introduced so far are normally implemented in volatile memory technologies. In order to provide persistent storage, some different memory technology must be used. For embedded systems, flash memory is frequently the best solution. In other cases, hard disks or Internet-based storage solutions (like the "cloud") may be used.

Memory hierarchies can be exploited in order to achieve a compromise between the design goals for the memory. Memory partitioning has been considered, for example, by A. Macii [342]. New memory technologies (including persistent memories) have the potential to change currently dominating hierarchies [370].

3.4.3   Register Files

The mentioned impact of the storage capacity on access times and energy consumption applies even to small memories such as register files. Figure 3.45 shows the cycle time and the power as a function of the size of memories used as register files [448]. The power needs to be considered due to frequent accesses to registers, as a result of which they can get very hot.

Fig. 3.45 Cycle time and power as a function of the register file size

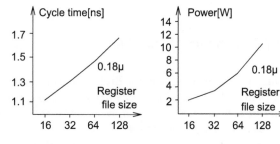

3.4.4   Caches

For caches it is required that the hardware checks whether or not the cache has a valid copy of the information associated with a certain address. This check involves comparing the tag fields of caches, containing a subset of the relevant address bits [205]. If the cache has no valid copy, the information in the cache is automatically updated.

Caches were initially introduced in order to provide good run-time efficiency. The name is derived from the French word *cacher* (to hide), indicating that programmers do not need to see or to be aware of caches, since updating information in caches is automatic. However, when large amounts of information need to be accessed, caches are not so invisible any more. This has been demonstrated very nicely by Drepper [134]. Drepper analyzed execution times of a program traversing a linear list of entries. Each entry contained NPAD 64-bit words. Execution times were measured for a Pentium P4 processor comprising a 16 kB level-1 cache requiring four processor cycles per access, a 1 MB level-2 cache requiring 14 processor cycles per access, and a main memory requiring 200 cycles per access. Figure 3.46 shows the average number of cycles per access to one list element as a function of the total size of the list for the case NPAD=0. For small sizes of the list, four cycles are required per list

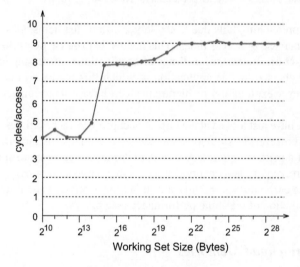

Fig. 3.46 Average number of cycles per access for NPAD=0

element. This means that we are almost always accessing the level-1 cache, since it is large enough for this size of the list. If we increase the size of the list, we need eight cycles per access on average. In this case, we are accessing the level-2 cache. However, since the cache block size is large enough to hold two list elements, only every second access is actually an access to the level-2 cache. For even larger lists, the access time increases to nine cycles. In these cases, the list is larger than the level-2 cache, but automatic prefetching of level-2 cache entries hides some of the access latency of the main memory.

Figure 3.47 shows the average number of cycles per access to one list element as a function of the total size of the list for cases NPAD=0, 7, 15, and 31. For NPAD=7,

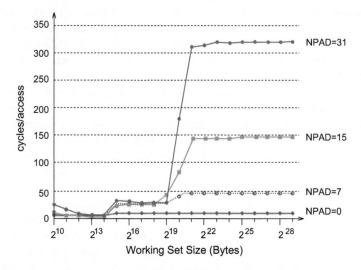

Fig. 3.47 Average number of cycles per access for NPAD=0, 7, 15, 31

15, and 31, prefetching fails due to the larger size of list items. Obviously, we see a dramatic increase of access times. This means that **the cache architecture has a strong impact on the execution times of applications**. Increasing cache size will only change the size of the application at which this increase in execution times happens. Clever exploitation of hierarchies can have a large impact on execution times.

So far, we have just looked at the impact of capacity on access times. In the context of Fig. 3.43 however, it is obvious that caches potentially also improve the energy efficiency of a memory system. Accesses to caches are accesses to small memories and, therefore, require less energy per access than large memories.

Predicting cache misses and hits at design time is difficult and is a burden for the accurate prediction of real-time performance (see p. 238).

3.4.5 Scratchpad Memories

Alternatively, small memories can be mapped into the address space (see Fig. 3.48).

Fig. 3.48 Memory map with scratchpad included

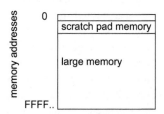

Such memories are called **scratchpad memories** (SPMs) or **tightly coupled memories** (TCM). SPMs are accessed by a proper selection of memory addresses. There is no need for checking tags, as for caches. Instead, the SPM is accessed whenever some simple address decoder is signaling an address to be in the address range of the SPM. SPMs are typically integrated together with processors on the same die. Hence, they are a special case of **on-chip memories**. For n-way set-associative caches, reads are usually reading n entries in parallel and select the right entry only afterward. These energy hungry parallel reads are avoided for SPMs. As a result, SPMs are very energy-efficient.

Figure 3.49 shows a comparison between the energy required per access to the scratchpad (SPM) and the energy required per access to the cache.

Fig. 3.49 Energy consumption per scratchpad and cache access

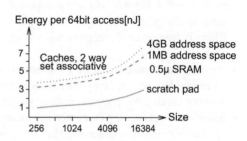

For a two-way set-associative cache, the two values differ by a factor of about three. The values in this example were computed using the energy consumption for RAM arrays as estimated by CACTI [561]. A detailed comparison between figures of merit for caches and scratchpads was published by Banakar et al. [36].

Frequently used variables and instructions should be allocated to the address space of SPMs. SPMs can improve the memory access times very predictably if the compiler is in charge of keeping frequently used variables in the SPM (see p. 350).

3.5 Communication

Information must be communicated before it can be processed in an embedded system. Communication is particularly important for the Internet of Things. Information can be communicated through various **channels**. Channels are abstract entities characterized by the essential properties of communication, such as maximum information transfer capacity and noise parameters. The probability of communication errors can be computed using communication theory techniques. The physical entities enabling communication are called communication **media**. Important media classes include: wireless media (radio frequency media, infrared), optical media (fibers), and wires.

There is a huge variety of communication requirements between the various classes of embedded systems. In general, connecting the different embedded hardware components is far from trivial. Some common requirements can be identified.

3.5.1 Requirements

The following list contains some of the requirements that must be met:

- **Real-time behavior**: This requirement has far-reaching consequences on the design of the communication system. Several low-cost solutions such as standard Ethernet fail to meet this requirement.
- **Efficiency**: Connecting different hardware components can be quite expensive. For example, point-to-point connections in large buildings are almost impossible. Also, it has been found that separate wires between control units and external devices in cars significantly add to the cost and the weight of the car. With separate wires, it is also very difficult to add new components. The need of providing cost-efficient designs also affects the way in which power is made available to external devices. There is frequently the need to use a central power supply in order to reduce the cost.
- **Appropriate bandwidth and communication delay**: bandwidth requirements of embedded systems may vary. It is important to provide sufficient bandwidth without making the communication system too expensive.
- **Support for event-driven communication**: polling-based systems provide a very predictable real-time behavior. However, their communication delay may be too large and there should be mechanisms for fast, event-oriented communication. For example, emergency situations should be communicated immediately and should not remain unnoticed until some central controller polls for messages.
- **Robustness**: Cyber-physical systems may be used at extreme temperatures, close to major sources of electromagnetic radiation, etc. Car engines, for example, can be exposed to temperatures less than -20 and up to $+180$ degrees Celsius (-4 to 356 degrees Fahrenheit). Voltage levels and clock frequencies could be affected due to this large variation in temperatures. Still, reliable communication must be maintained.
- **Fault tolerance**: Despite all the efforts for robustness, faults may occur. Cyber-physical systems should be operational even after faults, if at all feasible. Restarts, like the ones found in personal computers, cannot be accepted. This means that retries may be required after attempts to communicate failed. A conflict exists with the first requirement: if we allow retries, then it is difficult to meet strict real-time requirements.
- **Maintainability, diagnosability**: obviously, it should be possible to repair embedded systems within reasonable time frames.
- **Security/privacy**: Ensuring security/privacy of confidential information may require the use of encryption.

These communication requirements are a direct consequence of the general characteristics of embedded/cyber-physical systems mentioned in Chap. 1. Due to the conflicts between some of the requirements, compromises must be made. For example, there may be different communication modes: one high-bandwidth mode guaranteeing real-time behavior but no fault tolerance (this mode is appropriate for multimedia streams) and a second fault-tolerant, low-bandwidth mode for short messages that must not be dropped.

3.5.2 Electrical Robustness

There are some basic techniques for electrical robustness. Digital communication within chips is normally using so-called single-ended signaling. For single-ended signaling, signals are propagated on a single wire (see Fig. 3.50).

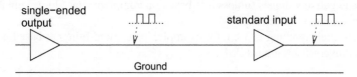

Fig. 3.50 Single-ended signaling

Such signals are represented by voltages with respect to a common ground (less frequently by currents). A single ground wire is sufficient for a number of single-ended signals. Single-ended signaling is very much susceptible to external noise. If external noise (originating from, for example, motors being switched on) affects the voltage, messages can easily be corrupted. Also, it is difficult to establish high-quality common ground signals between a large number of communicating systems, due to the resistance (and self inductance) on the ground wires. This is different for differential signaling. For differential signaling, each signal needs two wires (see Fig. 3.51).

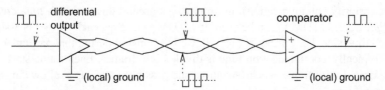

Fig. 3.51 Differential signaling

Using differential signaling, binary values are encoded as follows: if the voltage on the first wire with respect to the second is positive, then this is decoded as '1', otherwise values are decoded as '0'. The two wires will typically be twisted to form so-called **twisted pairs**. There will be local ground signals, but a nonzero voltage between the local ground signals does not hurt. Advantages of differential signaling include:

- Noise is added to the two wires in essentially the same way. The comparator, therefore, removes almost all the noise.

- The logic value depends just on the polarity of the voltage between the two wires. The magnitude of the voltage can be affected by reflections or because of the resistance of the wires; this has no effect on the decoded value.
- Signals do not generate any currents on the ground wires. Hence, the quality of the ground wires becomes less important.
- No common ground wire is required. Hence, there is no need to establish a high-quality ground wiring between a large number of communicating partners.
- As a consequence of the properties mentioned so far, differential signaling allows a larger throughput than single-ended signaling.

However, differential signaling requires two wires for every signal and it also requires negative voltages (unless it is based on complementary logic signals using voltages for single-ended signals).

Differential signaling is used, for example, in standard Ethernet-based networks and the universal serial bus (USB).

3.5.3 Guaranteeing Real-Time Behavior

For internal communication, computers may be using dedicated point-to-point communication or shared buses. Point-to-point communication can have a good real-time behavior, but requires many connections and there may be congestion at the receivers. Wiring is easier with common, shared buses. Typically, such buses use priority-based arbitration if several access requests to the communication media exist (see, for example, [205]). Priority-based arbitration comes with poor timing predictability, since conflicts are difficult to anticipate at design time. Priority-based schemes can even lead to "starvation" (low-priority communication can be completely blocked by higher priority communication). In order to get around this problem, *time division multiple access* (TDMA) can be used. In a TDMA-scheme, each partner is assigned a fixed time slot. The partner is only allowed to transmit during that particular time slot. Typically, communication time is divided into frames. Each frame starts with some time slot for frame synchronization, and possibly some gap to allow the sender to turn off (see Fig. 3.52, [291]).

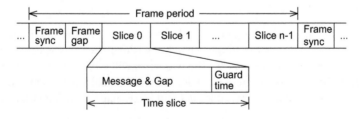

Fig. 3.52 TDMA-based communication

This gap is followed by a number of slices, each of which serves for communicating messages. Each slice also contains some gap and guard time to take clock speed variations of the partners into account. Slices are assigned to communication partners. Variations of this scheme exist. For example, truncation of unused slices or the assignment of partners to several slices are feasible. TDMA reduces the maximum amount of data available per frame and partner, but guarantees a certain bandwidth for all partners. Starvation can be avoided. The ARM AMBA-bus [18] includes TDMA-based bus allocation.

Communication between computers is frequently based on Ethernet standards. For 10 Mbit/s and 100 Mbit/s versions of Ethernet, there can be collisions between various communication partners. This means: Several partners are trying to communicate at about the same time and the signals on the wires are corrupted. Whenever this occurs, the partners must stop communications, wait for some time, and then retry. The waiting time is chosen at random, so that it is not very likely that the next attempt to communicate results in another collision. This method is called **carrier-sense multiple access/collision detect** (CSMA/CD). For CSMA/CD, communication time can become huge, since conflicts can repeat a large number of times, even though this is not very likely. Hence, CSMA/CD cannot be used when real-time constraints must be met.

This problem can be solved with CSMA/CA (**carrier-sense multiple access/ collision avoidance**). As the name indicates, collisions are completely avoided, rather than just detected. For CSMA/CA, priorities are assigned to all partners. Communication media are allocated to communication partners during **arbitration phases**, which follow communication phases. During arbitration phases, partners wanting to communicate indicate this on the media. Partners finding such indications of higher priority must immediately remove their indication.

Provided that there is an upper bound on the time between arbitration phases, CSMA/CA guarantees a predictable real-time behavior for the partner having the highest priority. For other partners, real-time behavior can be guaranteed if the higher priority partners do not continuously request access to the media.

Note that high-speed versions of Ethernet (≥ 1 Gbit/s) also avoid collisions. TDMA-schemes are also used for wireless communication. For example, mobile phone standards such as GSM, use TDMA for accesses to the communication medium.

3.5.4 Examples

- **Sensor/actuator buses**: Sensor/actuator buses provide communication between simple devices such as switches or lamps and the processing equipment. There may be many such devices, and the cost of the wiring needs special attention for such buses.
- **Field buses**: Field buses are similar to sensor/actuator buses. In general, they are supposed to support larger data rates than sensor/actuator buses. Examples of field buses include the following:

– **Controller Area Network (CAN)**: This bus was developed in 1981 by Bosch and Intel for connecting controllers and peripherals. It is popular in the automotive industry, since it allows the replacement of a large amount of wires by a single bus. Due to the size of the automotive market, CAN components are relatively cheap and are therefore also used in other areas such as smart homes and fabrication equipment. CAN has the following properties:
  · differential signaling with twisted pairs,
  · arbitration using CSMA/CA,
  · throughput between 10 kbit/s and 1 Mbit/s,
  · low and high-priority signals,
  · maximum latency of 134 μs for high-priority signals, and
  · coding of signals similar to that of serial (RS-232) lines of PCs, with modifications for differential signaling.

  CSMA/CA-based arbitration does not prevent starvation. This is an inherent problem of the CAN protocol.
– The **Time-Triggered-Protocol (TTP)** [293] for fault-tolerant safety systems such as air bags in cars.
– **FlexRay**™ [160, 244] is a TDMA protocol which has been developed by the FlexRay consortium (BMW, Daimler AG, General Motors, Ford, Bosch, Motorola, and Philips Semiconductors). FlexRay is a combination of a variant of the TTP and the byteflight [89] protocol.

  FlexRay includes a static as well as a dynamic arbitration phase. The static phase uses a TDMA-like arbitration scheme. It can be used for real-time communication, and starvation can be avoided. The dynamic phase provides a good bandwidth for non-real-time communication. Communicating partners can be connected to up to two buses for fault tolerance reasons. **Bus guardians** may protect partners against partners flooding the bus with redundant messages, so-called **babbling idiots**. Partners may use their own local clock periods. Periods common to all partners are defined as multiples of such local clock periods. Time slots allocated to partners for communication are based on these common periods.

  The levi simulation allows simulating the protocol in a laboratory environment [470].
– **LIN** (local interconnect network) is a low-cost communication standard for connecting sensors and actuators in the automotive domain [330].
– **MAP**: MAP is a bus designed for car factories.
– **EIB**: The European installation bus (EIB) is a bus designed for smart homes.

• The **Inter-Integrated Circuit** (I^2C) Bus [406] is a simple low-cost bus designed to communicate at short distances (meter range) with relatively low data rates. The bus needs only four wires: ground, SCL (clock), SDA (data), and a voltage supply line. Data and clock lines are open collector lines (see pp. 86 to 87). This means that connected devices pull these lines only toward ground. Separate resistors are needed to pull these lines up. The standard speed of I^2C is 100 kb/s, but versions for 10 kb/s and up to 3.4 Mb/s do also exist. The voltage on the supply voltage line may vary between interfaces. Only the standards for detecting high and low logic levels are defined relative to the supply voltage. The bus is supported on some microcontroller boards.

- **Wired multimedia communication**: for wired multimedia communication, larger data rates are required. Example: **MOST** (media-oriented systems transport) is a communication standard for multimedia and infotainment equipment in the automotive domain [385]. Standards such as IEEE 1394 (FireWire) may be used for the same purpose.
- **Wireless communication**: This kind of communication is becoming more popular. There are several standards for wireless communication, including the following:

  - **Mobile communication** is becoming available at increased data rates. 7 Mbit/s is obtained with HSPA (high-speed packet access). About ten times higher rates are available with **long-term evolution (LTE)**.
  - **Bluetooth** is a standard for connecting devices such as mobile phones and their headsets over short distances.
  - **Wireless local area networks** (WLANs) are standardized as IEEE standard 802.11, with several supplementary standards.
  - **ZigBee** (see http://www.zigbee.org) is a communication protocol designed to create personal area networks using low-power radios. Applications include home automation and the Internet of Things.
  - **Digital European Cordless Telecommunications (DECT)** is a standard used for wireless phones. It is being used throughout the world, except for different frequencies used in North America (see https://en.wikipedia.org/wiki/Digital_Enhanced_Cordless_Telecommunications).

3.6 Output

Output devices of embedded/cyber-physical systems include:

- **Displays**: Display technology is an area which is extremely important. Accordingly, a large amount of information [478] exists on this technology. Major research and development efforts lead to new display technology such as organic displays [328]. Organic displays are emitting light and can be fabricated with very high densities. In contrast to LCD displays, they do not need backlight and polarizing filters. Major changes are, therefore, expected in these markets.
- **Electro-mechanical devices**: These influence the environment through motors and other electro-mechanical equipment.

Analog as well as digital output devices are used. In the case of analog output devices, the digital information must first be converted by digital-to-analog converters (DACs). These converters can be found on the path from analog inputs of embedded systems to their outputs. Figure 3.53 shows the naming convention of signals along the path which we use. Purpose and function of the boxes will be explained in this section.

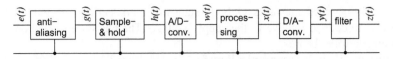

Fig. 3.53 Naming convention for signals between analog inputs and outputs

3.6.1 Digital-to-Analog Converters

Digital-to-analog converters (DACs) are not very complex. Figure 3.54 shows the schematic of a simple so-called weighted-resistor DAC.

Fig. 3.54 DAC

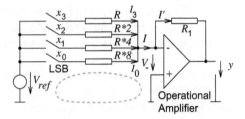

The key idea of the converter is to first generate a current which is proportional to the value represented by a digital signal x. Such a current can hardly be used by a following system. Therefore, this current is converted into a proportional voltage y. This conversion is done with an operational amplifier (depicted by a triangle in Fig. 3.54). Essential characteristics of operational amplifiers are described in Appendix B of this book.

How do we compute the output voltage y? Consider the four resistors on the left in Fig. 3.54. The current through any resistor is zero if the corresponding element of digital signal x is '0'. If it is '1', the current corresponds to the weight of that bit, since resistor values are chosen accordingly. Now, consider the loop indicated by the red dashed line in Fig. 3.54. We can apply Kirchhoff's Loop Rule (see Appendix B) to the loop turned on by the least significant bit x_0 of x. Let us start the loop traversal at the corresponding resistor and continue in a clockwise fashion. The second term is the voltage V_- between the inputs of the operational amplifier, counted as positive, since we proceed in the direction of the arrow. The third term is contributed by the constant voltage source, counted as negative, since we proceed against the direction of the arrow. Overall, we have

$$x_0 * I_0 * 8 * R + V_- - V_{ref} = 0 \qquad (3.22)$$

V_- is approximately 0 (see Appendix B, Eq. (B.14)). Therefore, we have

$$I_0 = x_0 * \frac{V_{ref}}{8 * R} \qquad (3.23)$$

Corresponding equations hold for the currents I_1 to I_3 through the other resistors. We can now apply Kirchhoff's Node Rule to the circuit node connecting all resistors. At this node, the outgoing current must be equal to the sum of the incoming currents. Therefore, we have

$$I = I_3 + I_2 + I_1 + I_0 \tag{3.24}$$

$$I = x_3 * \frac{V_{ref}}{R} + x_2 * \frac{V_{ref}}{2*R} + x_1 * \frac{V_{ref}}{4*R} + x_0 * \frac{V_{ref}}{8*R}$$

$$= \frac{V_{ref}}{R} * \sum_{i=0}^{3} x_i * 2^{i-3} \tag{3.25}$$

Now, we can apply Kirchhoff's Loop Rule to the loop comprising R_1, y, and V_-. Since V_- is approximately 0, we have:

$$y + R_1 * I' = 0. \tag{3.26}$$

Next, we can apply Kirchhoff's Node Rule to the node connecting I, I', and the inverting signal input of the operational amplifier. The current into this input is practically zero, and currents I and I' are equal: $I = I'$. Hence, we have:

$$y + R_1 * I = 0 \tag{3.27}$$

From Eqs. (3.25) and (3.27) we obtain:

$$y = -V_{ref} * \frac{R_1}{R} * \sum_{i=0}^{3} x_i * 2^{i-3} = -V_{ref} * \frac{R_1}{8*R} * nat(x) \tag{3.28}$$

nat denotes the natural number represented by digital signal x. Obviously, y is proportional to the value represented by x. Positive output voltages and bit vectors representing two's complement numbers require minor extensions.

From a DSP point of view, $y(t)$ is a function over a discrete time domain: It provides us with a **sequence** of voltage levels. In our running example, it is defined only over integer times. From a practical point of view, this is inconvenient, since we would typically observe the output of the circuit of Fig. 3.54 continuously. Therefore, DACs are frequently extended by a **"zero-order hold" functionality**. This means that the converter will keep the previous value until the next value is converted. Actually, the DAC of Fig. 3.54 will do exactly this if we do not change the settings of the switches until the next discrete time instant. Hence, the output of the converter is a step function $y'(t)$ corresponding to the sequence $y(t)$[16]. $y'(t)$ is a function over the continuous time domain.

---

[16]In practice, due to rise and fall times being > 0, transitions from one step to the next will not be ideal, but take some time.

As an example, let us consider the output resulting from the conversion of the signal of Eq. (3.3), assuming a resolution of 0.125. For this case, Fig. 3.55 shows $y'(t)$ instead of $y(t)$, since $y'(t)$ is a bit easier to visualize.

Fig. 3.55 $y'(t)$ (red) generated from signal $e_3(t)$ (blue) (Eq. (3.3)) sampled at integer times

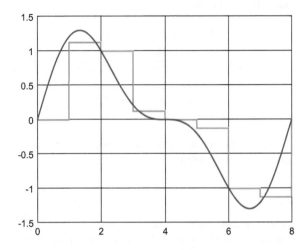

DACs enable a conversion from time- and value-discrete signals to signals in the continuous time and value domain. However, neither $y(t)$ nor $y'(t)$ reflect the values of the input signal in-between the sampling instances.

3.6.2   Sampling Theorem

Suppose that the processors used in the hardware loop forward values from ADCs unchanged to the DACs. We could also think of storing values $x(t)$ on a CD and aiming at generating an excellent analog audio signal. Would it be possible to reconstruct the original analog voltage $e(t)$ (see Figs. 3.8, 3.21, and 3.53) at the outputs of the DACs?

It is obvious that reconstruction is not possible if we have aliasing of the type described in Fig. 3.7 on p. 130[17]. So, we assume that the sampling rate is larger than twice the highest frequency of the decomposition of the input signal into sine waves (sampling criterion, see Eq. (3.8)). Does meeting this criterion allow us to reconstruct the original signal? Let us have a closer look!

Feeding DACs with a discrete sequence of digital values will result in a sequence of analog values being generated. Values of the input signal in-between the sampling instances are not generated by DACs. The simple zero-order hold functionality (if present) would generate only step functions. This seems to indicate that reconstruc-

---

[17]Reconstruction may be possible if additional information about the signal is available, i.e., if we restrict ourselves to certain signal types.

tion of $e(t)$ would require an infinitely large sampling rate, such that all intermediate values can be generated.

However, there could be some kind of smart interpolation computing values in-between the sampling instances from the values at sampling instances. And indeed, sampling theory [418] tells us that a corresponding time-continuous signal $z(t)$ can be constructed from the sequence $y(t)$ of analog values.

Let $\{t_s\}$, $s = \ldots, -1, 0, 1, 2, \ldots$ be the time points at which we sample our input signal. Let us assume a constant sampling rate of $f_s = \frac{1}{T_s}$ ($\forall s : T_s = t_{s+1} - t_s$). Then, sampling theory tells us that we can approximate $e(t)$ from $y(t)$ as follows:

$$z(t) = \sum_{s=-\infty}^{\infty} \frac{y(t_s)sin\frac{\pi}{T_s}(t - t_s)}{\frac{\pi}{T_s}(t - t_s)} \tag{3.29}$$

This equation is known as the **Shannon–Whittaker interpolation**. $y(t_s)$ is the contribution of signal y at sampling instance t_s. The influence of this contribution decreases the further t is away from t_s. The decrease follows a weighting factor, also known as the *sinc*-function:

$$sinc(t - t_s) = \frac{sin(\frac{\pi}{T_s}(t - t_s))}{\frac{\pi}{T_s}(t - t_s)} \tag{3.30}$$

which decreases non-monotonically as a function of $|t - t_s|$. This weighting factor is used to compute values in-between the sampling instances. Figure 3.56 shows the weighting factor for the case $T_s = 1$.

Fig. 3.56 Visualization of Eq. (3.30) used for interpolation

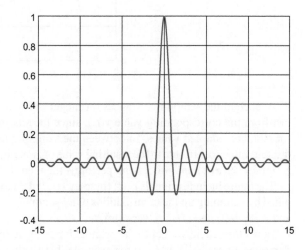

Using the *sinc*-function, we can compute the terms of the sum in Eq. (3.29). Figures 3.57 and 3.58 show the resulting terms if $e(t) = e_3(t)$ and processing performs the identify function ($x(t) = w(t)$).

Fig. 3.57 $y'(t)$ (red) and the first three terms of Eq. (3.29)

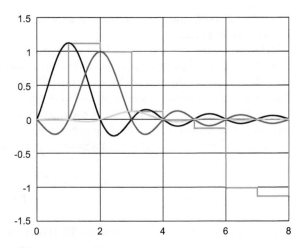

Fig. 3.58 $y'(t)$ (red) and the last three nonzero terms of Eq. (3.29)

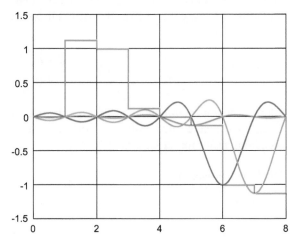

At each of the sampling instances t_s (integer times in our case), $z(t_s)$ is computed just from the corresponding value $y(t_s)$, since the *sinc*-function is zero in this case for all other sampled values. In between the sampling instances, all of the adjacent discrete values contribute to the resulting value of $z(t)$. Figure 3.59 shows the resulting $z(t)$ if $e(t) = e_3(t)$ and processing performs the identify function ($x(t) = w(t)$).

The figure includes signals $e_3(t)$ (blue), $y'(t)$ (red), and $z(t)$ (magenta). $z(t)$ is computed by summing up the contributions of all sampling instances shown in Figs. 3.57 and 3.58. $e_3(t)$ and $z(t)$ are very similar.

How close could we get to the original input signal by implementing Eq. (3.29)? Sampling theory tells us (see, for example, [418]) that **equation (3.29) computes an exact approximation** if the sampling criterion (Eq. (3.8)) is met. Therefore, let us see how we can implement Eq. (3.29).

Fig. 3.59 $e_3(t)$ (blue), $y'(t)$ (red), $z(t)$ (magenta)

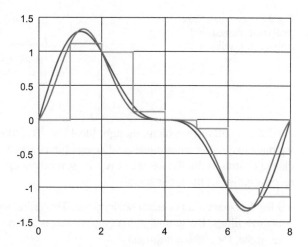

How do we compute Eq. (3.29) in an electronic system? We cannot compute this equation in the discrete time domain using a digital signal processor for this, since this computation has to generate a time-continuous signal. Computing such a complex equation with analog circuits seems to be difficult at first sight.

Fortunately, the required computation is a so-called *folding operation* between signal $y(t)$ and the *sinc*-function. According to the classical theory of Fourier transforms, a folding operation in the time domain is equivalent to a multiplication with frequency-dependent filter function in the frequency domain. This filter function is the Fourier transform of the corresponding function in the time domain. Therefore, Eq. (3.29) can be computed with some appropriate filter. Figure 3.60 shows the corresponding placement of the filter.

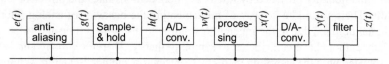

Fig. 3.60 Converting signal $e(t)$ from the analog time/value domain to the digital domain and back

The remaining question is: which frequency-dependent filter function is the Fourier transform of the *sinc*-function? Computing the Fourier transform of the *sinc*-function yields a low-pass filter function [418]. So, "all" we must do to compute Eq. (3.29) is to pass signal $y(t)$ through a low-pass filter, filtering frequencies as shown for the ideal filter in Fig. 3.61. Note that the representation of function $y(t)$ as a sum of sine waves would require very high-frequency components, making such a filtering non-redundant, even though we have already assumed an anti-aliasing filter to be present at the input.

Fig. 3.61 Low-pass filter: ideal (blue, dashed) and realistic (red, solid)

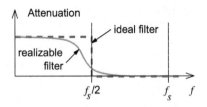

There is still one problem, though: Ideal low-pass filters do not exist. Therefore, we must live with compromises and design filters approximating the low-pass filter characteristics. Actually, we must live with several imperfections preventing a precise reconstruction of the input signals:

- Ideal low-pass filters cannot be designed. Therefore, we must use approximations of such filters. Designing good compromises is an art (performed extensively, for example, for audio equipment).
- For the same reason, we cannot completely remove input frequencies beyond the Nyquist frequency.
- The impact of value quantization is visible in Fig. 3.59. Due to value quantization, $e_3(t)$ is sometimes different from $z(t)$. Quantization noise, as introduced by ADCs, cannot be removed during output generation. Signal $w(t)$ from the output of the ADC will remain distorted by the quantization noise. However, this effect does not affect the signal $h(t)$ from the output of sample-and-hold circuits.
- Equation (3.29) is based on an infinite sum, involving also values at future instances in time. In practice, we can delay signals by some finite amount to know a finite number of "future" samples. Infinite delays are impossible. In Fig. 3.59, we did not consider contributions of sampling instances outside the diagram.

The functionality provided by low-pass filters demonstrates the power of analog circuits: There would be no way of implementing the behavior of analog filters in the digital domain, due to the inherent restriction to discretized time and values.

Many authors have contributed to sampling theory. Therefore, many names can be associated with the sampling theorem. Contributors include Shannon, Whittaker, Kotelnikov, Nyquist, and Küpfmüller. Therefore, the fact that the original signal can be reconstructed should simply be called the sampling theorem, since there is no way of attaching all names of relevant contributors to the theorem.

3.6.3 Pulse-width Modulation

In practice, the presented generation of analog signals has a number of disadvantages:

- DACs using an array of resistors are difficult to build. The precision of the resistors must be excellent. The deviation of the resistor handling the most significant bit from its nominal value must be less than the overall resolution of the converter. For example, this means that, for a 14-bit converter, the deviation of the real resistance from its nominal value must be in the order of 0.01%. This precision is difficult to achieve in practice, in particular over the full temperature range. If this precision is not achieved, the converter is not linear, possibly not even monotone.

- In order to generate a sufficient power for motors, lamps, loudspeakers, etc., analog outputs would need to be amplified in a power amplifier. Analog power amplifiers, such as so-called class A power amplifiers, are very power-inefficient, since they contain an always conducting path between the two rails of the power supply. This path results in a constant power consumption, irrespective of the actual output signal. For very small output signals, the ratio between the actually used power and the consumed power is, therefore, very small. As a result, the efficiency of audio power amplifiers for low-volume audio would be terribly bad.
- It is not easy to integrate analog circuitry on digital microcontroller chips. Adding external analog active components increases costs substantially.

Therefore, pulse-width modulation (PWM) is very popular. With PWM, we are using a digital output and generate a digital signal whose duty cycle corresponds to the value to be converted. Figure 3.62 shows digital signals with duty cycles of 25% and 75%. Such signals can be represented by Fourier series such as in Eq. (3.1). For applications of PWM, we try to eliminate effects of higher frequency components.

Fig. 3.62 Duty cycles

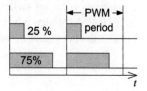

PWM signals can be generated by comparing a counter against a value stored in a programmable register (see Fig. 3.63). A high voltage is output whenever the value in the counter exceeds the value in the register. Otherwise, a voltage close to zero is generated. The clock signal of the counter must be programmable to select the basic frequency of the PWM signals. In our schematic, we have assumed that the PWM frequency is identical for all PWM outputs. Registers must be loaded with the values to be converted, typically at the sampling rate of the analog signals.

Fig. 3.63 Hardware for PWM output

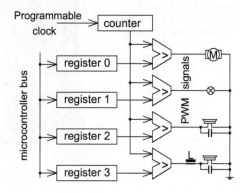

The effort required for filtering higher frequency components depends upon the application. For driving a motor, the averaging takes place in the motor, due to the mass of the moving parts in the motor and possibly also due to the self inductance of the motor. Hence, no external components are needed (see Fig. 3.63). For lamps, the averaging takes place in the human eye, as long as the frequencies are not too low. It may also be ok, to drive simple buzzers directly. In other cases, filtering out higher frequency components may be needed. For example, electromagnetic radiation caused by higher frequency components may be unacceptable or audio applications may be demanding filtered high-frequency signals. In Fig. 3.63, two capacitors and one inductor have been used to filter out high-frequency components for the loudspeakers. In our example, we are showing four PWM outputs. Having several PWM outputs is a common situation. For example, Atmel 32-Bit AVR microcontrollers in the AT32UC3A series have seven PWM outputs [25]. In practice, there are many options for the detailed behavior of PWM hardware.

The choice of the basic frequency (the reciprocal of the period) of the PWM signal and the filter is a matter of compromises. The basic frequency has to be higher than the highest frequency component of the analog signal to be converted. Higher frequencies simplify the design of the filter if any is present. Selecting a too high-frequency results in more electromagnetic radiation and in unnecessary energy consumption, since switching will consume energy. Compromises typically use a basic PWM frequency that is larger than the highest frequency of the analog signal by a factor between two and 10.

3.6.4 Actuators

There is a huge amount of actuators [145]. Actuators range from large ones that are able to move tons of weight to tiny ones with dimensions in the μm area, like the one shown in Fig. 3.64.

Fig. 3.64 Micro-system technology-based actuator motor (partial view; courtesy E. Obermeier, MAT, TU Berlin), ©TU Berlin

Figure 3.64 shows a tiny motor manufactured with micro-system technology. The dimensions are in the μm range. The rotating center is controlled by electrostatic forces.

As an example, we mention only a special kind of actuators which will become more important in the future: Micro-system technology enables the fabrication of tiny actuators, which can be put into the human body, for example. Using such tiny actuators, the amount of drugs fed into the body can be adapted to the actual need. This allows a much better medication than needle-based injections.

Actuators are important for the Internet of Things. It is impossible to provide a complete overview over actuators.

3.7 Electrical Energy: Energy Efficiency, Generation, and Storage

General constraints and objectives for the design of embedded and cyber-physical systems (see pp. 7 to 14 and Table 1.2) have to be obeyed for hardware design. Among the different objectives, we will focus on energy efficiency. Reasons for caring about the energy efficiency were listed in Table 1.1 on p. 12.

3.7.1 Energy Efficiency of Hardware Components

We will continue our discussion of energy efficiency by comparing the energy efficiency for the different technologies which we have at our disposal. Hardware components discussed in this chapter are quite different as far as their energy efficiency is concerned. A comparison between these technologies and changes over time (corresponding to a certain fabrication technology) shown in Fig. 3.65[18]. The figure reflects the conflict between efficiency and flexibility of currently available hardware technologies.

The diagram shows the energy efficiency GOP/J in terms of number of operations per unit of energy of various target technologies as a function of time and the target technology. In this context, operations could be 32-bit additions. Obviously, the number of operations per Joule is increasing as technology advances to smaller and smaller feature sizes of integrated circuits. However, for any given technology, the number of operations per Joule is largest for hardwired application-specific integrated circuits (ASICs). For reconfigurable logic usually coming in the form of field programmable gate arrays (FPGAs; see p. 186), this value is about one order of magnitude less. For programmable processors, it is even lower. However, processors offer the largest amount of flexibility, resulting from the flexibility of software.

---

[18]The figure approximates information provided by H. De Man [347] and is based on information provided by Philips.

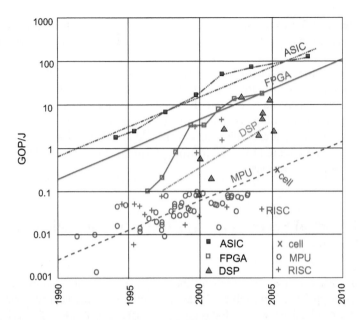

Fig. 3.65   Hardware efficiency (©De Man and Philips)

There is also some flexibility for reconfigurable logic, but it is limited to the size of applications that can be mapped to such logic. For hardwired designs, there is no flexibility. The trade-off between flexibility and efficiency also applies to processors: For processors optimized for an application domain, such as processors optimized for digital signal processing (DSP), power efficiency values approach those of reconfigurable logic. For general standard microprocessors, the values for this figure of merit are the worst. This is shown from Fig. 3.65, comprising values for microprocessors such as x86-like processors (see "MPU" entries), RISC processors, and the cell processor designed by IBM, Toshiba, and Sony.

Figure 3.65 does not identify exactly the applications which are compared, and it does not allow us to study the type of application mapping that has been performed.

More detailed and more recent comparisons have been made, enabling us to study the assumptions and the approach of these comparisons in a more comprehensive manner. A survey of comparisons involving GPUs has been published by Mittal et al. [381]. The survey includes a list of 28 publications for which GPUs have been found to be more energy-efficient than CPUs and two publications for which the reverse was true. Also, the survey comprises a list of 26 publications for which FPGAs have been found to be more energy-efficient than GPUs and one for which the reverse was true. For example, Hamada et al. [194] found for a gravitational n-body simulation that the number of operations per Watt was by a factor of 15 higher for FPGAs than for GPUs. For a comparison against CPUs, the factor was 34. The exact factors certainly depend on the application, but as a rule of THUMB, we can state the following: If we aim at top power- and energy-efficient designs, we should use ASICs. If we cannot afford ASICs, we should go for FPGAs. If FPGAs are also not an option, we should select GPUs. Also, we have already seen that heterogeneous processors are in general more energy-efficient than homogeneous processors. More detailed information can be computed for particular application areas.

3.7.1.1 The Case of Mobile Phones

Among the different applications of embedded systems (see pp. 4 to 7), we are now looking at telecommunication and smart phones. For smart phones, computational requirements are increasing at a rapid rate, especially for multimedia applications. De Man and Philips estimated that advanced multimedia applications need about 10–100 billion operations per second. Figure 3.65 demonstrates that advanced hardware technologies provided us more or less with this number of operations per Joule (= Ws) in 2007. This means that the most power-efficient platform technologies hardly provided the efficiency which was needed. Standard processors (entries for MPU and RISC) were hopelessly inefficient. It also meant that all sources of efficiency improvements needed to be exploited. More recently, the power efficiency has been improved. However, all such improvements are typically compensated by trends to provide a higher quality, e.g., by an increase of the resolution of still and moving images as well as a higher bandwidth for communication.

A detailed analysis of the power consumption has been published by Berkel [50] and by Carroll et al. [85]. A more recent analysis including LTE mobile phones has been published by Dusza et al. [138]. A power consumption of up to around four Watts has been observed. The display itself caused a consumption of up to around one Watt, depending on the display brightness.

Improving battery technology would allow us to consume power over longer periods, but the thermal limitation prevents us from going significantly beyond the current consumption in the near future. Due to thermal issues, it has become standard to design mobile phones with temperature sensors and to throttle devices in case of overheating. Of course, a larger power consumption would be feasible for larger devices. Nevertheless, environmental concerns also result in the need to keep the power consumption low.

Technology forecasts have been published as so-called International Technology Roadmap for Semiconductors. In the ITRS edition of 2013 [249], it is explicitly stated that mobile phones are driving technological development: *"System integration has shifted from a computational, PC-centric approach to a highly diversified mobile communication approach. The heterogeneous integration of multiple technologies in a limited space (e.g., GPS, phone, tablet, mobile phones, etc.) has truly revolutionized the semiconductor industry by shifting the main goal of any design from a performance driven approach to a reduced power driven approach. In few words, in the past performance was the one and only goal; today minimization of power consumption drives IC design"*.

3.7.1.2   Sensor Networks

Sensor networks used for the Internet of Things are another special case. For sensor networks, there may be even much less energy available than for mobile phones. Hence, energy efficiency is of utmost importance, comprising of course energy-efficient communication [518].

3.7.2   Sources of Electrical Energy

For plugged devices (i.e., for those connected to the power grid), energy is easily available. For all others, energy must be made available via other techniques. In particular, this applies to sensor networks used in IoT systems where energy can be a very scarce resource. Batteries store energy in the form of chemical energy. Their main limitation is that they must be carried to the location where the energy is required. If we would like to avoid this limitation, we have to use **energy harvesting**, also called **energy scavenging**. A large amount of techniques for energy harvesting is available [545, 551], but the amount of energy is typically much more limited:

- **Photovoltaics** allows the conversion of light into electrical energy. The conversion is usually based on the photovoltaic effect of semiconductors. Panels of photovoltaic material are in widespread use. Examples are shown in Fig. 3.66.

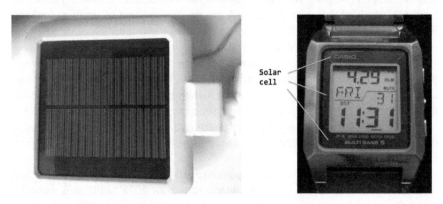

Fig. 3.66 Photovoltaic material: **left**: panel; **right**: solar-powered watch

- The **piezoelectric effect** can be used to convert mechanical strain into electrical energy. Piezoelectric lighters exploit this effect.
- **Thermoelectric generators** (TEGs) allow turning temperature gradients into electrical energy. They can be used even on the human body.
- **Kinetic energy** can be turned into electrical energy. This is exploited, for example, for some watches. Also, wind energy falls into this category.
- **Ambient electromagnetic radiation** can be turned into electrical energy as well.
- There are many other physical effects allowing us to convert other forms of energy into electrical energy.

3.7.3 Energy Storage

For many applications of embedded systems, power sources are not guaranteed to provide power whenever it is needed. However, we may be able to store electrical energy. Methods for storing electrical energy include the following:

1. **Non-rechargeable batteries** can be used only once and will not be considered further.
2. **Capacitors**: capacitors are a very convenient means of storing electrical energy. Their advantages include a potentially fast charging process, very high output currents, close to 100% efficiency, low leakage currents (for high-quality capacitors), and a large number of charge/discharge cycles. The limited amount of energy that can be stored is their main disadvantage.
3. **Rechargeable batteries** allow storing and using electrical energy, very much such as capacitors. Storing electrical energy is based on certain chemical processes and using this energy is based on reversing these chemical processes.

Due to their importance for embedded systems, we will discuss rechargeable batteries. If we want to include sources of electrical energy in our system model, we will need models of rechargeable batteries. Various models can be used. They differ in the amount of details that are included, and there is not a single model that fits all needs [443]. The following models are popular:

- **Chemical and physical models**: They describe the chemical and/or physical operation of the battery in detail. Such models may include partial differential equations, including many parameters. These models are beneficial for battery manufacturers, but typically too complex for designers of embedded systems (who will typically not know the parameters).
- **Simple empirical models**: such models are based on simple equations for which some parameter fitting has been performed. Peukert's law [428] is a frequently cited empirical model. According to this law, the lifetime of a battery is

$$\text{lifetime} = C/I^\alpha \tag{3.31}$$

where $\alpha > 1$ is the result of some empirical fitting process. Peukert's law reflects the fact, that higher currents will typically lead to an effective decrease of the battery capacity. Other details of battery behavior are not included in this model.
- **Abstract models** provide more details than the very simple empirical models, but do not refer to chemical processes. We would like to present two such models:

  - The model proposed by Chen and Ricón [97]. The model is an electrical model, as shown in Fig. 3.67. According to this model, a charging current I_{Batt} controls a current source in the left part of the schematic. The current generated by the current source is equal to the charging current entering on the right. This current will charge the capacitor $C_{Capacity}$. The amount of charge on the capacitor is called **state of charge** (SoC). The state of charge is reflected by the voltage

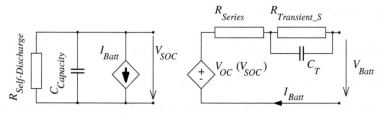

Fig. 3.67 Battery model according to Chen et al. (simplified)

V_{SOC} on the capacitor, since the charge on the capacitor can be computed as $Q = C_{Capacity} * V_{SOC}$. Resistor $R_{Self-Discharge}$ models the self-discharge (leakage) of this capacitor which happens even when no current is drawn at the terminal pins of the battery.

Let us consider the voltage which is available at the battery terminals when the current through these terminals is zero. The voltage at the battery terminals will typically nonlinearly depend on V_{SOC}. This dependency can be modeled by a nonlinear function $V_{OC}(V_{SOC})$, representing the **open terminal output voltage** of the battery. This voltage decreases when the battery provides some current. For a constant discharging current, $R_{Series} + R_{Transient\_s}$ models the corresponding voltage drop. For short current spikes, the decrease is determined by the value of R_{Series} only, since C_T will act as a buffer. When the current consumption increases, *time constant* $R_{Transient\_s} * C_T$ determines the speed for the transition from only R_{Series} causing the voltage drop to $R_{Series} + R_{Transient\_s}$ causing the voltage drop. The original proposal by Chen et al. includes a second resistor/capacitor pair in order to model transient output voltage behavior more precisely. Overall, this model captures the impact of high output currents on the voltage, the nonlinear dependency of the output voltage, and self-discharge reasonably well. Simpler versions of this model exist, i.e., ones that do not model all three effects.

- Actual batteries exhibit the so-called charge recovery effect: whenever the discharge process of batteries is paused for some time interval, the battery recovers, i.e., more charge becomes available and the voltage is typically also increased. This effect is not considered in Chen's model. However, it is the focus of the so-called kinetic battery model (KiBaM) of Manwell et al. [348]. The name reflects the analogy upon which this model is based. The model assumes two different bins of charge, as shown in Fig. 3.68. The right bin contains the charge

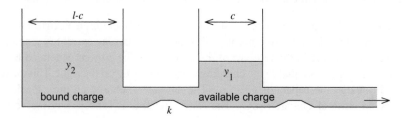

Fig. 3.68 Kinetic battery model

y_1 which is immediately available. The left bin contains charge y_2 which exists in the battery, but which needs to flow into the right bin to become available. An interval of heavy usage of the battery may almost empty the right bin. It will then take some time for charge to become available again. The speed of the recovery process is determined by parameter k, the width of the pipe connecting the two bins. The details of the model (such as the amount of charge flowing) reflect the physical situation of the bins. This model describes the charge recovery process with some reasonable precision, but fails to describe transients and self-discharge as captured in Chen's model. The kinetic model has an impact on how embedded systems should be used. For example, it has been demonstrated that it is beneficial to plan for intervals, during which wireless transmission is turned off [138].

Overall, the two models demonstrate nicely that models must be selected to reflect the effects that should be taken into account.

- There may be **mixed models** which are partially based on abstract models and partially on chemical and physical models.

3.8 Secure Hardware

The general requirements for embedded systems can often include security (see p. 7). In particular, security is important for the Internet of Things. If security is a major concern, special secure hardware may need to be developed. Security may need to be guaranteed for communication and for storage [296]. Security has to be provided despite possible attacks. **Attacks** can be partitioned into the following [289]:

- **Software attacks** which do not require physical access: the deployment of software Trojans is an example of such an attack. Also, software defects can be exploited. Buffer overflows are a frequent cause of security hazards.
- Attacks which require physical access and which can be classified into the following:
  - **Physical attacks**: These try to physically tamper with the system. For example, silicon chips can be opened and analyzed. The first step in this procedure is de-packaging (removing the plastic covering the silicon). Next, micro-probing or optical analysis can be used. Such attacks are difficult, but they reveal many details of the chip.
  - **Side-channel attacks**: These try to exploit additional sources of information complementing the specified interfaces.

    For example, **timing analysis** may reveal which data is being processed. This is especially true if execution times of software are data-dependent. Security-relevant algorithms should be designed such that their execution time does not depend on data values. This requirement also affects the implementation of computer arithmetic: Instructions should not have data-dependent execution times.

Power analysis is a second class of attacks. Power analysis techniques include simple power analysis (SPA) and differential power analysis (DPA). In some cases, SPA may be sufficient to compute encryption keys directly from simple power measurements. In other cases, advanced statistical methods may be needed to compute keys from small statistical fluctuations of measured currents.

Analysis of electromagnetic radiation is a third class of side-channel attack.

Different classes of people might try these attacks, and different classes of people may have an interest in blocking these attacks. The attacker may actually be the user of an embedded device trying to obtain unauthorized network access or unauthorized access to protected media such as music.

We can distinguish between the following **counter measures**:

- A security-aware software development process is required as a shield against software attacks.
- Tamper-resistant devices include special mechanisms for physical protection (shielding, or sensors to detect tampering with the modules).
- Devices can be designed such that processed data patterns have very little impact on the power consumption. This requires special devices which are typically not used in complex chips.
- Logical security, typically provided by cryptographic methods: Encryption can be based on either symmetric or asymmetric ciphers.

  – For symmetric ciphers, sender and receiver are using the same secret key to encrypt and decrypt messages. DES, 3DES, and AES are examples of symmetric ciphers.
  – For asymmetric ciphers, messages are encrypted with a public key and decrypted with a private key. RSA and Diffie–Hellman are examples of asymmetric ciphers.
  – Also, hash codes can be added to messages, allowing the detection of message modifications. MD5 and SHA are examples of hashing algorithms.

Due to the performance gap, some processors may support encryption and decryption with dedicated instructions. Also, specialized solutions such as ARM's Trust-Zone computing exist. "*At the heart of the TrustZone approach is the concept of secure and non-secure worlds that are hardware separated, with non-secure software blocked from accessing secure resources directly. Within the processor, software either resides in the secure world or the non-secure world; a switch between these two worlds is accomplished via software referred to as the secure monitor (Cortex-A) or by the core logic (Cortex-M). This concept of secure (trusted) and non-secure (non-trusted) worlds extends beyond the processor to encompass memory, software, bus transactions, interrupts and peripherals within an SoC*" (see https://www.arm.com/products/security-on-arm/trustzone).

The Kalray MPPA2®-256 multi-core processor chip contains as many as 128 specialized crypto coprocessors connected to a matrix of 288 "regular" cores (see http://www.kalrayinc.com/kalray/products/). Cores are 64-bit VLIW processors.

The following **challenges** exist for the design of counter measures [289]:

1. **Processing gap**: Due to the limited performance of embedded systems, advanced encryption techniques may be too slow, in particular if high data rates have to be processed.
2. **Battery gap**: Advanced encryption techniques require a significant amount of energy. This energy may be unavailable in a portable system. Smart cards are a special case of hardware that must run using a very small amount of energy.
3. **Flexibility**: Frequently, many different security protocols are required within one system and these protocols may have to be updated from time to time. This hinders using special hardware accelerators for encryption.
4. **Tamper resistance**: Mechanisms against malicious attacks need to be built in. Their design is far from trivial. For example, it may be difficult if not impossible to guarantee that the current consumption is independent of the cryptographic keys that are processed.
5. **Assurance gap**: The verification of security requires extra efforts during the design.
6. **Cost**: Higher security levels increase the cost of the system.

Ravi et al. have analyzed these challenges in detail for a secure socket layer (SSL) protocol [289].

More information on secure hardware is available, for example, in a book by Gebotys [174] and in proceedings of a workshop series dedicated to this topic (see [177] for the most recent edition).

3.9 Problems

We suggest solving the following problems either at home or during a flipped classroom session:

3.1: It is suggested that locally available small robots are used to demonstrate hardware in the loop, corresponding to Fig. 3.2. The robots should include sensors and actuators. Robots should run a program implementing a control loop. For example, an optical sensor could be used to let a robot follow a black line on the ground. The details of this assignment depend on the availability of robots.

3.2: Define the term "signal"!

3.3: Which circuit do we need for the transition from continuous time to discrete time?

3.4: What does the sampling theorem tell us?

3.5: Assume that we have an input signal x consisting of the sum of sine waves of 1.75 and 2 kHz. We are sampling x at a rate of 3 kHz. Will we be able to reconstruct the original signal after discretization of time? Please explain your result!

3.6: Discretization of values is based on ADCs. Develop the schematic of a flash-based ADC for positive and negative input voltages! The output should be encoded as 3-bit two's complement numbers, allowing to distinguish between eight different voltage intervals.

3.7: Suppose that we are working with a successive approximation-based 4-bit ADC. The input voltage range extends from V_{min} =1 V (="0000") to V_{max} =4.75 V (="1111"). Which steps are used to convert voltages of 2.25, 3.75, and 1.8 V? Draw a diagram similar to Fig. 3.12 which depicts the successive approximation to these voltages!

3.8: Compare the complexity of flash-based and successive approximation-based ADC. Assume that you would like to distinguish between n different voltage intervals. Enter the complexity into the Table 3.2, using the O-notation.

Table 3.2 Complexity of ADCs	Flash-based converter	Successive approximation converter
Time complexity		
Space complexity		

3.9: Suppose a sine wave is used as an input signal to the converter designed in Problem 3.6. Depict the quantization noise signal for this case!

3.10: Create a list of features of DSP processors!

3.11: Which components do FPGA comprise? Which of these are used to implement Boolean functions? How are FPGAs configured? Are FPGAs energy-efficient? Which kind of applications are FPGAs good for?

3.12: What is the key idea of VLIW processors?

3.13: What is a "single-ISA heterogeneous multi-core architecture"? Which advantages do you see for such an architecture?

3.14: Explain the terms "GPU" and "MPSoC"!

3.15: Some FPGAs support an implementation of all Boolean functions of six variables. How many such functions exist? We ignore that some functions differ only by a renaming of variables.

3.16: In the context of memories, we are sometimes saying "small is beautiful". What could be the reason for this?

3.17: Some levels of the memory hierarchy may be hidden from the application programmer. Why should such a programmer nevertheless care about the architecture of such levels?

3.18: What is a "scratchpad memory" (SPM)? How can we ensure that some memory object is stored in the SPM?

3.19: Develop the following FlexRay™cluster: The cluster consists of the five nodes A, B, C, D, and E. All nodes should be connected via two channels. The cluster uses a bus topology. The nodes A, B, and C are executing a safety critical task, and, therefore, their bus requests should be guaranteed at the time of 20 macroticks. The following is expected from you:

- Download the levi FlexRay simulator [470]. Unpack the .zip file and install!
- Start the training module by executing the file leviFRP.jar.
- Design the described FlexRay cluster within the training module.
- Configure the communication cycle such that the nodes A, B, and C have a guaranteed bus access within a maximal delay of 20 macroticks. The nodes D and E should use only the dynamic segment.
- Configure the node bus requests. The node A sends a message every cycle. The nodes B and C send a message every second cycle. The node D sends a message of the length of 2 minislots every cycle and the node E sends every second cycle a message of the length of 2 minislots.
- Start the visualization and check if the bus requests of the nodes A, B, and C are guaranteed.
- Swap the positions of nodes D and E in the dynamic segment. What is the resulting behavior?

3.20: Develop the schematic of a 3-bit DAC! The conversion should be done for a 3-bit vector x encoding positive numbers. Prove that the output voltage is proportional to the value represented by the input vector x. How would you modify the circuit if x represented two's complement numbers?

3.21: The circuit shown in Fig. B.4 in Appendix B is an amplifier, amplifying input voltage V_1:

$$V_{out} = g_{closed} * V_1$$

Compute the gain g_{closed} for the circuit of Fig. B.4 as a function of R and R_1!

3.22: How do different hardware technologies differ with respect to their energy efficiency?

3.23: The computational efficiency is sometimes also measured in terms of billions of operations per second per Watt. How is this different from the figure of merit used in Fig. 3.65?

3.24: Why is it so important to optimize embedded systems? Compare different technologies for processing information in an embedded system with respect to their efficiency!

3.25: Suppose that your mobile phone uses a lithium battery rated at 720 mAh. The nominal voltage of the battery is 3.7 V. Assuming a constant power consumption of 1 W, how long would it take to empty the battery? All secondary effects such as decreasing voltages should be ignored in this calculation.

3.26: Which challenges do you see for the security of embedded systems?

3.27: What is a "side-channel attack"? Please provide examples of side-channel attacks!

Chapter 4
System Software

Not all components of embedded systems need to be designed from scratch. Instead, there are standard components that can be reused. These components comprise knowledge from earlier design efforts and constitute **intellectual property** (IP). IP reuse is one key technique in coping with the increasing complexity of designs. The term "IP reuse" frequently denotes the reuse of hardware. However, reusing hardware is not enough. Sangiovanni-Vincentelli pointed out that software components need to be reused as well. Therefore, the platform-based design methodology advocated by Sangiovanni-Vincentelli [452] (see p. 302) comprises the reuse of hardware and software IP.

Standard software components that can be reused include system software components such as embedded operating systems (OSs) and **middleware**. The last term denotes software that provides an intermediate layer between the OS and application software. We include libraries for communication as a special case of middleware. Such libraries extend the basic communication facilities provided by operating systems. Also, we consider real-time databases (see Sect. 4.7) to be a second class of middleware. Calls to standard software components may already need to be included in the specification. Therefore, information about the application programming interface (API) of these standard components may already be needed for completing executable specifications of the SUD.

Consistent with the design information flow, we will describe embedded operating systems and middleware in this chapter (see also Fig. 4.1).

© Springer International Publishing AG 2018
P. Marwedel, *Embedded System Design*, Embedded Systems,
DOI 10.1007/978-3-319-56045-8_4

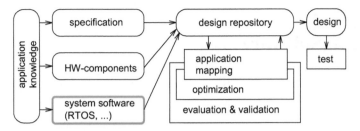

Fig. 4.1 Simplified design information flow

4.1 Embedded Operating Systems

4.1.1 General Requirements

Except for very simple systems, scheduling, context switching, and I/O require the support of an operating system suited for embedded applications. Context switching algorithms multiplex processors such that each process seems to have its own processor.

For systems with virtual addressing[1], we can distinguish between different address spaces and between processes and threads. **Each process has its own address space, whereas several threads may share an address space.** Context switches which change the address space require more time than those which do not. Threads sharing an address space will typically communicate via shared memory. Operating systems must provide communication and synchronization methods for threads and processes. More information about the just touched standard topics in system software can be found in textbooks on operating systems, such as the book by Tanenbaum [502][2].

The following are essential features of embedded operating systems:

- Due to the large variety of embedded systems, there is also a large variety of requirements for the functionality of embedded OSs. Due to efficiency requirements, it is not possible to work with OSs which provide the union of all functionalities. For most applications, the OS must be small. Hence, we need operating systems which can be **flexibly tailored** toward the application at hand. **Configurability** is therefore one of the main characteristics of embedded OSs. There are various techniques of implementing configurability, including[3]:

  - **Object orientation**, used for a derivation of proper subclasses: For example, we could have a general scheduler class. From this class, we could derive schedulers having particular features. However, object-oriented approaches

---

[1] See Appendix C.

[2] Students who have not attended a course on operating systems may have to browse through one of these textbooks before proceeding any further.

[3] This list is sorted by the position of the technique in the development process or tool chain.

typically come with an additional overhead. For example, dynamic binding of methods does create run-time overhead. Ideas for reducing this overhead exist (see, for example, https://github.com/lefticus/cppbestpractices/blob/master/08-Considering_Performance.md). Nevertheless, remaining overhead and potential timing unpredictability may be unacceptable for performance-critical system software.

- **Aspect-oriented programming** [334]: With this approach, orthogonal aspects of software can be described independently and then can be added automatically to all relevant parts of the program code. For example, some code for profiling can be described in a single module. It can then be automatically added to or dropped from all relevant parts of the source code. The CIAO family of operating systems has been designed in this way [335].
- **Conditional compilation**: In this case, we are using some macropreprocessor and we are taking advantage of #if and #ifdef preprocessor commands.
- **Advanced compile-time evaluation**: Configurations could be performed by defining constant values of variables before compiling the OS. The compiler could then propagate the knowledge of these values as much as possible. Advanced compiler optimizations may also be useful in this context. For example, if a particular function parameter is always constant, this parameter can be dropped from the parameter list. Partial evaluation [264] provides a framework for such compiler optimizations. In a sophisticated form, dynamic data might be replaced by static data [24]. A survey of operating system specialization was published by McNamee et al. [369].
- **Linker-based removal of unused functions**: At link-time, there may be more information about used and unused functions than during earlier phases. For example, the linker can figure out, which library functions are used. Unused library functions can be accordingly dropped, and specializations can take place [93].

These techniques are frequently combined with a rule-based selection of files to be included in the operating system. Tailoring the OS can be made easy through a graphical user interface hiding the techniques employed for achieving this configurability. For example, VxWorks [562] from Wind River is configured via a graphical user interface.

Verification is a potential problem of systems with a large number of derived tailored OSs. Each and every derived OS must be tested thoroughly. Takada mentions this as a potential problem for eCos (an open-source RTOS, see http://ecos.sourceware.org, and Massa [364]), comprising 100–200 configuration points [500]. For Linux, this problem is even larger [503]. Software product line engineering [431] can contribute toward solving this problem.

- There is a large variety of peripheral devices employed in embedded systems. Many embedded systems do not have a hard disk, a keyboard, a screen, or a mouse. There is effectively **no device that needs to be supported by all variants of the OS**, except maybe the system timer. Frequently, applications are designed to handle particular devices. In such cases, devices are not shared between applications, and hence, there is no need to manage the devices by the OS. Due to the large variety of devices, it would also be difficult to provide all required device drivers together with the OS. Hence, it makes sense to decouple OS and drivers by using special processes instead of integrating their drivers into the kernel of the OS. Due to the limited speed of many embedded peripheral devices, there is also no need for an integration into the OS in order to meet performance requirements. This may lead to a different stack of software layers. For PCs, some drivers, such as disk drivers, network drivers, or audio drivers, are implicitly assumed to be present. They are implemented at a very low level of the stack. The application software and middleware are implemented on top of the application programming interface, which is standard for all applications. For an embedded OS, device drivers are implemented on top of the kernel. Applications and middleware may be implemented on top of appropriate drivers, not on top of a standardized API of the OS (see Fig. 4.2). Drivers may even be included in the application itself.

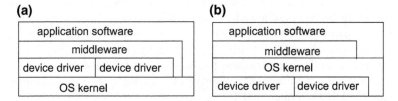

Fig. 4.2 Device drivers implemented on top of (**a**) or below (**b**) the OS kernel

- **Protection mechanisms are sometimes not necessary**, since embedded systems are sometimes designed for a single purpose (they are not supposed to support so-called multiprogramming). Untested programs have traditionally hardly ever been loaded. After the software has been tested, it could be assumed to be reliable. This also applies to input/output. In contrast to desktop applications, it is possibly not always necessary to implement I/O instructions as privileged instructions and processes can sometimes be allowed to do their own I/O. This matches nicely with the previous item and reduces the overhead of I/O operations.

Example: Let switch correspond to the (memory-mapped) I/O address of some switch which needs to be checked by some program. We can simply use a

```
load register,switch
```

instruction to query the switch. There is no need to go through an OS service call, which would create overhead for saving and restoring the context (registers etc.).

However, there is a trend toward more dynamic embedded systems. Also, safety and security requirements might make protection necessary. Special memory protection units (MPUs) have been proposed for this (see Fiorin [158] for an example).

For systems with a mix of critical and non-critical applications (**mixed-criticality systems**), configurable memory protection [336] may be a goal.

- **Interrupts can be connected to any thread or process**. Using OS service calls, we can request the OS to start or stop them if certain interrupts happen. We could even store the start address of a thread or process in the interrupt vector address table, but this technique is very dangerous, since the OS would be unaware of the thread or process actually running. Also composability may suffer from this: If a specific thread is directly connected to some interrupt, then it may be difficult to add another thread which also needs to be started by some event. Application-specific device drivers (if used) might also establish links between interrupts and threads and processes. Techniques for establishing safe links have been studied by Hofer et al. [210].
- Many embedded systems are real-time (RT) systems, and hence, the OS used in these systems **must be a real-time operating system** (RTOS).

Additional information about embedded operating systems can be found in a chapter written by Bertolotti [54]. This chapter comprises information about the architecture of embedded operating systems, the POSIX standard, open-source real-time operating systems and virtualization.

4.1.2 Real-Time Operating Systems

Definition 4.1: (A) *"real-time operating system is an operating system that supports the construction of real-time systems"* [500].

What does it take to make an OS an RTOS? There are four key requirements[4]:

- **The timing behavior of the OS must be predictable**. For each service of the OS, an upper bound on the execution time must be guaranteed. In practice, there are various levels of predictability. For example, there may be sets of OS service calls for which an upper bound is known and for which there is not a significant variation of the execution time. Calls like "get me the time of the day" may fall into this class. For other calls, there may be a huge variation. Calls like "get me 4MB of free memory" may fall into this second class. In particular, the scheduling policy of any RTOS must be deterministic.

  There may also be times during which interrupts must be disabled to avoid interferences between components of the OS. Less importantly, they can also be disabled to avoid interferences between processes. The periods during which interrupts are disabled must be quite short in order to avoid unpredictable delays in the processing of critical events.

  For RTOSs implementing file systems still using hard disks, it may be necessary to implement contiguous files (files stored in contiguous disk areas) to avoid unpredictable disk head movements.

---

[4]This section includes information from Hiroaki Takada's tutorial [500].

- **The OS must manage the scheduling of threads and processes**. Scheduling can be defined as mapping from sets of threads or processes to intervals of execution time (including the mapping to start times as a special case) and to processors (in case of multiprocessor systems). Also, the OS possibly has to be aware of deadlines so that the OS can apply appropriate scheduling techniques. There are, however, cases in which scheduling is done completely off-line, and the OS only needs to provide services to start threads or processes at specific times or priority levels. Scheduling algorithms will be discussed in detail in Chap. 6.
- **Some systems require the OS to manage time**. This management is mandatory if internal processing is linked to an absolute time in the physical environment. Physical time is described by real numbers. In computers, discrete time standards are typically used instead. The precise requirements may vary:

1. In some systems, synchronization with global time standards is necessary. In this case, **global clock synchronization** is performed. Two standards are available for this:
   - **Universal Time Coordinated (UTC)**: UTC is defined by astronomical standards. Due to variations regarding the movement of the earth, this standard has to be adjusted from time to time. Several seconds have been added during the transition from one year to the next. The adjustments can be problematic, since incorrectly implemented software could get the impression that the next year starts twice during the same night.
   - **International atomic time** (in French: *temps atomic internationale*, or TAI). This standard is free of any artificial artifacts.
     Some connection to the environment is used to obtain accurate time information. External synchronization is typically based on wireless communication standards such as the global positioning system (GPS) [395] or mobile networks.
2. If embedded systems are used in a network, it is frequently sufficient to synchronize time information within the network. Local clock synchronization can be used for this. In this case, connected embedded systems try to agree on a consistent view of the current time.
3. There may be cases in which provision for precise local delays is all that is needed.

For several applications, precise time services with a high resolution must be provided. They are required for example in order to distinguish between original and subsequent errors. For example, they can help to identify the power plant(s) that are responsible for blackouts (see [405]). The precision of time services depends on how they are supported by a particular execution platform. They are very imprecise (with precisions in the millisecond range) if they are implemented through processes at the application level and very precise (with precisions in the microsecond range) if they are supported by communication hardware. More information about time services and clock synchronization is contained in the book by Kopetz [292].

- **The OS must be fast**. An operating system meeting all the requirements mentioned so far would be useless if it were very slow. Therefore, the OS must obviously be fast.

Each RTOS includes a so-called real-time OS **kernel**. This kernel manages the resources which are found in every real-time system, including the processor, the memory, and the system timer. Major functions in the kernel include the process and thread management, inter-process synchronization and communication, time management, and memory management.

While some RTOSs are designed for general embedded applications, others focus on a specific area. For example, OSEK/VDX-compatible operating systems focus on automotive control. Operating systems for a selected area can provide a dedicated service for that particular area and can be more compact than operating systems for several application areas.

Similarly, while some RTOSs provide a standard API, others come with their own, proprietary API. For example, some RTOSs are compliant with the standardized POSIX RT-extension [195] for UNIX, with the OSEK ISO 17356-3:2005 standard, or with the ITRON specification developed in Japan (see http://www.ertl.jp/ITRON/). Many RT-kernel type of OSs has their own API. ITRON, mentioned in this context, is a mature RTOS which employs link-time configuration.

Available RTOSs can further be classified into the following categories [187]:

- **Fast proprietary kernels**: According to Gupta, *"for complex systems, these kernels are inadequate, because they are designed to be fast, rather than to be predictable in every respect."* Examples include QNX, PDOS, VCOS, VTRX32, and VxWorks.
- **Real-time extensions to standard OSs**: In order to take advantage of comfortable mainstream operating systems, hybrid systems have been developed. For such systems, there is an RT-kernel running all RT-processes. The standard operating system is then executed as one of these processes (see Fig. 4.3).

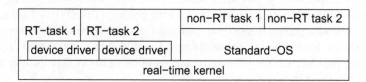

Fig. 4.3 Hybrid OSs

This approach has some advantages: The system can be equipped with a standard OS API and can have graphical user interfaces (GUIs), file systems, etc., and enhancements to standard OSs become quickly available in the embedded world as well. Also, problems with the standard OS and its non-RT-processes do not negatively affect the RT-processes. The standard OS can even crash, and this would not affect the RT-processes. On the downside, and this is already visible from Fig. 4.3, there may be problems with device drivers, since the standard OS will have its own device drivers. In order to avoid interference between the drivers for RT-processes and those for the other processes, it may be necessary to partition devices into those handled by RT-processes and those handled by the standard OS. Also, RT-processes cannot use the services of the standard OS. So all the nice features such as file system access and GUIs are normally not available to those processes, even though some attempts may be made to bridge the gap between the two types of processes without losing the RT-capability. RT-Linux is an example of such hybrid OSs.

According to Gupta [187], trying to use a version of a standard OS is *"not the correct approach because too many basic and inappropriate underlying assumptions still exist such as optimizing for the average case (rather than the worst case), ... ignoring most if not all semantic information, and independent CPU scheduling and resource allocation."* Indeed, dependencies between processes are not very frequent for most applications of standard operating systems and are therefore frequently ignored by such systems. This situation is different for embedded systems, since dependencies between processes are quite common and they should be taken into account. Unfortunately, this is not always done if extensions to standard operating systems are used. Furthermore, resource allocation and scheduling are rarely combined for standard operating systems. However, integrated resource allocation and scheduling algorithms are required in order to guarantee meeting timing constraints.

- There is a number of **research systems** which aim at avoiding the above limitations. These include Melody [544] and (according to Gupta [187]) MARS, Spring, MARUTI, Arts, Hartos, and DARK.

Takada [500] mentions low overhead memory protection, temporal protection of computing resources (targeting at preventing processes from computing for longer periods of time than initially planned), RTOSs for on-chip multiprocessors (especially for heterogeneous multiprocessors and multi-threaded processors), and support for continuous media and quality of service control as research issues.

Due to the potential growth in the Internet of Things (IoT) system market, vendors of standard OSs are offering variations of their products (like Windows Embedded [377]) and obtain market shares from traditional vendors such as Wind River Systems [563]. Due to the increasing connectedness, Linux and its derivative Android® are becoming popular. Advantages and limitations of using Linux in embedded systems will be described in Sect. 4.4.

4.1.3 Virtual Machines

In certain environments, it may be useful to emulate several processors on a single real processor. This is possible with **virtual machines** executed on the bare hardware. On top of such a virtual machine, several operating systems can be executed. Obviously, this allows several operating systems to be run on a single processor. For embedded systems, this approach has to be used with care since the temporal behavior of such an approach may be problematic and timing predictability may be lost. Nevertheless, sometimes this approach may be useful. For example, we may need to integrate several legacy applications using different operating systems on a single hardware processor. A full coverage of virtual machines is beyond the scope of this book. Interested readers should refer to books by Smith et al. [477] and Craig [113]. PikeOS is an example of a virtualization concept dedicated toward embedded systems [497]. PikeOS allows the system's resources (e.g., memory, I/O devices, CPU-time) to be divided into separate subsets. PikeOS comes with a small micro-kernel. Several operating systems, application programming interfaces (APIs), and run-time environments (RTEs) can be implemented on top of this kernel (see Fig. 4.4).

Fig. 4.4 PikeOS virtualization (©SYSGO)

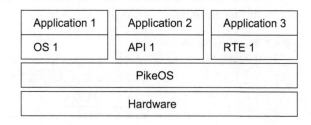

4.2 Resource Access Protocols

In the following, we will use the term **job** to denote either a thread or a process.

4.2.1 Priority Inversion

There are cases in which jobs must be granted exclusive access to resources such as global shared variables or devices in order to avoid non-deterministic or otherwise unwanted program behavior. Such exclusive access is very important for embedded systems, e.g., for implementing shared memory-based communication or exclusive access to some special hardware device. Program sections during which such exclusive access is required are called **critical sections**. Critical sections should be short. Operating systems typically provide primitives for requesting and releasing exclusive access to resources, also called **mutex primitives**. Jobs not being granted exclusive access must wait until the resource is released. Accordingly, the release operation has to check for waiting processes and resume the job of highest priority.

In this book, we will call the request operation or lock operation P(S) and the release or unlock operation V(S), where S corresponds to the particular resource requested. P(S) and V(S) are so-called **semaphore** operations. Semaphores allow up to n (with n being a parameter) threads or processes to use a particular resource protected by S concurrently. S is a data structure maintaining a count on how many resources are still available. P(S) checks the count and blocks the caller if all resources are in use. Otherwise, the count is modified, and the caller is allowed to continue.

V(S) increments the number of available resources and makes sure that a blocked caller (if it exists) is unblocked. The names P(S) and V(S) are derived from the Dutch language. We will use these operations only in the form of binary semaphores with $n = 1$; i.e., we will allow only a single caller to use the resource.

For embedded systems, dependencies between processes are the rule, rather than an exception. Also, the effective job priority of real-time applications is more important than for non-real applications. Mutually exclusive access can lead to priority inversion, an effect which changes the effective priority of processes. Priority inversion exists on non-embedded systems as well. However, due to the reasons just listed, the priority inversion problem can be considered a more serious problem in embedded systems.

A first case of the consequences resulting from the combination of "mutual exclusion" with "no preemption" is shown in Fig. 4.5.

Fig. 4.5 Blocking of a job
by a lower priority job

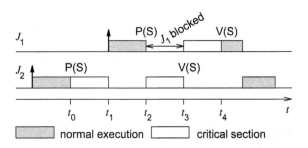

Bold upward pointing arrows indicate the times at which jobs are released, or "ready." At time t_0, job J_2 enters a critical section after requesting exclusive access to some resource via an operation P. At time t_1, job J_1 becomes ready and preempts J_2. At time t_2, J_1 fails getting exclusive access to the resource in use by J_2 and becomes blocked. Job J_2 resumes and after some time releases the resource. The release operation checks for pending jobs of higher priority and preempts J_2. During the time J_1 has been blocked, a lower priority job has effectively blocked a higher priority job. The necessity of providing exclusive access to some resources is the main reason for this effect. Fortunately, in the particular case of Fig. 4.5, the duration of the blocking cannot exceed the length of the critical section of J_2. This situation is problematic, but difficult to avoid.

In more general cases, the situation can be even worse. This can be seen, for example, from Fig. 4.6.

Fig. 4.6 Priority inversion
with potentially large delay

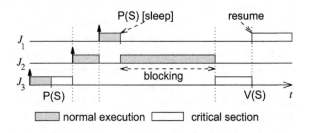

We assume that jobs J_1, J_2, and J_3 are given. J_1 has the highest priority, J_2 has a medium priority, and J_3 has the lowest priority. Furthermore, we assume that J_1 and J_3 require exclusive use of some resource via operation P(S). Now, let J_3 be in its critical section when it its preempted by J_2. When J_1 preempts J_2 and tries to use the same resource that J_3 is having exclusive access of, it blocks and lets J_2 continue. As long as J_2 is continuing, J_3 cannot release the resource. Hence, J_2 is effectively blocking J_1 even though the priority of J_1 is higher than that of J_2. In this example, the blocking of J_1 continues as long as J_2 executes. J_1 is blocked by a job of lower priority, which is not in its critical section. This effect is called **priority inversion**[5]. In fact, priority inversion happens even though J_2 is unrelated to J_1 and J_3. The duration of the priority inversion situation is not bounded by the length of any critical section. This example and other examples can be simulated with the levi simulation software [472].

One of the most prominent cases of priority inversion happened in the Mars Pathfinder, where an exclusive use of a shared memory area led to priority inversion on Mars [263].

4.2.2   Priority Inheritance

One way of dealing with priority inversion is to use the **priority inheritance protocol** (PIP). This protocol is a standard protocol available in many real-time operating systems. It works as follows:

- Jobs are scheduled according to their active priorities. Jobs with the same priorities are scheduled on a first-come, first-served basis.
- When a job J_1 executes P(S) and exclusive access is already granted to some other job J_2, then J_1 will become blocked. If the priority of J_2 is lower than that of J_1, J_2 inherits the priority of J_1. Hence, J_2 resumes execution. In general, every job inherits the highest priority of jobs blocked by it.
- When a job J_2 executes V(S), its priority is decreased to the highest priority of the jobs blocked by it. If no other job is blocked by J_2, its priority is reset to the original value. Furthermore, the highest priority job so far blocked on S is resumed.
- Priority inheritance is transitive: If J_x blocks J_y and J_y blocks J_z, then J_x inherits the priority of J_z.

This way, high-priority jobs being blocked by low-priority jobs propagate their priority to the low-priority jobs such that the low-priority jobs can release semaphores as soon as possible.

In the example of Fig. 4.6, J_3 would inherit the priority of J_1 when J_1 executes P(S). This would avoid the problem mentioned since J_2 could not preempt J_3 (see Fig. 4.7).

---

[5] Some authors do already consider the case of Fig. 4.5 as a case of priority inversion. This was also done in earlier versions of this book.

Fig. 4.7 Priority inheritance
for the example of Fig. 4.6

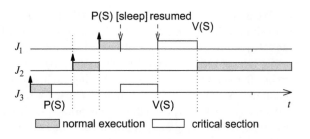

Figure 4.8 shows an example of nested critical sections [82]. Note that the priority of job J_3 is not reset to its original value at time t_0. Instead, its priority is decreased to the highest priority of the jobs blocked by it; in this case, it remains at priority p_1 of J_1.

Fig. 4.8 Nested critical
sections

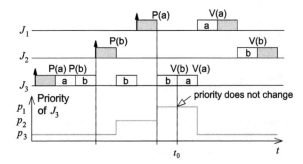

Transitiveness of priority inheritance is shown in Fig. 4.9 [82].

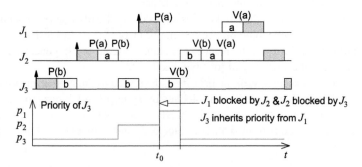

Fig. 4.9 Transitiveness of priority inheritance

At time t_0, J_1 is blocked by J_2 which in turn is blocked by J_3. Therefore, J_3 inherits the priority p_1 of J_1.

Priority inheritance is also used by Ada: During a rendez-vous, the priority of two threads is set to their maximum.

Priority inheritance also solved the Mars Pathfinder problem: the VxWorks operating system used in the Pathfinder implements a flag for the calls to mutex primitives. This flag allows priority inheritance to be set to "on." When the software was shipped, it was set to "off." The problem on Mars was corrected by using the debugging facilities of VxWorks to change the flag to "on," while the Pathfinder was already on Mars [263]. Priority inheritance can be simulated with the levi simulation software [472].

While priority inheritance solves some problems, it does not solve others. For example, there may be a large number of jobs having a high priority.

There may also be deadlocks. The possible existence of deadlocks can be shown by means of an example [82]. Suppose that we have two jobs J_1 and J_2. For job J_1, we assume a code sequence of the form ...; P(a); P(b); V(b); V(a); ...;. For job J_2, we assume a code sequence of the form ...; P(b); P(a); V(a); V(b); ...;. A possible execution sequence is shown in Fig. 4.10.

Fig. 4.10 Priority inheritance deadlock

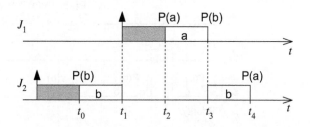

We assume that the priority of J_1 is higher than that of J_2. Hence, J_1 preempts J_2 at time t_1 and runs until it calls P(b), while b is held by J_2. Hence, J_2 resumes. However, it runs into a deadlock when it calls P(a). Such a deadlock would also exist if we were not using any resource access protocol.

4.2.3 Priority Ceiling Protocol

Deadlocks can be avoided with the **priority ceiling protocol** [460] (PCP)[6]. PCP requires jobs to be known at design time. With PCP, a job is not allowed to enter a critical section if there are already locked semaphores which could block it eventually. Hence, once a job enters a critical section, it can not be blocked by lower priority jobs until its completion. This is achieved by assigning a priority ceiling. Each semaphore S is assigned a priority ceiling $C(S)$. It is the static priority of the highest priority job that can lock S.

PCP works as follows:

- Let us assume that some job J is running and wants to lock semaphore S. Then, J can lock S only if the priority of J exceeds the priority ceiling $C(S')$ of semaphore S' where S' is the semaphore with the highest priority ceiling among all the semaphores which are currently locked by jobs other than J. If such a semaphore

---

[6]We adopt the example presented on previously available slides from Linköping University at http://www.ida.liu.se/~unmbo/RTS_CUGS_files/Lecture3.pdf.

exists, then J is said to be blocked by S' and the job currently holding S'. When J gets blocked by S', the job currently holding S' inherits the priority of J.

- When some job J leaves a critical section guarded by S, it unlocks S and the highest priority job, if any, which is blocked by S is awakened. The priority of J is set to the highest priority among all the jobs that which are still blocked by some semaphore which J is still holding. If J is not blocking any other job, then the priority of J is set to its normal priority.

Figure 4.11 shows an example. In this example, semaphores a, b, and c are used. The highest priority of a and b is p_1, and the highest priority of c is p_2.

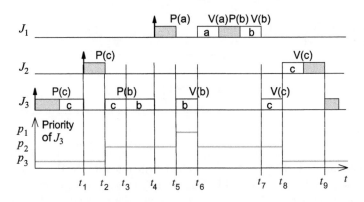

Fig. 4.11 Locking with the priority ceiling protocol

At time t_2, J_2 wants to lock c, but c is already locked. Furthermore, the priority of J_2 does not exceed the ceiling of c. Nevertheless, the attempt to lock c results in an increase of the priority of J_3 to p_2.

At time t_5, J_1 tries to lock a. a is not yet locked, but J_3 has locked b and the ceiling for b does not exceed the current priority of J_1. So, J_1 gets blocked. This is the key property of PCP: This blocking avoids potential later deadlocks. J_3 inherits the priority of J_1, reflecting that J_1 is waiting for the semaphore b to be released by J_3.

At time t_6, J_3 unlocks b. J_1 is the highest priority job so far blocked by b and now awakened. The priority of J_3 drops to p_2. J_1 locks and unlocks a and b and runs to completion. At time t_7, J_2 is still blocked by c, and for all jobs with priority p_2, J_3 is the only one that can be resumed. At time t_8, J_3 unlocks c and its priority drops to p_3. J_2 is no longer blocked, and it preempts J_3 and locks c. J_3 is only resumed after J_2 has run to completion.

Let us consider a second example, to be used later for comparison with an extended PCP protocol. Figure 4.12 shows this second example[7].

---

[7]This example was adopted from http://www.ida.liu.se/~unmbo/RTS_CUGS_files/Lecture3.pdf.

Fig. 4.12 Second PCP
example

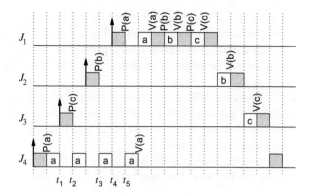

The highest priority of all semaphores is the priority of J_1. At time t_2, there is a request by J_3 for semaphore c, but the priority of J_3 is lower than the ceiling for the already locked semaphore a, and J_4 inherits the priority of J_3. At time t_3, there is a request for b, but the priority of J_2 is again lower than for the ceiling of the already locked semaphore a, and J_4 inherits the priority of J_2. At time t_5, there is a request for a, but the priority of J_1 is not exceeding the ceiling for a and J_4 inherits the priority of J_1. When J_4 releases a, no semaphore is blocked and its priority drops to its normal priority. At this time, J_1 has the highest priority and executes until it terminates. Remaining executions are determined by the regular priorities.

It can be proven that PCP prevents deadlocks (see [82], Theorem 7.3).

There are certain variants of PCP with different times at which the priority is changed. The Distributed Priority Ceiling Protocol (DPCP) [441] and the Multiprocessor Priority Ceiling Protocol (MPCP) [440] are extensions of PCP for multiprocessors.

4.2.4 Stack Resource Policy

In contrast to PCP, the stack resource policy (SRP) supports dynamic priority scheduling; i.e., SRP can be used with dynamic priorities as computed by EDF scheduling (see Sect. 6.2.1.2 on p. 294). For SRP, we have to distinguish between jobs and tasks. Tasks may be describing repeating computations. Each computation is a job in the sense the term has been used so far. The notion of tasks captures features that apply to a set of jobs, e.g., the same code which needs to be executed periodically. Accordingly, for each task τ_i, there is a corresponding set of jobs. See also Definition 6.1 on p. 284. SRP does not just consider each job of a task separately, but defines properties which apply to tasks globally. Furthermore, SRP supports multi-unit resources like, for example, memory buffers. The following values are defined:

- The **preemption level** l_i of a task τ_i provides information about which tasks can be preempted by jobs of τ_i. A task τ_i can preempt some other task τ_j only if $l_i > l_j$. We require that if task τ_i arrives after τ_j and τ_i has a higher priority, then τ_i must have a higher preemption level than τ_j. For sporadic EDF scheduling (see p. 304), this means that the preemption levels are ordered inversely with respect to the relative deadlines. The larger the deadline, the easier it is to preempt the job. l_i is a static value.

- The **resource ceiling** of a resource is the highest preemption level of the tasks that could be blocked by issuing their maximum request for units of this resource. The resource ceiling is a dynamic value which depends on the number of currently available resource units.
- The **system ceiling** is the highest resource ceiling of all the resources which are currently blocked. This value is dynamic and changes with resource accesses.

SRP blocks the job at the time it attempts to preempt, instead of the time at which it tries to lock: A job can preempt another job if it has the highest priority, and its preemption level is higher than the system ceiling. A job is not allowed to start until the resources currently available are sufficient to meet the maximum requirement of every job that could preempt it.

Figure 4.13 demonstrates the difference between PCP and SRP by means of the example[8] shown in Fig. 4.12.

Fig. 4.13 SRP example

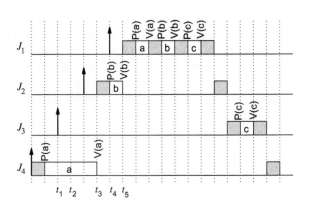

For SRP, at time t_1 there is no preemption since the preemption level is not higher than the ceiling. The same happens at t_4. Overall, SRP has significantly less preemptions than PCP. This property has made SRP a popular protocol.

SRP is called *stack* resource policy, since jobs cannot be blocked by jobs with a lower l_i and can resume only when the job completes. Hence, jobs on the same level l_i can share stack space. With many jobs at the same level, a substantial amount of space can be saved.

SRP is also free of deadlocks (see Baker [33]). For more details about SRP, refer also to Buttazzo [82].

---

[8]This example was adopted from http://www.ida.liu.se/~unmbo/RTS_CUGS_files/Lecture3.pdf.

PIP, PCP, and SRP protocols have been designed for single processors. A first overview of resource access protocols for multiprocessors was published by Rajkumar et al. [442]. At the time of writing this book, there is not yet a standard resource access protocol for multi-cores (see Baruah et al. [40], Chap. 23).

4.3   ERIKA

Several embedded systems (such as automotive systems and home appliances) require the entire application to be hosted on small microcontrollers[9]. For that reason, the operating system services provided by the firmware on such systems must be limited to a minimal set of features allowing multi-threaded execution of periodic and aperiodic jobs, with support for shared resources to avoid the priority inversion phenomenon.

Such requirements have been formalized in the 1990s by the OSEK/VDX Consortium [17], which defined the minimal services of a multi-threaded real-time operating system allowing implementations of 1–10 kilobytes code footprint on 8-bit microcontrollers. The OSEK/VDX API has been recently extended by the AUTOSAR Consortium [26] which provided enhancements to support time protection, scheduling tables for time triggered systems, and memory protection to protect the execution of different applications hosted on the same microcontroller. This section briefly describes the main features and requirements of such systems, considering as a reference implementation the open-source ERIKA Enterprise real-time kernel [150].

The first feature that distinguishes an OSEK kernel from other operating systems is that all kernel objects are *statically* defined at compile time. In particular, most of these systems do not support dynamic memory allocation and dynamic creation of jobs. To help the user in configuring the system, the OSEK/VDX standard provides a configuration language, named OIL, to specify the objects that must be instantiated in the application. When the application is compiled, the OIL compiler generates the operating system data structures, allocating the exact amount of memory needed. This approach allows allocating only the data really needed by the application, to be put in flash memory (which is less expensive than RAM memory on most microcontrollers).

The second feature distinguishing an OSEK/VDX system is the support for *Stack Sharing*. The reason for providing stack sharing is that RAM memory is very expensive on small microcontrollers. The possibility of implementing a stack sharing system is related to how the code is written.

In traditional real-time systems, we consider the repetitive execution of code. A job corresponds to a single execution of the code. The code to be executed repeatedly is called a **task**. In particular, tasks may be periodically causing the execution of a job. The typical implementation of such a periodic task is structured according to the following scheme:

---

[9]This section was contributed by G. Buttazzo and P. Gai (Pisa).

```
task(x) {
  int local;
  initialization();
  for (;;) {
    do_instance();
    end_instance();
}}
```

Such a scheme is characterized by a forever loop containing an instance (job) of the periodic task that terminates with a blocking primitive (end_instance()), which has the effect of blocking the task until the next activation. When following such a programming scheme (called *extended task* in OSEK/VDX), the task is always present in the stack, even during waiting times. In this case, the stack cannot be shared and a separate stack space must be allocated for each task.

The OSEK/VDX standard also provides support for *basic tasks*, which are special tasks that are implemented in a way more similar to functions, according to the following scheme:

```
int local;
Task x() {
  do_instance();
}
System_initialization() {
  initialization();
  ...}
```

With respect to extended tasks, in basic tasks, the persistent state that must be maintained between different instances is not stored in the stack, but in global variables. Also, the initialization part is moved to system initialization, because tasks are not dynamically created, but they exist since the beginning. Finally, no synchronization primitive is needed to block the task until its next period, because the task is activated every time a new instance starts. Also, the task cannot call any blocking primitive; therefore, it can either be preempted by higher priority tasks or execute until completion. In this way, the task behaves like a function, which allocates a frame on the stack, runs, and then cleans the frame. For this reason, the task does not occupy stack space between two executions, allowing the stack to be shared among all tasks in the system. ERIKA Enterprise supports stack sharing, allowing all basic tasks in the system to share a single stack, so reducing the overall RAM memory used for this purpose.

Concerning task management, OSEK/VDX kernels provide support for fixed priority scheduling with Immediate Priority Ceiling to avoid the priority inversion problem. The usage of Immediate Priority Ceiling is supported through the specification of the resource usage of each task in the OIL configuration file. The OIL compiler computes the resource ceiling of each task based on the resource usage declared by each task in the OIL file.

OSEK/VDX systems also support non-preemptive scheduling and preemption thresholds to limit the overall stack usage. The main idea is that limiting the preemption between tasks reduces the number of tasks allocated on the system stack at the same time, further reducing the overall amount of required RAM. Note that reducing preemptions may degrade the schedulability of the tasks set, and hence, the degree of preemption must be traded off with the system schedulability and the overall RAM memory used in the system.

Another requirement for operating systems designed for small microcontrollers is *scalability*, which means supporting reduced versions of the API for smaller footprint implementations. In mass production systems, in fact, the footprint significantly impacts on the overall cost. In this context, scalability is provided through the concept of *Conformance Classes*, which define specific subsets of the operating system API. Conformance classes are also accompanied by an upgrade path between them, with the final objective of supporting partial implementation of the standard with reduced footprint. The conformance classes supported by the OSEK/VDX standard (and by ERIKA Enterprise) are as follows:

- BCC1: This is the smallest conformance class, supporting a minimum of 8 tasks with different priority and 1 shared resource.
- BCC2: Compared to BCC1, this conformance class adds the possibility to have more than one task at the same priority. Each task can have pending activations; that is, the operating system records the number of instances that have been activated but not yet executed.
- ECC1: Compared to BCC1, this conformance class adds the possibility to have extended tasks that can wait for an event to appear.
- ECC2: This conformance class adds both multiple activations and extended tasks.

ERIKA Enterprise further extends these conformance classes by providing the following two conformance classes:

- EDF: This conformance class does not use a fixed priority scheduler but an Earliest Deadline First (EDF) scheduler (see Sect. 6.2.1.2) optimized for the implementation on small microcontrollers.
- FRSH: This conformance class extends the EDF scheduler class by providing a resource reservation scheduler based on the IRIS scheduling algorithm [363].

Another interesting feature of OSEK/VDX systems is that the system provides an API for controlling interrupts. This is a major difference when compared to POSIX-like systems, where interrupts are an exclusive domain of the operating system and are not exported to the operating system API. The rationale for this is that on small microcontrollers users often want to directly control interrupt priorities, and hence, it is important to provide a standard way to deal with interrupt disabling/enabling. Moreover, the OSEK/VDX standard specifies two types of Interrupt Service Routines (ISRs):

- Category 1: Simpler and faster does not implement a call to the scheduler at the end of the ISR;

- Category 2: This ISR can call some primitives that change the scheduling behavior. The end of the ISR is a rescheduling point. ISR1 has always a higher priority of ISR2.

An important feature of OSEK/VDX kernels is the possibility to fine-tune the footprint by removing error-checking code from the production versions, as well as to define hooks that will be called by the system when specific events occur. These features allow for a fine-tuning of the application footprint that will be larger (and safer) when debugging and smaller in production when most bugs will be found and removed from the code.

To support a better debugging experience, the OSEK/VDX standard defines a textual language, named ORTI, which describes where the various objects of the operating system are allocated. The ORTI file is typically generated by the OIL compiler and is used by debuggers to print detailed information about operating system objects defined in the system (e.g., the debugger could print the list of the tasks in an application with their current status).

All the features defined by the OSEK/VDX standard have been implemented in the open-source ERIKA Enterprise kernel [150], for a set of embedded microcontrollers, with a final footprint ranging between 1 and 5 kilobytes of object code. ERIKA Enterprise also implements additional features, like the EDF scheduler, providing an open and free of charge operating system that can be used to learn, test, and implement real applications for industrial and educational purposes.

4.4   Embedded Linux

Increasing requirements to the functionality of embedded systems, such as Internet connectivity (in particular for the Internet of Things) or sophisticated graphics displays, demand that a large amount of software is added to a typical embedded system's simple operating system. It has been shown that it is possible to add some of this functionality to small embedded real-time operating systems, e.g., by integrating a small Internet Protocol (IP) network stack [136]. However, integrating a number of different additional software components is a complex task and may lead to functional as well as security deficiencies.

A different approach, enabled by the exponential growth of semiconductor densities according to Moore's Law, is the adaptation of a well-tested code base with the required functionality to run in an embedded context. Here, Linux[10] has become the OS of choice for a large number of complex embedded applications following this approach, such as Internet routers, GPS satellite navigation systems, network-attached storage devices, smart television sets, and mobile phones. These applications benefit from easy portability—Linux has been ported to more than 30 processor architectures, including the popular embedded ARM, MIPS, and PowerPC architectures—as well as the system's open-source nature, which avoids the licensing costs arising for commercial embedded operating systems.

---

[10]This section on embedded Linux was contributed by M. Engel (Coburg).

Adapting Linux to typical embedded environments poses a number of challenges due to its original design as a server and desktop OS. Below, we detail solutions available in Linux to tackle the most common problems that arise in its use in embedded systems.

4.4.1 Embedded Linux Structure and Size

Strictly speaking, the term "Linux" denotes only the kernel of a Linux-based operating system. To create a complete, working operating system, a number of additional components are required that run on top of the Linux kernel. A configuration for a typical Linux system, including system-level user mode components, is shown in Fig. 4.14. On top of the Linux kernel reside a number of—commonly dynamically linked—libraries, which form the basis for system-level tools and applications. Device drivers in Linux are usually implemented as loadable kernel modules; however, restricted user mode access to hardware is also possible.

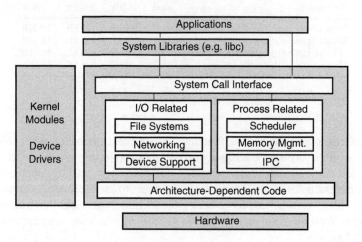

Fig. 4.14 Structure of typical Linux-based system

The open-source nature of Linux allows to tailor the kernel and other system components to the requirements of a given application and platform. This, in turn, results in a small system which enables the use of Linux in systems with restricted memory sizes.

One of the essential components of a Unix-like system is the C library, which provides basic functionality for file I/O, process synchronization and communication, string handling, arithmetic operations, and memory management. The libc variant commonly used in Linux-based systems is GNU libc (glibc). However, glibc was designed with server and desktop systems in mind and, thus, provides much more functionality than typically required in embedded applications. Linux-based Android® systems replace glibc with Bionic, a libc version derived from BSD Unix. Bionic is specifically designed to support systems running at lower clock speeds, e.g., by providing a tailored version of the pthreads multi-threading library to efficiently support Android's Dalvik Java VM. Bionic's size is estimated to be about half the size of a typical glibc version[11].

Several significantly smaller implementations of libc exist such as newlib, musl, uClibc, PDCLib, and dietlibc. Each of these is optimized for a specific use case; e.g., musl is optimized for static linking, uClibc was originally designed for MMU-less[12] Linux systems (see below), whereas newlib is a cross-platform libc also available for a number of other OS platforms. Sizes of the related shared library binary files range from 185 kB (dietlibc) to 560 kB (uClibc), whereas the glibc binary is 7.9 MB in size (all numbers taken from x86 binaries) according to a comprehensive comparison of different libc implementation features and sizes, compiled by Eta Labs[13]. Figure 4.15 gives an overview of the sizes of various libc variants and programs built using the different libraries.

libc version	musl	uClibc	dietlibc	glibc
Static library size	426 kB	500 kB	120 kB	2.0 MB
Shared library size	527 kB	560 kB	185 kB	7.9 MB
Minimal static C program size	1.8 kB	5 kB	0.2 kB	662 kB
Minimal static "Hello, World" size	13 kB	70 kB	6 kB	662 kB

Fig. 4.15 Size comparison of different Linux libc configurations

In addition to the C library, the functionality, size, and number of utility programs bundled with the OS can be adapted according to application requirements. These utilities are required in a Linux system to control system start-up, operation, and monitoring; examples are tools to mount file systems, to configure network interfaces, or to copy files. As is the case for glibc, a typical Linux system includes a set of tools appropriate for a large number of use cases, most of which are not required on an embedded system.

An alternative to a traditional set of diverse tools is BusyBox, software that provides a number of simplified essential Unix utilities in a single executable file. It was specifically created for embedded operating systems with very limited resources. BusyBox reduces the overhead introduced by the executable file format and allows code to be shared between multiple applications without requiring a library. A comparison of BusyBox with alternative approaches to provide a small user mode tool set can be found in [508].

---

[11] The glibc-shared library size includes internationalization support.

[12] See Appendix C for an introduction to MMUs.

[13] Available online at http://www.etalabs.net/compare_libcs.html.

4.4.2 Real-Time Properties

Achieving real-time guarantees in a system based on a general-purpose operating system kernel is one of the most complex challenges in adapting an OS to run in an embedded context. As shown above in Fig. 4.3, one common approach is to run the Linux kernel and all Linux user mode processes as a dedicated task of an underlying RTOS, only to be activated when no real-time task needs to run. In Linux, competing approaches exist that follow this design pattern. RTAI (real-time application interface) [133] is based on the Adeos hypervisor[14], which is implemented as a Linux kernel extension. Adeos enables multiple prioritized domains (one of which is the Linux kernel itself) to exist simultaneously on the same hardware. On top of this, RTAI provides a service API, for example, to control interrupts and system timers. Xenomai [176] was codeveloped with RTAI for several years, but became an independent project in 2005. It is based on its own abstract "nucleus" RTOS core, which provides real-time scheduling, timer, memory allocation, and virtual file handling services. Both projects differ in their aims and implementations. However, they share the support for the Real-Time Driver Model (RTDM), a method to unify interfaces for developing device drivers and related applications in real-time Linux systems. The third approach using an underlying real-time kernel is RT-Linux [580], developed as a project at the New Mexico Institute of Mining and Technology and then commercialized at the company FSMLabs, which was acquired by WindRiver in 2007. The related product was discontinued in 2011. The use of RT-Linux in products was controversial, since its initiators vigorously defended their intellectual property, for which they obtained a software patent [579]. The decision to patent the RT-Linux methods was not well received by the Linux developer community, leading to spin-offs resulting in the above-mentioned RTAI and Xenomai projects.

A more recent approach to add real-time capabilities to Linux, integrated into the kernel as of version 3.14 (2014), is SCHED_DEADLINE, a CPU scheduling policy based on the Earliest Deadline First (EDF) and Constant Bandwidth Server (CBS) [3] algorithms and supporting resource reservations. The SCHED_DEADLINE policy is designed to coexist with other Linux scheduling policies. However, it takes precedence before all other policies to guarantee real-time properties.

Each task τ_i scheduled under SCHED_DEADLINE is associated with a run-time budget C_i and a period T_i, indicating to the kernel that C_i time units are required by that task every T_i time units, on any processor. For real-time applications, T_i corresponds to the minimum time elapsing between subsequent activations (releases) of the task, and C_i corresponds to the worst-case execution time needed by each execution of the task. On addition of a new task to this scheduling policy, a schedulability test is performed and the task is only accepted if the test succeeds. During scheduling, a task is suspended when it tries to run for longer than the preallocated budget and deferred to its next execution period. This non-work conserving

---

[14]See http://home.gna.org/adeos/.

strategy[15] is required to guarantee temporal isolation between different tasks. Thus, on single processor or partitioned multiprocessor systems (with tasks pinned to a specific CPU), all accepted SCHED_DEADLINE tasks are guaranteed to be scheduled for an overall time equal to their budget in every time window as long as their period.

In the general case of tasks which are free to migrate on a multiprocessor, as SCHED_DEADLINE implements global EDF (as described in detail in Sect. 6.3.3), the general tardiness bound for global EDF applies [123]. Benchmarks performed in [322] give an amount of missed deadlines of less than 0.2% when running SCHED_DEADLINE on a four-processor system with a utilization of 380 and 0.615% with a utilization of 390%. The numbers cited for a six-processor system are of similar magnitude. Of course, no deadline misses occur on single processor systems or multi-core systems with processes pinned to a fixed processor core.

4.4.3 Flash Memory File Systems

Embedded systems pose different requirements to permanent storage than server or desktop environments. Often, there is a large amount of static (read-only) data, whereas the amount of varying data is in many cases quite limited.

Accordingly, file system storage can benefit from these special conditions. Since most of the read-only data in current embedded SoCs is implemented as flash ROM, optimization for this storage is an important aspect for the use of Linux in embedded systems. Accordingly, a number of different file systems specifically designed for using NAND-based flash storage have been developed.

One of the most stable flash-specific file systems available is the log-structured Journaling Flash File System version 2 (JFFS2) [568]. In JFFS2, changes to files and directories are "logged" to flash memory in so-called nodes. Two types of nodes exist, inodes (shown in Fig. 4.16), which consist of a header with file metadata followed by an optional payload of file data, and dirent nodes, which are directory entries each holding a name and an inode number. Nodes start out as valid when they are created and become obsolete when a newer version has been created in a different place in flash memory. JFFS2 supports transparent data compression by storing compressed data as inode payloads.

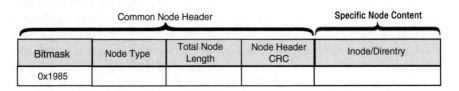

Fig. 4.16 Structure of the JFFS2 inode content

---

[15]This means that the processor may be idle even when tasks could be executed. A definition of the term can be found in Chap. 6 on p. 298.

However, compared to other log-structured file systems such as Berkeley lfs [449], there is no circular log. Instead, JFFS2 uses blocks, a unit the same size as the erase segment of the flash medium. Blocks are filled with nodes in a bottom-up manner one at a time, as shown in Fig. 4.17.

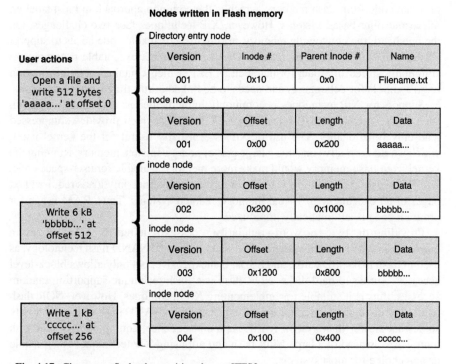

Nodes written in Flash memory

User actions

Directory entry node			
Version	Inode #	Parent Inode #	Name
001	0x10	0x0	Filename.txt

Open a file and write 512 bytes 'aaaaa...' at offset 0

inode node			
Version	Offset	Length	Data
001	0x00	0x200	aaaaa...

inode node			
Version	Offset	Length	Data
002	0x200	0x1000	bbbbb...

Write 6 kB 'bbbbb...' at offset 512

inode node			
Version	Offset	Length	Data
003	0x1200	0x800	bbbbb...

inode node			
Version	Offset	Length	Data
004	0x100	0x400	ccccc...

Write 1 kB 'ccccc...' at offset 256

Fig. 4.17 Changes to flash when writing data to JFFS2

Clean blocks contain only valid nodes, whereas dirty blocks contain at least one obsolete node. In order to reclaim memory, a background garbage collector collects dirty blocks and frees them. Valid nodes from dirty blocks are copies into a new block, whereas obsolete blocks are skipped. After copying, the dirty block is marked as free. The garbage collector is also able to consume clean blocks in order to even out the flash memory wear-leveling and prevent localized erasure of blocks in a mostly static file system, as is common in many embedded systems.

4.4.4 Reducing RAM Usage

Traditionally, Unix-like operating systems treat main memory (RAM) as a cache for secondary storage on disk, i.e., swap space [367]. While this is a useful assumption for desktop and server systems with large disks and equally large memory requirements, it results in a waste of resources for embedded systems, since programs which exist in a system's nonvolatile memory have to be loaded into volatile memory for execution. This commonly includes the rather large operating system kernel.

To eliminate this duplication of memory requirements, a number of Execute-in-Place (XiP) techniques have been developed which allow the direct execution of program code from flash memory, which is the common approach in most smaller, microcontroller-based systems. However, XiP techniques face two challenges. On the one hand, the nonvolatile memory storing the executable code needs to support accesses in byte or word granularity. On the other hand, executable programs are commonly stored in a data format such as ELF, which contains meta information (e.g., symbols for debugging) and needs to be linked at run-time before execution.

Support for XiP techniques is commonly implemented as a special file system, such as the Advanced XiP Filesystem (AXFS) [44], which provides compressed read-only functionality. The use of XiP is especially useful for the kernel itself, which would normally consume a large part of non-swappable memory. Running the kernel from flash memory would make more memory available for user-space code. XiP for user mode code itself is less useful, since the kernel only loads required text pages of an executable in virtual memory enabled systems. Thus, RAM usage for program code is automatically minimized.

Providing the byte- or word-granularity accesses required for XiP is mostly a question of cost in current systems. The commonly used NAND flash technology, as used in flash disks, SD cards, and SSDs, is inexpensive, but only allows block-level accesses, similar to hard disks. NOR flash is a flash technique supporting random accesses, thus is it suitable for implementing XiP techniques. However, NOR flash tends to be an order of magnitude more expensive than NAND flash and is commonly somewhat slower than system RAM. As a consequence, equipping a system with more RAM instead of a large NOR flash and not using XiP techniques is a sensible design choice for most systems.

4.4.5   uClinux—Linux for MMU-Less Systems

One final resource restriction is apparent in low-end microcontroller systems, such as ARM's Cortex-M series. The processor cores in these SoCs were developed for typical real-time OS scenarios, which often use a simple library OS approach, as described for Erika above. Thus, they lack crucial OS support hardware such as a paging memory management unit (see Appendix C). However, the large address space and relatively high clock speeds of these microcontrollers enable running a Linux-like operating system with some restrictions. Thus, uClinux was created as a derivative of the Linux kernel for MMU-less systems. Since kernel version 2.5.46, uClinux support is available in the mainstream kernel source tree for a number of architectures including ARM7TDMI, ARM Cortex-M3/4/7/R, MIPS, M68k/ColdFire, as well as FPGA-based softcores such as Altera Nios II, Xilinx MicroBlaze, and Lattice Mico32.

The lack of memory management hardware in uClinux-supported platforms comes with a number of disadvantages. An obvious drawback is the lack of memory protection, so any process is able to read and write other processes' memory. The lack of an MMU also has consequences for the traditional Unix process creation approach. Commonly, processes in Unix are created as a copy of an existing process using the fork() system call [447]. Instead of creating a physical copy in memory, which would require copying potentially large amounts of data, only the page table entries of the process executing fork() are replicated and point to physical page frames of the parent process. When the newly created process memory starts to differ from its parent due to data writes, only the affected page frames are copied on demand using a copy-on-write strategy. The lack of hardware support for copy-on-write semantics and the overhead involved in actually copying pages result in the fork() system call being unavailable in uClinux.

Instead, uClinux provides the vfork() system call. This system call makes use of the fact that most Unix-style processes immediately call exec() after a fork to start a different executable file by overloading their memory image with text and data segments of that different binary:

```
pid_t childPID;
childPID = vfork();
  if (childPID == 0) { // in child process
    execl("/bin/sh","sh", 0);
  }
printf("Parent program running again, child PID is %d", childPID);
```

The direct calling of exec() after vfork() implies that the complete address space of the newly created process will be replaced in any case and only a small part of the executable calling vfork() is actually used. In contrast to standard Unix behavior, vfork guarantees that the parent process is stopped after calling fork until the child process has called the exec() system call. Thus, the parent process is unable to interfere with the execution of the child process until the new program image has been loaded. However, some restrictions have to be observed to guarantee safe operation of vfork(). It is not permitted to modify the stack in the created child process, i.e., no function calls may be executed before exec. As a consequence, returning from vfork in case of an error, e.g., insufficient memory or inability to execute the new program, is impossible, since this would modify the stack. Instead, it is recommended to exit() from the child process in case of a problem.

To summarize, uClinux is a way to use some Linux functionality on low-end, microcontroller-style embedded systems. However, the on-chip memory even in high-end microcontrollers is restricted to several hundred kB. A minimal uClinux version, however, requires about 8 MB RAM, so the addition of an external RAM chip is essential. For systems offering a smaller memory footprint, more traditional RTOS systems are still the more feasible solution.

4.4.6 Evaluating the Use of Linux in Embedded Systems

In addition to technical criteria, the decision whether to base an embedded system on Linux also has to consider legal and business questions.

On the technical side, Linux includes support for a large number of CPU architectures, SoCs, and peripheral devices as well as communication protocols commonly used in embedded applications, such as Internet Protocol TCP/IP, CAN, Bluetooth® or IEEE802.15.4/ZigBee®. It provides a POSIX-like API that enables easy porting of existing code, not only written in C or C++, but also in scripting languages such as Python or Lua, and even more specialized languages like Erlang. Linux development tools are available free of charge and can easily be integrated into development toolflows utilizing IDEs such as Eclipse and continuous integration testing services such as Jenkins. While in general, the Linux code base is well tested, the quality of support varies with the targeted platform. When utilizing a less common hardware platform, it is recommended to thoroughly investigate the stability of CPU and driver support. One drawback of using Linux is the inherent complexity of the large code base, requiring a good insight into and experience with the system to debug problems. However, a number of semiconductor manufacturers and third-party companies offer commercial support for embedded Linux, including the provisioning of complete board support software packages (BSPs) for a number of reference designs.

From a business perspective, the obvious benefit of using Linux is the availability of its source code free of cost. However, the GPL License version 2[16] governing the kernel source code also requires that the source code for modifications to the existing code base is provided along with the binary code. This might jeopardize trade secrets of hardware components or violate non-disclosure agreements with hardware intellectual property owners. For some hardware, such as GPU drivers, this is circumvented by the inclusion of binary code "blobs" which are loaded by an open-source device driver stub. However, this approach is being actively discouraged by the Linux kernel developers.

An increasingly serious problem is the security of embedded systems built on Linux, especially in the context of the Internet of Things. Many security problems affecting the Linux kernel also apply to embedded Linux. Inexpensive consumer devices, such as Internet-based cameras, routers, and mobile phones, rarely receive software updates, but may be in active use for many years. This exposes them to security vulnerabilities which are already being actively exploited, e.g., for distributed denial-of-service attacks (DDOS) emanating from thousands of hijacked embedded Linux devices. As a consequence, the cost of continually updating devices in production as well as legacy devices in the field has to be considered in order to provide secure systems.

---

[16] See http://www.gnu.org/licenses/gpl-2.0.html.

4.5 Hardware Abstraction Layers

Hardware abstraction layers (HALs) provide a way for accessing hardware through a hardware-independent application programming interface (API). For example, we could come up with a hardware-independent technique for accessing timers, irrespective of the addresses to which timers are mapped. Hardware abstraction layers are used mostly between the hardware and operating system layers. They provide software intellectual property (IP), but they are neither part of operating systems nor can they be classified as middleware. A survey over work in this area is provided by Ecker, Müller, and Dömer [139].

4.6 Middleware

Communication libraries provide a means for adding communication functionality to languages lacking this feature. They add communication functionality on top of the basic functionality provided by operating systems. Due to being added on top of the OS, they can be independent of the OS (and obviously also of the underlying processor hardware). As a result, we will obtain cyber-physical systems in the sense that we place emphasis on communications. Also, such communication is needed for the Internet of Things (IoT). There is a trend toward supporting communication within some local system as well as communication over longer distances. The use of Internet Protocols in general is becoming more popular. Frequently, such protocols enable **secure communication**, based on en- and decryption (see p. 192). The corresponding algorithms are a special case of middleware.

4.6.1 OSEK/VDX COM

OSEK/VDX® COM is a special communication standard for the OSEK automotive operating systems [419][17]. OSEK COM provides an "Interaction Layer" as an application programming interface (API) through which internal communication (communication within one ECU) and external communication (communication with other ECUs) can be performed. OSEK COM specifies just the functionality of the Interaction layer. Conforming implementations must be developed separately.

The Interaction layer communicates with other ECUs via a "Network Layer" and a "Data Link" layer. Some requirements for these layers are specified by OSEK COM, but these layers themselves are not part of OSEK COM. This way, communication can be implemented on top of different network protocols.

---

[17]OSEK is a trademark of Continental Automotive GmbH.

OSEK COM is an example of communication middleware dedicated toward embedded systems. In addition to middleware tailored for embedded systems, many communication standards developed for non-embedded applications can be adopted for embedded systems as well.

4.6.2 CORBA®

CORBA® (Common Object Request Broker Architecture) [407] is one example of such adopted standards. CORBA facilitates the access to remote services. With CORBA, remote objects can be accessed through standardized interfaces. Clients are communicating with local stubs, imitating the access to the remote objects. These clients send information about the object to be accessed as well as parameters (if any) to the Object Request Broker (ORB, see Fig. 4.18). The ORB then determines the location of the object to be accessed and sends information via a standardized protocol, e.g., the IIOP protocol, to where the object is located. This information is then forwarded to the object via a skeleton, and the information requested from the object (if any) is returned using the ORB again.

Fig. 4.18 Access to remote objects using CORBA

Standard CORBA does not provide the predictability required for real-time applications. Therefore, a separate real-time CORBA (RT-CORBA) standard has been defined [408]. A very essential feature of RT-CORBA is to provide *end-to-end predictability of timeliness in a fixed priority system*. This involves *respecting thread priorities between client and server for resolving resource contention* and bounding the latencies of operation invocations. One particular problem of real-time systems is that thread priorities might not be respected when threads obtain mutually exclusive access to resources. The priority inversion problem (see p. 205) has to be addressed in RT-CORBA. RT-CORBA includes provisions for bounding the time during which such priority inversion can happen. RT-CORBA also includes facilities for thread priority management. This priority is independent of the priorities of the underlying operating system, even though it is compatible with the real-time extensions of the POSIX standard for operating systems [195]. The thread priority of clients can be propagated to the server side. Priority management is also available for primitives providing mutually exclusive access to resources. The priority inheritance protocol just described must be available in implementations of RT-CORBA. Pools of preexisting threads avoid the overhead of thread creation and thread construction.

4.6.3 POSIX Threads (Pthreads)

The POSIX thread (Pthread) library is an application programming interface (API) to threads at the operating system level [37]. Pthreads are consistent with the IEEE POSIX 1003.1c operating system standard. A set of threads can be run in the same address space. Therefore, communication can be based on shared memory communication. This avoids the memory copy operations typically required for MPI (see Sect. 2.8.4.1 on p. 110). The library is therefore appropriate for programming multi-core processors sharing the same address space, and it includes a standard API with mechanisms for mutual exclusion. Pthreads use completely explicit synchronization [530]. The exact semantics depends on the memory consistency model used. Synchronization is hard to program correctly. The library can be employed as a back-end for other programming models.

4.6.4 UPnP, DPWS, and JXTA

Universal Plug-and-Play (UPnP) is an extension of the plug-and-play concept of PCs toward devices connected within a network. Connecting network printers, storage space and switches in homes and offices easily can be seen as the key target [415]. Due to security concerns, only data is exchanged. Code cannot be transferred.

Devices Profile for Web Services (DPWS) aims at being more general than UPnP. *"The Devices Profile for Web Services (DPWS) defines a minimal set of implementation constraints to enable secure Web Service messaging, discovery, description, and eventing on resource-constrained devices"* [569]. DPWS specifies services for discovering devices connected to a network, for exchanging information about available services, and for publishing and subscribing to events.

In addition to libraries designed for high-performance computing (HPC), several comprehensive network communication libraries can be used. These are typically designed for a loose coupling over Internet-based communication protocols. JXTA™[265] is an open-source peer-to-peer protocol specification. It defines a protocol by a set of XML messages that allow any device connected to a network peer to exchange messages and collaborate independently of the network topology. JXTA creates a virtual overlay network, allowing a peer to interact with other peers even when some of the peers and resources are behind firewalls. The name is derived from the word "juxtapose."

CORBA, MPI, Pthreads, OpenMP, UPnP, DPWS, and JXTA are special cases of communication middleware (software to be used at a layer between the operating system and applications). Initially, they were essentially designed for communication between desktop computers. However, there are attempts to leverage the knowledge and techniques also for embedded systems. In particular, MPI (Message Passing Interface) is designed for message passing based communication and it is rather popular. It has recently been extended to also support shared memory-based communication.

For mobile devices like smartphones, using standard middleware may be appropriate. For systems with hard time constraints (see Definition 1.8 on p. 9), their overhead, their real-time capabilities, and their services may be inappropriate.

4.7 Real-Time Databases

Databases provide a convenient and structured way of storing and accessing information. Accordingly, databases provide an API for writing and reading information. A sequence of read and write operations is called a **transaction**. Transactions may have to be aborted for a variety of reasons: There could be hardware problems, deadlocks, problems with concurrency control, etc. A frequent requirement is that transactions do not affect the state of the database unless they have been executed to their very end. Hence, changes caused by transactions are normally not considered to be final until they have been **committed**. Most transactions are required to be **atomic**. This means that the end result (the new state of the database) generated by some transaction must be the same as if the transaction has been fully completed or not at all. Also, the database state resulting from a transaction must be **consistent**. Consistency requirements include, for example, that the values from read requests belonging to the same transaction are consistent (do not describe a state which never existed in the environment modeled by the database). Furthermore, to some other user of the database, no intermediate state resulting from a partial execution of a transaction must be visible (the transactions must be performed as if they were executed in **isolation**). Finally, the results of transactions should be persistent. This property is also called **durability** of the transactions. Together, the four properties printed in bold are known as ACID properties (see the book by Krishna and Shin [297], Chap. 5).

For some databases, there are soft real-time constraints. For example, time constraints for airline reservation systems are soft. In contrast, there may also be hard constraints. For example, automatic recognition of pedestrians in automobile applications and target recognition in military applications must meet hard real-time constraints. The above requirements make it very difficult to guarantee hard real-time constraints. For example, transactions may be aborted various times before they are finally committed. For all databases relying on demand paging and on hard disks, the access times to disks are hardly predictable. Possible solutions include main memory databases and predictable use of flash memory. Embedded databases are sometimes small enough to make this approach feasible. In other cases, it may be possible to relax the ACID requirements. For further information, see the book by Krishna and Shin as well as Lam and Kuo [305].

4.8 Problems

We suggest solving the following problems either at home or during a flipped classroom session:

4.1: Which requirements must be met for a embedded operating system?

4.2: Which techniques can be used to customize an embedded operating system in
the necessary way?

4.3: Which requirements must be met for a real-time operating system? How do
they differ from the requirements of a standard OS? Which features of a standard
OS like Windows or Linux could be missing in an RTOS?

4.4: How many seconds have been added at New Year's Eve to compensate for the
differences between UTC and TAI since 1958? You may search the Internet for an
answer to this question.

4.5: Find processors for which memory protection units are available! How are
memory protection units different from the more frequently used memory management units (MMUs)? You may search the Internet for an answer to this question.

4.6: Describe classes of embedded systems for which protection should definitely
be provided! Describe classes of systems, for which we would possibly not need
protection!

4.7: Provide an example demonstrating priority inversion for a system comprising
three jobs!

4.8: Download the levi learning module leviRTS from the levi Web site [472]. Model
a job set as described in Table 4.1.

Table 4.1 Set of jobs requesting exclusive use of resources

Job	Priority	Arrival	Run time	Printer		Comm line	
				$t_{P,P}$	$t_{V,P}$	$t_{P,C}$	$t_{V,C}$
J_1	1 (high)	3	4	1	4	–	–
J_2	2	10	3	–	–	1	2
J_3	3	5	6	–	–	4	6
J_4	4 (low)	0	7	2	5	–	–

$t_{P,P}$ and $t_{P,C}$ are the times relative to the start times, at which a job requests exclusive
use of the printer or the communication line, respectively (called ΔtP in levi). $t_{V,P}$
and $t_{V,C}$ are the times relative to the start times at which these resources are released.
Use priority-based, preemptive scheduling! Which problem occurs? How can it be
solved?

4.9: Which resource access protocols prevent deadlocks caused by exclusive access to resources?

4.10: How is the use of the system stack optimized in ERIKA?

4.11: Which problems have to be solved if Linux is used as an operating system for an embedded system?

4.12: Which impact does the priority inversion problem have on the design of network middleware?

4.13: How could flash memory have an influence on the design of real-time databases?

Chapter 5
Evaluation and Validation

5.1 Introduction

5.1.1 Scope

Specification, hardware platforms and system software provide us with the basic ingredients which we need for designing embedded systems. During the design process, we must validate and evaluate designs rather frequently. Therefore, we will describe validation and evaluation before we talk about design steps. Validation and evaluation, even though different from each other, are very much linked.

Definition 5.1: **Validation** is the process of checking whether or not a certain (possibly partial) design is appropriate for its purpose, meets all constraints and will perform as expected.

Definition 5.2: Validation with mathematical rigor is called **(formal) verification**.

Validation is important for any design procedure, and hardly any system would work as expected, had it not been validated during the design process. Validation is extremely important for safety-critical embedded systems. In theory, we could try to design verified tools which always generate correct implementations from the specification. In practice, this verification of tools does not work, except in very simple cases. As a consequence, each and every design has to be validated. In order to minimize the number of times that we must validate a design, we could try to validate it at the very end of the design process. Unfortunately, this approach normally does not work either, due to the large differences between the level of abstraction used for the specification and that used for the implementation. Therefore, validation is required at various phases during the design procedure (see Fig. 5.1). Validation and design should be intertwined and not be considered as two completely independent activities.

© Springer International Publishing AG 2018
P. Marwedel, *Embedded System Design*, Embedded Systems,
DOI 10.1007/978-3-319-56045-8_5

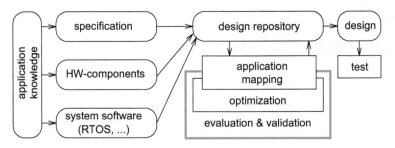

Fig. 5.1 Context of the current chapter

It would be nice to have a single validation technique applicable to all validation problems. In practice, none of the available techniques solves all the problems, and a mix of techniques has to be applied. In this Chapter, starting in Sect. 5.7, we will provide a brief overview of key techniques which are available. This material will be preceded by an overview of evaluation techniques.

Definition 5.3: **Evaluation** is the process of computing quantitative information of some key characteristics (or "objectives") of a certain (possibly partial) design.

5.1.2   Multi-objective Optimization

Design evaluations will, in general, lead to a characterization of the design by several criteria, such as average and worst case execution time, energy consumption, code size, dependability and safety. Merging all these criteria into a single objective function (e.g. by using a weighted average) is usually not advisable, as this would hide some of the essential characteristics of designs. Rather, it is recommended to return to the designer a set of designs among which the designer can then select an appropriate design. Such a set should, however, only contain "reasonable" designs. Finding such sets of designs is the purpose of **multi-objective optimization techniques**.

In order to perform multi-objective optimization, we do consider an m-dimensional space X of possible solutions of the optimization problem. These dimensions could, for example, reflect the number of processors, the sizes of memories, as well as the number and types of buses. For this space X, we define an n-dimensional function

$$f(x) = (f_1(x), \ldots, f_n(x)) \text{ where } x \in X$$

which evaluates designs with respect to several criteria or objectives (e.g. cost and performance). Let F be the n-dimensional space of values of these objectives (the so-called **objective space**). Suppose that, for each of the objectives, some total order $<$ and the corresponding $\leq$-order are defined. In the following, we assume that the goal is to **minimize** our objectives.

Definition 5.4: Vector $u = (u_1, \ldots, u_n) \in F$ **dominates** vector $v = (v_1, \ldots, v_n) \in F$ iff u is "better" than v with respect to at least one objective and not worse than v with respect to all other objectives:

$$\forall i \in \{1, \ldots, n\} : u_i \leq v_i \ \wedge \tag{5.1}$$

$$\exists j \in \{1, \ldots, n\} : u_j < v_j \tag{5.2}$$

Definition 5.5: Vector $u \in F$ is called **indifferent** with respect to vector $v \in F$ iff neither u dominates v nor v dominates u.

Definition 5.6: A design $x \in X$ is called **Pareto-optimal** with respect to X iff there is no design $y \in X$ such that $u = f(x)$ is dominated by $v = f(y)$.

The previous definition defines Pareto-optimality in the solution space. The next definition serves the same purpose in the objective space.

Definition 5.7: Let $S \subseteq F$ be a subset of vectors in the objective space. $v \in F$ is called a **non-dominated solution** with respect to S iff v is not dominated by any element $\in S$. v is called Pareto-optimal iff v is non-dominated with respect to all solutions F.

Figure 5.2 highlights the different areas in an objective space with objectives $O1$ and $O2$, relative to design point (1).

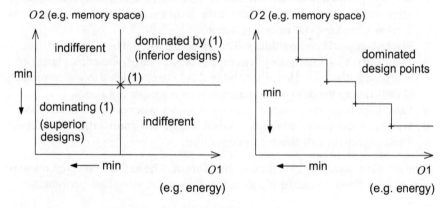

Fig. 5.2 Pareto optimality: **left**: Pareto point; **right**: Pareto front

The upper right area corresponds to designs that would be dominated by design (1), since they would be "worse" with respect to both objectives. Designs in the lower left rectangle would dominate design (1), since they would be "better" with respect to both objectives. Designs in the upper left and the lower right area are indifferent: they are "better" with respect to one objective and "worse" with respect to the other. Figure 5.2 (**right**) shows a set of Pareto points, i.e., the so-called **Pareto front**.

Definition 5.8: Design space exploration (DSE) based on Pareto points is the process of finding and returning a set of Pareto-optimal solutions to the designer, enabling the designer to select the most appropriate implementation.

5.1.3 Relevant Objectives

For servers and PCs, the average performance plays a dominating role. For embedded and cyber-physical systems, multiple objectives need to be considered. The following list explains if and where this objective is discussed in this book:

1. **Worst case performance/real-time behavior**: Some fundamental techniques for computing the worst-case execution time (WCET) will be presented in Sect. 5.2.2. This will be complemented by an introduction to real-time calculus in Sect. 5.2.3.
2. **Quality metrics**: Quality metrics will be presented in Sect. 5.3. In addition, transformations between number systems are discussed in Sect. 7.1.5.
3. **Energy/power consumption**: A brief overview of techniques for evaluating this objective will be presented in Sect. 5.4.
4. **Thermal models**: An introduction to this topic will be presented in Sect. 5.5.
5. **Dependability**: Dependability is the topic of Sect. 5.6, with subsections on safety, security and reliability.
6. **Average performance**: Information on this objective is available in Sect. 5.2. Moreover, an analysis of this objective is frequently based on simulations. Section 5.7 presents issues in simulation.
7. **Electromagnetic compatibility**: This objective will not be considered here.
8. **Testability**: Costs for testing systems can be very large, sometimes larger even than production costs. Hence, testability should be considered as well, preferably already during the design. Testability will be discussed in Chap. 8.
9. **Cost**: Cost in terms of silicon area or real money will not be considered here.
10. **Weight, robustness, usability, extendability, environmental friendliness**: These objectives will also not be considered.

There are more objectives than the ones listed above. The next Section presents some approaches for performance evaluation, focusing on the worst case performance.

5.2 Performance Evaluation

Performance evaluation aims at predicting the performance of systems. This is a major challenge (especially for cyber-physical systems) since we might need worst case information, rather than just average case information. Such information is necessary in order to guarantee real-time constraints.

5.2.1 Early Phases

Two different classes of techniques have been proposed for obtaining performance information already during early design phases:

- **Estimated cost and performance values**: Quite a number of estimators have been developed for this purpose. Examples include the work by Jha and Dutt [262] for hardware, Jain et al. [254], and Franke [162] for software. Generating sufficiently precise estimates requires considerable efforts.
- **Accurate cost and performance values**: We can also use the real software code (in the form of some binary) on a close-to-real hardware platform. This is only possible if interfaces to compilers exist. This method can be more precise than the previous one, but may be significantly (and sometimes prohibitively) more time consuming.

In order to obtain sufficiently precise information, communication needs to be considered as well. Unfortunately, it is typically difficult to compute communication cost already during early design phases.

Formal performance evaluation techniques have been proposed by many researchers. For embedded systems, the work of Thiele et al., Henia and Ernst et al., and Wilhelm et al. is particularly relevant (see, for example, [203, 513, 559]). These techniques require some knowledge of architectures. They are less appropriate for very early design phases, but some of them can still be used without knowing all the details about target architectures. These approaches model real, physical time.

5.2.2 WCET Estimation

Scheduling of tasks requires knowledge about the duration of task executions, especially if meeting time constraints has to be guaranteed, as in real-time (RT) systems. The **worst case execution time** (WCET) is the basis for most scheduling algorithms. Some definitions related to the WCET are shown in Fig. 5.3.

Fig. 5.3 WCET-related terms

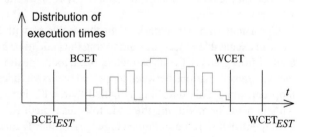

Definition 5.9: The **worst case execution time** (WCET) is the largest execution time of a program for any input and any initial execution state.

Unfortunately, the WCET is extremely difficult to compute. In general, it is undecidable whether or not the WCET is finite. This is obvious from the fact that it is undecidable whether or not a program terminates. Hence, the WCET can only be computed for certain programs/tasks. For example, for programs without recursion, without while loops and with loops having statically known iteration counts, decidability is not an issue. But even with such restrictions, it is usually practically impossible to compute the WCET exactly. The effect of modern processor architectures' pipelines with their different kinds of hazards and memory hierarchies with limited predictability of hit rates is difficult to precisely predict at design time. Computing the WCET for systems containing interrupts, virtual memory and multiple processors is an even greater challenge. As a result, we must be happy if we are able to compute good **upper bounds** on the WCET.

Such upper bounds are usually called **estimated worst case execution time**s, or $WCET_{EST}$. Such bounds should have at least two properties:

1. The bounds should be safe ($WCET_{EST} \geq WCET$).
2. The bounds should be tight ($WCET_{EST}\text{-}WCET \ll WCET$)

Note that the term "estimated" does not mean that the resulting times are unsafe.

Sometimes, architectural features which reduce the average execution time but cannot guarantee to reduce $WCET_{EST}$ are completely omitted from the real-time designs (see p. 150). Computing tight upper bounds on the execution time may still be difficult. The architectural features described above also present problems for the computation of $WCET_{EST}$. The computation of such bounds is extremely difficult for multi-cores. In fact, potential conflicts might even cause multi-cores to have larger worst-case bounds than the corresponding single cores.

Accordingly, the **best-case execution time** (BCET) and the corresponding estimate $BCET_{EST}$ are defined in an analogous manner. The $BCET_{EST}$ is a safe and tight lower bound on the execution time.

Computing tight bounds from a program written in a high-level language such as C without any knowledge of the generated assembly code and the underlying architectural platform is impossible. Therefore, a safe analysis must start from real machine code. Any other approach would lead to unsafe results.

In the following, we will study WCET estimation more closely. The presentation is based on the description of the tool aiT by R. Wilhelm [559]. The architecture of aiT is shown in Fig. 5.4.

Consistent with our remark about the problems with high-level code, aiT starts from an executable object file comprising the code to be analyzed. From this code, a control-flow graph (CFG) is extracted. Next, loop transformations are applied. These include transformations between loops and recursive function calls as well as virtual loop unrolling. This unrolling is called "virtual" since it is performed internally, without actually modifying the code to be executed. Results are represented in the CRL (control flow representation language) format. The next phase employs different static analyses. Static analyses read the AIP-file comprising designer's annotations. These annotations contain information which is difficult or impossible to extract

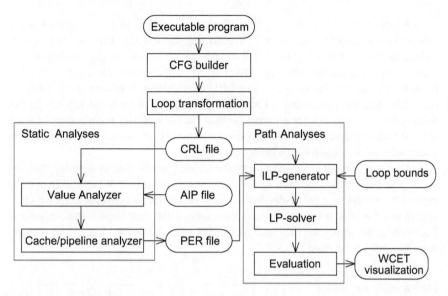

Fig. 5.4 Architecture of the aiT timing analysis tool

automatically from the program (for example, bounds of complex loops). Static analyses include value analysis, cache analysis, and pipeline analyses.

A **value analysis** computes enclosing intervals for possible values in registers and local variables. The resulting information can be used for control-flow analysis and for data-cache analysis. Frequently, values such as addresses are precisely known (especially for "clean" code) and this helps in predicting accesses to memories.

The next step is **cache** and **pipeline analysis**. We will present a few details about the cache analysis. Suppose using an n-way set associative cache (see Fig. 5.5)[1].

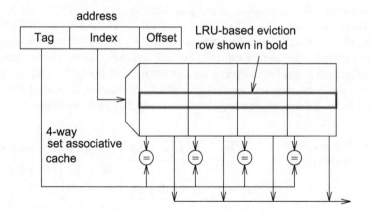

Fig. 5.5 Set associative cache (for $n = 4$)

---

[1] We assume that students are familiar with concepts of caches.

We consider that part of the cache (the **row**) corresponding to a certain index (shown in bold and blue in Fig. 5.5). We assume that eviction from the row is controlled by the least recently used (LRU) strategy[2]. This means that among all references for a particular index, the last n referenced memory blocks are stored in the row. We assume that the necessary LRU management hardware is available for each index and that each index is handled independently of other indexes. Under this assumption, all evictions for a particular index are completely independent of decisions for other indexes. This independence is extremely important, since it allows us to consider each of the indexes independently.

Let us now consider a row and a particular index. Suppose that we have information about potential entries for each of the cache ways (columns). What will happen in case of an access to a particular index? First of all, let us consider the case of an access to a variable e known to be in the cache. After that access, that variable is known to be the youngest (see Fig. 5.6). Entries on the left are assumed to be younger than the ones on the right.

Fig. 5.6 Access to variable e makes it the youngest
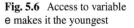 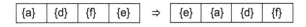

Now, assume that we have an access to some variable (say c) which is not yet in the cache. This access will remove the oldest entry from the cache (see Fig. 5.7).

Fig. 5.7 Access to variable c causes eviction of f

Furthermore, consider control flow joins. What do we know about the content of the partial cache after the join? We must distinguish between *may-* and *must-*information and the corresponding analysis. Must-analysis reveals the entries which **must** be in the cache. This information is useful for computing the WCET. May-analysis identifies the entries which **may** be in the cache. This information is typically used to conclude that certain information will definitely not be in the cache. This knowledge is then exploited during the computation of the BCET. As an example of must- and may-analysis, we consider must information at control flow joins. Figure 5.8 shows the corresponding situation. In Fig. 5.8, memory object c is assumed

Fig. 5.8 Must-analysis at program joins for LRU-caches

{c}	{e}	{a}	{d}

Intersection+maximum age ⇒

{}	{}	{a,c}	{d}

{a}	{}	{c,f}	{d}

to be the youngest object for one path to the join and a is assumed to be the youngest object for the other path. The age of the other entries is defined accordingly. What

---

[2]Unfortunately, this strategy is typically not available for processors.

do we know about the "worst" case after the join? A certain entry is guaranteed to be in the cache only if it is guaranteed to be in the cache for both paths. This means that the **intersection** of the memory objects defines the result of the must-analysis after the join. As a worst case, we must assume the **maximum of the ages** along the two paths. Figure 5.8 shows the result. This analysis uses sets of entries for each cache way.

Let us now consider may-analysis for control flow joins. Figure 5.9 depicts the situation. Some object being in the cache on either of the two paths to the join *may*

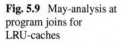

Fig. 5.9 May-analysis at program joins for LRU-caches

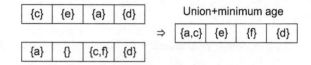

be in the cache after the join. Hence, the set of objects which may be in the cache after the join consists of the **union** of the objects that were in the cache before the join. As a best case, we use the **minimum of the ages** before the join. Figure 5.9 shows the result.

Static analyses also comprise pipeline analysis. Pipeline analysis has to compute safe bounds on the number of cycles required to execute code in the machine pipeline. Details of pipeline analysis are explained by Hahn et al. [190] and Thesing [511]. The result of static analyses consists of bounds on the execution times for each of the basic blocks of a program. Results are written to the PER-file shown in Fig. 5.4.

aiT's next phase exploits these bounds to derive $WCET_{EST}$ values for the entire program, using an **integer linear programming** (ILP) model (see p. 379), comprising two types of information:

- **The objective function**: In our application of ILP modeling, this function represents the overall execution time. This time is calculated as $\sum_{\text{basic blocks}} e_i * f_i$ where e_i is the worst case execution time of basic block i (as computed during static analysis) and f_i its worst case execution count. Only some of these counts can be determined automatically and additional designer-provided information, e.g. about loop bounds, may be required.
- **Linear constraints**: in our application of ILP modeling, these reflect the structure of the control flow graph.

Example 5.1: Let us consider the simple code shown below:

```
int main()  {  int i,j=0;
   _Pragma("loopbound min 100 max 100")     /* hint for bound analysis */
   for (i=0; i <100; i++) {
      if (i<50) j+=i;
      else j+=(i*13) % 42    }
   return j; }
```

Figure 5.10 (**left**) shows the control flow graph (CFG) corresponding to this small program. This graph is extended by additional start and exit nodes. Node _L1 reflects

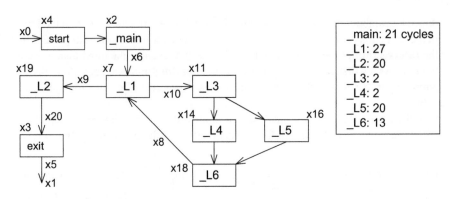

Fig. 5.10 Sample program: **left**: extended control flow graph; **right**: WCET$_{EST}$ of basic blocks

the for-testing, _L3 the if-testing, _L4 and _L5 the two cases of the if-statement and _L6 its join operation. Variables x0 to x20 denote the number of executions of the blocks and the number of transitions between blocks. For example, we are transitioning from node main into node _L1 x6 times and are executing the target node x7 times. We assume that the analysis of the WCET for each of the basic blocks has resulted in the list shown on the right of Fig. 5.10. The following is a partial list of the ILP-constraints:

```
01: 21 x2 + 27 x7 + 2 x11 + 2 x14 + 20 x16 + 13 x18 + 20 x19;/*objective*/
02: x7 - x8 - x6 = 0;        /* Constraint for flow entering CFG node _L1 */
03: x7 - x9 - x10 = 0;       /* Constraint for flow leaving CFG node _L1 */
04: x7 - 101 x9 >= 0;        /* Constraint for lower loop bound of _L1 */
05: x7 - 101 x9 <= 0; ...    /* Constraint for upper loop bound of _L1 */
06: x0 - x4 = 0;                         /* CFG Start Constraint */
07: x2 - x4 = 0;        /* Constraint for flow entering function _main */
08: x2 - x6 = 0;        /* Constraint for flow leaving CFG node _main */
09: ...
```

Line 01 contains the cost function. All other lines model constraints reflecting the structure of the graph. Consider, for example, node _L1. Constraints for this node are shown in lines 02 and 03. The number of times that we are branching into the node (x6+x8) is equal to its number of executions (x7). The number of times that we are leaving from the node (x9+x10) is also equal to its number of executions. Lines 04 and 05 reflect the number of loop iterations. This number is taken from the pragma in the code. Line 06 describes the fact that node start is executed exactly as many times as we are branching into the code. The other lines are reflecting the structure in a similar way.   ∇

The ILP problem can be solved with some standard ILP solver. Maximizing the objective function yields a safe upper bound on the WCET.

This technique for modeling execution time is called **implicit path enumeration**, since the problem of enumerating the potentially large number of execution paths is avoided.

aiT visualizes the results as annotated control flow graphs. The designer could optimize the system under design by exploiting these graphs.

Only few approaches exist for the WCET-analysis of multi-cores [252, 253, 273]. New probabilistic approaches [2] aim at complementing available methods. They are usually based on Extreme Value Theory [189].

5.2.3 Real-Time Calculus

WCET-estimates allow us to predict the execution of some algorithm for a single input event. However, the overall goal is more comprehensive. Overall, we should make sure that our hardware platform is capable of processing streams of events in a timely manner (which may be important for some parts of the Internet of Things).

This can be checked with Thiele's **real-time calculus (RTC)**. This calculus (**RTC**) is based on the description of the rate of incoming events[3]. This description also includes fluctuations of this rate. Toward this end, the timing characteristics of a sequence (or stream) of events are represented by a tuple of *arrival curves*:

$$\overline{\alpha}^{\,u}(\Delta), \overline{\alpha}^{\,l}(\Delta) \in \mathbb{R} \geq 0, \Delta \in \mathbb{R} \geq 0$$

These curves represent the maximal resp. the minimal number of events arriving within a time interval of length Δ. There are at most $\overline{\alpha}^{\,u}(\Delta)$ and at least $\overline{\alpha}^{\,l}(\Delta)$ events arriving within the time interval $(t, t + \Delta)$ for all $t \geq 0$. Figure 5.11 shows the number of possibly arriving events for some possible models of arriving events. For

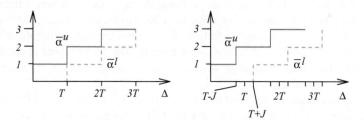

Fig. 5.11 Arrival curves: **left**: periodic stream; **right**: periodic stream with jitter J

example, in the case of periodic event streams with period T, there is a maximum of a single event happening in time interval $(0, T)$[4]. Similarly, there is an upper bound of two events within time interval $(T, 2T)$. Now, let us consider the lower bound for time interval $(0, T)$. There is possibly not a single event in this interval. Hence, the bound is zero. For time interval $(T, 2T)$, there has to be at least one event. Therefore, the bound is one. So, for $\Delta = 0.5T$, there will be at least zero and at most one incoming

---

[3]Our presentation of the real-time calculus is based on Thiele's presentation in the book edited by Zurawski [513]. Resulting considerations at the system level have been called *modular performance analysis* (MPA).

[4]We leave out the subtle discussion of dis-continuities at $\Delta = n * T$.

event (see Fig. 5.11 (**left**)). In the case of periodic event streams with jitter J, these curves are shifted by this amount (see Fig. 5.11 (**right**)). The upper bound is shifted to the left, the lower bound is shifted to the right. The jitter is assumed not to be accumulating. We are using bars on top of symbols (like $\bar{\alpha}$) for all entities referring to incoming events.

Available computational and communication service capacity can be described by *service functions*:

$$\beta^u(\Delta), \beta^l(\Delta) \in \mathbb{R} \geq 0, \Delta \in \mathbb{R} \geq 0$$

These functions allow us to model situations in which the available service capacity is fluctuating. Figure 5.12 shows the communication capacity of some *time division*

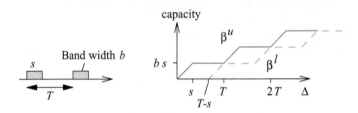

Fig. 5.12 Service functions for a TDMA bus

multiple access (TDMA) bus (see p. 173). Allocation is done periodically with a period of T. Bus arbitration allocates this bus during a time window s time units long. During this window, the bus achieves a bandwidth of b. The upper bound is obtained if the bus is allocated exactly at the time we are starting our observation. The transferred amount is then increasing linearly. The lower bound is obtained if the bus was just deallocated when we started our observation of length Δ. Then we must wait $T - s$ time units until the bus gets allocated again.

Separate methods are required to determine $\bar{\alpha}$ and β for streams of ("external") events arriving at the system to be modeled. Their computation is not part of RTC. In contrast, bounds for events generated within the system are derived by the calculus (see below).

Up till now, there is no information about the **workload** required by each of the incoming events. This workload is represented by additional functions $\gamma^u(e), \gamma^l(e) \in \mathbb{R} \geq 0$ for each sequence e of incoming events. This information can be derived from bounds on the execution time of code required for each of the events. Figure 5.13 shows an example of such functions. This example is based on the assumption that between three and four time units are required for processing a single event. Accordingly, the workload for a single event varies between three and four time units, the work load for two events varies between six and eight time units, etc. The dashed lines are not part of the function, since it is defined only for an integer number of events. The work load resulting from an incoming stream of

events can now be easily computed. Upper and lower bounds are characterized by
the functions

$$\alpha^u(\Delta) = \gamma^u(\overline{\alpha}^u(\Delta)) \text{ and} \tag{5.3}$$

$$\alpha^l(\Delta) = \gamma^l(\overline{\alpha}^l(\Delta)) \tag{5.4}$$

Fig. 5.13 Work load characterization (WCET$_{EST}$ may be used instead of WCET)

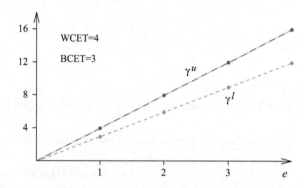

There should be enough computational or communication capacity to handle this
work load. The number of events which can be processed with the available computational capacity can be computed as

$$\overline{\beta}^u(\Delta) = (\gamma^l)^{-1}(\beta^u(\Delta)) \text{ and} \tag{5.5}$$

$$\overline{\beta}^l(\Delta) = (\gamma^u)^{-1}(\beta^l(\Delta)) \tag{5.6}$$

Equations (5.5) and (5.6) use the inverse of functions γ^u and γ^l to convert bounds
on the available capacity (measured in real time units) into bounds measured in terms
of the number of events that can be processed.

Based on this information, it is possible to derive the properties of outgoing
streams of events from incoming streams of events. Suppose the incoming stream
is characterized by bounds $[\overline{\alpha}^l, \overline{\alpha}^u]$. We can then compute characteristics of the
outgoing streams such as the corresponding bounds $[\overline{\alpha}^{l\prime}, \overline{\alpha}^{u\prime}]$ of the outgoing stream
of events and the remaining service capacity, available for other tasks. This remaining
capacity is derived by transforming *service curves* $[\overline{\beta}^l, \overline{\beta}^u]$ into *service curves*
$[\overline{\beta}^{l\prime}, \overline{\beta}^{u\prime}]$ (see Fig. 5.14). This remaining service capacity can be employed for
lower priority tasks to be executed on the same processor.

According to Thiele et al., outgoing streams and remaining service capacities are
bounded by the following functions [513]:

$$\overline{\alpha}^{u\prime} = [(\overline{\alpha}^u \underline{\otimes}\overline{\beta}^u)\overline{\oslash}\overline{\beta}^l] \wedge \overline{\beta}^u \tag{5.7}$$

$$\overline{\alpha}^{l\prime} = [(\overline{\alpha}^l \overline{\oslash}\overline{\beta}^u)\underline{\otimes}\overline{\beta}^l] \wedge \overline{\beta}^l \tag{5.8}$$

$$\overline{\beta}^{u\prime} = (\overline{\beta}^u - \overline{\alpha}^l)\underline{\oslash}0 \tag{5.9}$$

$$\overline{\beta}^{l\prime} = (\overline{\beta}^l - \overline{\alpha}^u)\overline{\otimes}0 \tag{5.10}$$

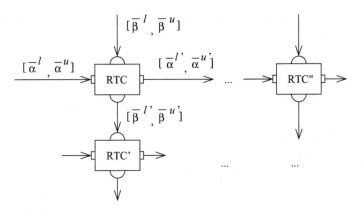

Fig. 5.14 Transformation of event stream and service capacities by real-time components

Operators used in these equations are defined as follows:

$$(f \otimes g)(t) = inf_{0 \leq u \leq t}\{f(t - u) + g(u)\} \tag{5.11}$$

$$(f \overline{\otimes} g)(t) = sup_{0 \leq u \leq t}\{f(t - u) + g(u)\} \tag{5.12}$$

$$(f \overline{\oslash} g)(t) = sup_{u \geq 0}\{f(t + u) - g(u)\} \tag{5.13}$$

$$(f \underline{\oslash} g)(t) = inf_{u \geq 0}\{f(t + u) - g(u)\} \tag{5.14}$$

$\wedge$ denotes the minimum operator.

In essence, these equations characterize outgoing streams and capacities. These equations have been adopted from communications theory. Proofs regarding these equations are provided by Network Calculus [313]. The easiest way of using these equations is to download a MATLAB® toolbox [536].

The same theory also allows to compute the delay caused by the real-time components as well as the size of the buffer required to temporarily store incoming/outgoing events. This way, performance and other characteristics of the system can be computed from information about the components.

A second performance analysis method has been proposed by Henia, Ernst et al. In this so-called SymTA/S approach [203], the different curves in Thiele's approach are replaced by standard models of event streams such as periodic event streams, periodic event streams with random jitter and periodic event streams with bursts. SymTA/S explicitly supports the combination and integration of different kinds of analysis techniques known from real-time research.

5.3 Quality Metrics

5.3.1 Approximate Computing

Sometimes, computing the best possible output of some algorithm requires a significant amount of resources (in terms of computing time, energy, thermal headroom etc.). For some applications, the best possible output is not actually needed, since minor degradations will possibly not even be recognized by users. This can be exploited in a resource-constrained environment in order to trade-off the quality of the output against needed resources. A certain deviation of the actual output from the best possible output is accepted, for example, for lossy audio, video and image encoding. This leads us to consider **approximate computing**.

Definition 5.10: Computing which tolerates a certain deviation of generated output of some algorithm from the best possible result is called **approximate computing** [380].

With approximate computing, it is necessary to consider the quality of the generated output as one of the objectives. Unfortunately, it is not easy to evaluate the quality of some generated result and several metrics can be used.

5.3.2 Simple Criteria of Quality

Some simple metrics can be applied whenever the true (or the best possible) output is known. Suppose that $x_1, \ldots, x_n$ are n samples of some signal x in discrete time. Furthermore, suppose that instead of the real (or the best possible) values $x_1, \ldots, x_n$ we measure or compute approximate values $y_1, \ldots, y_n$.

Then, our first metric, the Mean-Squared Error (MSE), is defined as follows:

Definition 5.11: The **Mean-Squared Error** (MSE) is defined as

$$MSE(x, y) = \frac{1}{n} \sum_{i=1}^{n} (x_i - y_i)^2 \tag{5.15}$$

The second metric is the root-mean-squared error.

Definition 5.12: The **Root-Mean-Squared Error** (RMSE) is defined as

$$RMSE(x, y) = \sqrt{\frac{1}{n} \sum_{i=1}^{n} (x_i - y_i)^2} \tag{5.16}$$

RMSE has the same dimension as the difference between the actual and the real value, but it should not be confused with the "average error" which is defined next:

Definition 5.13: The **Mean-Absolute Error** (MAE) is defined as

$$MAE(x, y) = \frac{1}{n} \sum_{i=1}^{n} |x_i - y_i| \tag{5.17}$$

For identical deviations of the measured signal y from real values x, the MAE is equal to the RMSE. However, the RMSE emphasizes large deviations between real and measured values (so-called outliers).

The signal-to-noise ratio (SNR) was already defined on p. 139.

Next, we define the Peak-Signal-to-Noise Ratio, which is similar to the SNR. Let x_{max} be the maximum value for signal x.

Definition 5.14: The **Peak-Signal-to-Noise ratio** (PSNR) is defined as

$$PSNR(x, y) = 10 \log_{10} \left(\frac{x_{max}^2}{MSE(x, y)} \right) \tag{5.18}$$

$$= 20 \log_{10} \left(\frac{x_{max}}{RMSE(x, y)} \right) \tag{5.19}$$

The PSNR, just like the SNR, is measured in decibels (dB).

The above values are easy to compute, but they are agnostic of the impression which humans might have of certain errors [278]. It is known that certain deviations between real and computed signal values are hardly noticed by humans. This is the foundation of lossy coding techniques such as MP3, JPEG or digital TV standards. None of the metrics presented so far reflects the impression of deviations by humans. Next, we will present the **Universal Image Quality Index** (UIQI) [539]. This index tries to capture changes in the structure of images, since the human eye is very sensitive to it. We will present the computation of this index for gray-scale images. Several values need to be computed [278]:

$$\mu_x = \frac{1}{n} \sum_{i=1}^{n} x_i \tag{5.20}$$

$$\mu_y = \frac{1}{n} \sum_{i=1}^{n} y_i \tag{5.21}$$

$$\ell(x, y) = \frac{2\mu_x \mu_y}{\mu_x^2 + \mu_y^2} \tag{5.22}$$

Equations (5.20) and (5.21) compute the average brightness of each of the images and these averages are used to compute $\ell(x, y)$. For images of the same average brightness, $\ell(x, y)$ will be equal to 1. Otherwise, this value will be less than 1.

Furthermore, we consider variances. Equations (5.23) and (5.24) compute the contrast of each of the images and these averages are used to compute $c(x, y)$. For images of the same average contrast, $c(x, y)$ will be equal to 1. Otherwise, this value will be less than 1:

$$\sigma_x = \sqrt{\frac{1}{(n-1)} \sum_{i=1}^{n} (x_i - \mu_x)^2} \tag{5.23}$$

$$\sigma_y = \sqrt{\frac{1}{(n-1)} \sum_{i=1}^{n} (x_i - \mu_y)^2} \tag{5.24}$$

$$c(x, y) = \frac{2\sigma_x \sigma_y}{\sigma_x^2 + \sigma_y^2} \tag{5.25}$$

Equation (5.26) computes the cross-correlation of the two images:

$$\sigma_{x,y} = \frac{1}{n-1} \sum_{i=1}^{n} (x_i - \mu_x)(y_i - \mu_y) \tag{5.26}$$

$$s(x, y) = \frac{\sigma_{x,y}}{\sigma_x \sigma_y} \tag{5.27}$$

Positive values of $s(x, y)$ as computed from Eq. (5.27) correspond to a good correlation of the two images, negative values correspond to an inverse correlation.

An overall quality index is then computed by Eq. (5.28). $Q = 1$ for identical images, and Q will be negative for inversely correlated images:

$$Q(x, y) = \frac{2\mu_x \mu_y}{\mu_x^2 + \mu_y^2} * \frac{2\sigma_x \sigma_y}{\sigma_x^2 + \sigma_y^2} * \frac{\sigma_{x,y}}{\sigma_x \sigma_y} \tag{5.28}$$

It does not make sense to consider the correlation of images globally, since some inverse correlation in a particular block will already provide a negative impression about the image. Hence, Eq. (5.28) is computed only for blocks of pixels. The global UIQI-value takes the values of Q for the different blocks into account.

The Structural Similarity Index (SSIM) [540] is an extension of the UIQI objective.

Kühn compared the different metrics and found that none of these is really superior to others [278]. He recommends that several of these metrics should be computed and a careful comparison should be performed in practice. An overview over some useful objectives is also provided by Mittal [380].

In digital communications, the bit error ratio (BER) is an important metric.

Definition 5.15: The **bit error ratio (BER)** is ratio of the number of bit errors divided by total number of communicated bits.

5.3.3 Criteria for Data Analysis

In data analysis, the output of algorithms is statistical anyway. In a way, we are dealing with approximate computing even though this term was not used in this context.

For data analysis, classification of objects is a very frequent goal. Let X be a set of objects which we would like to classify. Suppose that we restrict ourselves to binary classification. For example, consider the case of searching for amber at a beach. Unfortunately, white phosphorus as a leftover from bombs found e.g. at the Baltic ocean looks very much like amber, but starts to suddenly burn at $1300\,C$ when it dries. Classifying some found objects as either amber or phosphorus is thus a very delicate task (and hence, inexperienced people should not touch such objects anyway).

In this context, four cases are possible

- **True positives** (TP): we classify some object as amber and it is actually valuable amber.
- **False positive** (FP): we classify some object as amber and it is actually dangerous.
- **True negative** (TN): we classify some object as dangerous and it is actually dangerous.
- **False negative** (FN): we classify some object as dangerous and it is actually valuable amber.

Absolute numbers have to be related to each other. Hence, the following metrics have been defined:

Definition 5.16: The **precision** p is defined as the fraction

$$p = \frac{TP}{TP + FP} \qquad (5.29)$$

In the case of searching for amber, we aim at a precision of 1, since we do not want to get burnt.

Definition 5.17: The **recall** r (or **sensitivity**) is defined as the fraction

$$r = \frac{TP}{TP + FN} \qquad (5.30)$$

In order to obtain a good precision, we will have to accept some false negatives (e.g. amber classified as phosphorus).

Definition 5.18: The **accuracy** acc is defined as the fraction

$$acc = \frac{TP + TN}{TP + FP + TN + FN} \qquad (5.31)$$

In the case of searching for amber, we might tolerate a non-optimal accuracy, due to the importance of keeping false positives as close to zero as possible and, hence, we might have several false negatives.

Definition 5.19: The **specificity** is defined as the fraction

$$\text{specificity} = \frac{TN}{TN + FP} \qquad (5.32)$$

Definition 5.20: The **F1 score** or **F-measure** is defined as the harmonic mean of precision and recall:

$$F1 = 2\,\frac{p * r}{p + r} \tag{5.33}$$

In a more general context, the **Quality-of-Service** (QoS) is another well-known metric. Frequently, it is related to the quality of communication channels, where bit-error rates, latency and bandwidth are indicators of quality.

In an even wider sense, we may also consider not just those technical parameters, but also the overall experience for the user. This is captured in the **Quality of Experience** (QoE) metric, which refers to the overall user experience including all aspects which might be considered by a user. There is a number of metrics which can be used to estimate the overall quality of experience [383].

5.4 Energy and Power Models

5.4.1 General Properties

Energy models and **power models** are essential for evaluating the corresponding objectives. Such models are needed for optimizations aiming at a reduction of power and energy consumptions. They are also required for optimizations trying to reduce operating temperatures and to improve reliability. Power estimation is used in **power management** algorithms (see p. 361).

Energy and power models are closely related, as can be seen from Eq. (3.13). In general, we can use:

1. **Measurements on real hardware**: measurements can be very precise, but they apply only to the hardware at hands. Measuring voltages is typically rather easy and does not require complex procedures.

   Measuring currents can be done with a current clamp or a shunt resistor.

   Current clamps have to enclose one of the wires of the power supply cable. They measure the magnetic field caused by the current flowing through the cable. The advantage of this approach is that no power wires have to be broken and power will remain connected unchanged to the device being analyzed. The disadvantage is that current clamps do not allow precise measurements.

   A typical circuit containing a **shunt** is shown in Fig. 5.15 (**left**). The advantage of using a shunt resistor over using a simple ammeter is that the shunt can be integrated into the power wires, whereas the ammeter's cables would possibly make the setup susceptible to noise.

   Due to the shunt resistor, currents flowing into the device under test will cause a voltage drop across the shunt and this voltage can be measured and used to compute the current from Ohm's law. Finding the right resistance of the shunt

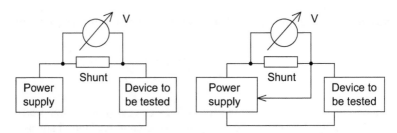

Fig. 5.15 Measuring current: **left**: 2-wire connection; **right**: feedback into voltage regulator

is an issue. If the resistance is too large, the device under test will be powered with a voltage lower than the original voltage and might even fail to work. If the resistance is too small, the voltage across the shunt will be too small to be reliably measured and will be subject to a substantial amount of noise. Selecting the right resistance depends on the current flowing into the device under test. If this current varies substantially, it may even be necessary to employ several shunt resistors and switch between them, depending on the current actually flowing. The problem regarding the voltage drop can be partially avoided when regulated power supplies are used and the regulator feedback input can be connected to the voltage actually powering the device (see Fig. 5.15 (**right**)). The power supply would then try to keep the voltage at the device at its nominal level. However, the voltage across the shunt is affected by the current flowing back into the voltage regulator input.

Unfortunately, there will not be a separate power pin or wire for every component within the device and we can compute only a lumped sum of currents drawn by the device. We may have to stimulate the device in a particular way in order to get information about the consumption of the different components.

2. **Models**: models can be used even when real hardware is not available, but they can be very imprecise. Models have to be validated, otherwise they would remain very questionable. Two validation methods can be found for many of the available power and energy models: either models are validated against more detailed models at a lower level of abstraction or they are compared with measurement for real devices, resulting in a hybrid model. Validation against measurements requires a method for selecting model parameters. Frequently, linear models are selected and parameters are selected with using the least square method (minimizing the MSE as per Eq. (5.15)). Curve fitting with this method is typically available in mathematical tool boxes such as MATLAB®.

There is no one-approach-fits-all solution for energy consumption modeling. Instead, the usual approach is to combine ideas for modeling to fit the needs at hand. Therefore, we will present representative examples of power models, and hope that the reader will identify the combination of methods which fits his/her constraints best.

5.4.2 Analysis of Memories

As described in the section on memory hardware (see p. 164), the energy consumption of caches and other memories can be computed with CACTI [391, 561]. CACTI assumes an abstract layout of the memory, extracts capacitances from the layout, and computes access times, cycle times, area, leakage and dynamic power consumption from this information. CACTI has been validated against models of the same memories at a more detailed level, employing SPICE [496] as the solver at that level. Currently (in 2017), the most recent version of CACTI (version 6.5) is available from http://www.hpl.hp.com/research/cacti/. Recent enhancements include detailed modeling of the interconnect and modeling of non-uniform memory accesses. Models of transmitters and sense amplifiers have been included. Also, used architectural and technological parameters can be specified. A web interface and a modifiable C++-version are available. Only the C++-version is currently up to date.

5.4.3 Analysis of Instructions and Inter-Instruction Effects

One of the first power models was proposed by Tiwari [519]. The model includes so-called base costs and inter-instruction costs. Base costs of an instruction correspond to the energy consumed per instruction execution if an infinite sequence of instances of that instruction is executed. Base costs have been computed by running programs consisting of 120 identical instructions and a branch back to the beginning of this sequence. Programs are designed such that no stall cycles appear. This may require the adding of no-operation instructions and some simple calculations to eliminate their contribution to the energy consumption.

Inter-instruction costs model the additional energy consumed by the processor if instructions change. This additional energy is required, for example, due to switching functional units on and off. Inter-instruction costs reflect the impact of the initial circuit state on the overall energy consumption of an instruction. These costs can be computed by running programs containing an alternating sequence of instructions pairs.

Base costs and inter-instruction costs are computed for a program not generating any cache misses. The effect of cache misses has to be added to these two costs. This requires the knowledge of the cache miss ratio and the memory access energy. The memory energy depends on the addresses accessed. No attempt is made to statically predict memory addresses. Hence, this contribution can only be determined dynamically, during the execution of the program.

The model has been applied to two real systems, an Intel 486 DX2 and a Fujitsu SPARClite 934. Measurements of the currents have been used to calibrate the model.

5.4.4   Analysis of Major Functional Processor Units

The Wattch power estimation tool [72] estimates the power consumption of micro-processor systems at the architectural level. Wattch uses the SimpleScalar simulator to simulate processors. SimpleScalar can be configured to model the processor at hand as closely as possible. The number of pipeline stages and functional units is typically correctly modeled, whereas some more specialized features are possibly not. Wattch is based on detailed information on the energy consumption of the different components which we could find in a microprocessor. While running, SimpleScalar keeps track of invoked functional units. Wattch exploits this information in order to compute an overall energy consumption.

Wattch requires much more information about the architecture than Tiwari's instruction set level approach. For example, Wattch includes its own detailed model of the energy consumption in memories. Also, clocking is taken explicitly into account, including conditional clocking if clock gating is used.

In the original paper [72], results have been validated for three different processors.

5.4.5   Analysis of Processor and Memory Energy Consumption

The level of details of the model by Steinke et al. [487] lies between that of Tiwari and that of Wattch. First of all, the model considers the sum of the energies consumed in the CPU and the memory for instructions and for data:

$$E_{total} = E_{cpu\_instr} + E_{cpu\_data} + E_{mem\_instr} + E_{mem\_data} \tag{5.34}$$

Each of the four terms is then computed from detailed equations. The following notation is used in these equations: m is the number of instructions considered, $w(b)$ returns the number of ones in its argument (either code or data), $h(b_1, b_2)$ returns the Hamming distance between its two arguments, dir denotes the direction of data transfer, and α_i and β_i ($i \in \{1..10\}$) are constants computed from curve fitting of measured energies. Using this notation, E_{cpu_data} can be computed as follows:

$$E_{cpu\_data} = \sum_{i=1}^{m} \{\alpha_5 * w(DAddr_i) + \beta_5 * h(DAddr_{i-1}, DAddr_i)$$
$$+ \alpha_{6,dir} * w(Data_i) + \beta_{6,dir} * h(Data_{i-1}, Data_i)\} \tag{5.35}$$

where $Data_i$ is the data value used in instruction i, and $DAddr_i$ is its address.

Furthermore, consider E_{mem_data}, a term which is relevant only when the data is actually loaded from the main memory:

$$E_{mem\_data} = \sum_{i=1}^{m} \{BaseMem(DataMem, dir, Word\_width)$$
$$+ \alpha_9 * w(DAddr_i) + \beta_9 * h(DAddr_{i-1}, DAddr_i) \tag{5.36}$$
$$+ \alpha_{10,dir} * w(Data_i) + \beta_{10,dir} * h(Data_{i-1}, Data_i)\}$$

where *BaseMem* is the base cost for accessing a memory object of a particular width in direction *dir*.

E_{mem_instr} can be computed as follows:

$$E_{mem\_instr} = \sum_{i=1}^{m} \{BaseMem(InstrMem, Word\_width_i)$$
$$+ \alpha_7 * w(IAddr_i) + \beta_7 * h(IAddr_{i-1}, IAddr_i) \tag{5.37}$$
$$+ \alpha_8 * w(IData_i) + \beta_8 * h(IData_{i-1}, IData_i)\}$$

where *BaseMem* is the base cost for accessing a memory word of a particular width from the instruction memory, $IAddr_i$ is the address of the instruction, and $IData_i$ is instruction *i* itself.

E_{cpu_instr} can be computed from the following equation:

$$E_{cpu\_instr} = \sum_{i=1}^{m} \{BaseCPU(Opcode_i) + FUChange(Instr_{i-1}, Instr_i)$$
$$+ \alpha_4 * w(IAddr_i) + \beta_4 * h(IAddr_{i-1}, IAddr_i)$$
$$+ \sum_{j=1}^{s} (\alpha_1 * w(Imm_{i,j}) + \beta_1 * h(Imm_{i-1,j}, Imm_{i,j})) \tag{5.38}$$
$$+ \sum_{k=1}^{t} (\alpha_2 * w(Reg_{i,k}) + \beta_2 * h(Reg_{i-1,k}, Reg_{i,k}))$$
$$+ \sum_{k=1}^{t} (\alpha_3 * w(RegVal_{i,k}) + \beta_3 * h(RegVal_{i-1,k}, RegVal_{i,k}))\}$$

where *BaseCPU* is the base cost for $Opcode_i$, $FUChange(..)$ reflects the costs caused by the transition from instruction $i - 1$ to *i*, *Imm* reflects the impact of up to *s* immediate values per instruction, *Reg* reflects the register numbers of up to *t* registers per instruction, and *RegVal* reflects up to *t* register values per instruction.

To determine constants, dedicated code sequences have to be designed in order to attribute energy consumption to particular terms of the equations.

Example 5.2: The following code sequence allows measuring the energy required for executing a load word instruction:

```
start: lw R1, address
       ...                          /* about 50-100 identical instructions */
       bra start                    /* back to the start */
```

The impact of the branch back to the beginning on the energy consumption can be neglected. The impact of different addresses, register numbers and register content can be studied by varying these values. For example, we can initially set all these values to zero and then incrementally study the impact of additional ones.   ∇

In our own experiments, constants were determined by running a linear regression method on the data. A significant impact of the number of ones in the data was found, which would have been unnoticed for Tiwari's model.

5.4.6 Analysis of an Entire Application

The Odroid XU3 [196] platform (see Fig. 5.16) comprises several current sensors. The sensors enable precise measurement of the consumed power during the execution of applications, measuring the consumption of ARM® big cores, little cores, GPU and DRAM individually. This possibility is exploited by several researchers. For example, Neugebauer et al. [397] have integrated Odroid XU3 processors into their design space exploration for one application. Hence, design space exploration is based on a realistic analysis of the consumed energy. This approach eliminates the use of models of unknown precision. The overall approach for design space exploration enabled by the XU3 is shown in Fig. 5.17.

Fig. 5.16 Odroid XU3

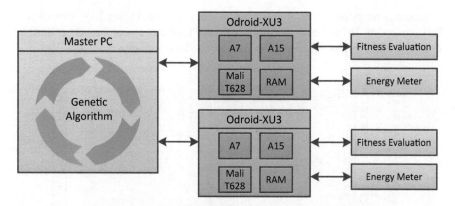

Fig. 5.17 Evolutionary algorithm, fitness estimation based on real measurements

The design space exploration is based on an evolutionary algorithm. The evaluation of a particular solution is based on real execution of the code on an XU3. Unfortunately, the Odroid XU3 has been discontinued and replaced by the XU4 not including current sensors.

5.4.7 Analysis of Multiple Applications with Multithreading

Kerrison and Eder analyzed the energy consumption of the XMOS XS1-L multi-threaded processor design for real-time applications [277]. One of the particular features of that processor is its hardware-supported multithreading: it performs fast context switching between four threads in hardware. One of the research questions was: how much does the hardware-context switching between threads cost? Due to the availability of real hardware, this question could be answered with real measurements. The power consumed by the XMOS XS1-L was measured with a shunt resistor inserted into its power cable and the resistor was connected to an INA219 power measurement chip (see http://www.ti.com/product/ina219). The software running on the processor was controlled from a second processor. It turned out that the best energy efficiency was reached when all four hardware threads are used. However, hardware multi-threading leads to many charging/discharging operations and a corresponding energy consumption. The interesting experimental results include an analysis of the impact of executed instructions on the energy consumption, as shown in Fig. 5.18 for the case of 8 bit data. Figure 5.19 displays the corresponding information for the case of 16 bit data. The two dimensions of the diagrams encode the applications which are run in the odd and even threads, respectively. In these figures, a change in the number of operands is indicated by dashed lines. Instructions with three or more operands are shown at the top and at the right end of each diagram. Obviously, the

consumed energy increases with the number of operands. Figure 5.19 demonstrates
that processing 16 bit data requires more energy than processing 8 bit data. Kerrison
et al. use these results in order to optimize embedded software.

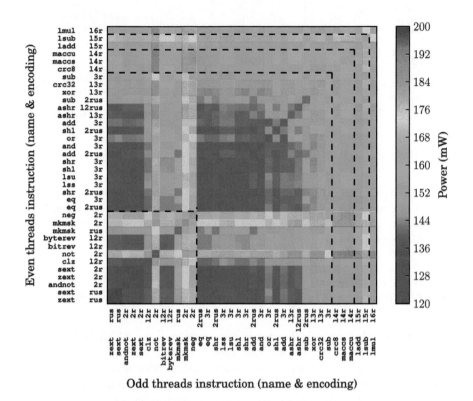

Fig. 5.18 Power analysis for multithreading for 8 bit data, © Kerrison, Eder

5.4.8 Analysis for Communication in an Android Phone

Zhang et al. [583] describe a power model construction technique for an HTC Android
phone, called PowerBooter. Their technique uses the following equation:

$$
\begin{aligned}
E = {} & (\beta_{uh} * freq_h + \beta_{ul} * freq_l) * util + \beta_{CPU} * CPU_{on} \\
& + \beta_{br} * brightness + \beta_{Gon} * GPS\_on + \beta_{Gsl} * GPS\_sl \\
& + \beta_{WiFi\_l} * WiFi_l + \beta_{WiFi\_h} * WiFi_h + \beta_{3G\_idle} * 3G_{idle} \\
& + \beta_{3G\_FACH} * 3G_{FACH} + \beta_{3G\_DCH} * 3G_{DCH}
\end{aligned}
\tag{5.39}
$$

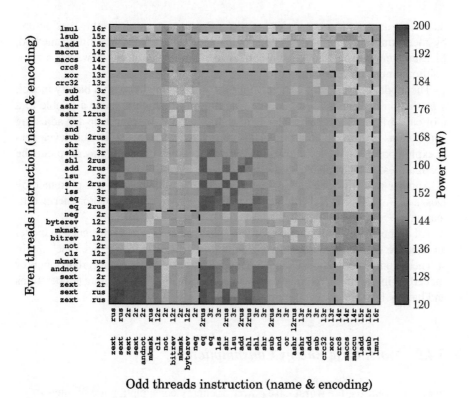

Fig. 5.19 Power analysis for multithreading for 16 bit data, © Kerrison, Eder

where

$\beta_{..}$: Constants to be determined

$freq_i$: CPU frequencies

$util$: CPU utilization

CPU_{on} : refers to processor utilization

$brightness$: takes illumination into account

$GPS_{..}$: Relates to GPS usage

$WiFi_l$: Amount of time, Wi-Fi is in low-speed mode

$WiFi_h$: Amount of time, Wi-Fi is in high-speed mode

$3G_{3G\_idle}$: Amount of time, 3G is idle

$3G_{FACH}$: Amount of time, a shared 3G channel is used

$3G_{DCH}$: Amount of time, a dedicated 3G channel is used

Obviously, PowerBooter is abstracting much more from the details of the hardware implementation. Note that PowerBooter also includes communication, which was not taken into account in our previous models. Parameters are determined, as before, by measuring currents in dedicated setups and using some curve fitting method. Measurements are based on a Monsoon power monitor (see http://www.msoon.com/LabEquipment/PowerMonitor/).

The model construction technique allows, in combination with a battery model, a prediction of battery life time. The resulting information is made available to a tool called PowerTutor. PowerTutor is intended to provide some help for adjusting applications to different hardware platforms and as an aid for application developers to exploit power saving techniques in their application without digging deep into the peculiarities of the available hardware.

Another model for the energy consumption in mobile phones was presented by Dusza et al. [138]. Several commercial tools also provide power and/or energy estimation.

All of the energy consumption models considered so far were designed to model an **average case** power or energy consumption, where term "average case" might still need some clarification. Computed models might apply only for certain inputs or for certain initial states. Average case results are valuable for predicting temperatures and battery life time for certain time intervals.

5.4.9   Worst Case Energy Consumption

In certain contexts, the **worst case** power or energy consumption are of interest.

Definition 5.21: The **worst case energy consumption (WCEC)** of an embedded system is defined as the largest energy consumption, computed as the maximum of the energy consumption for all inputs and initial states.

Definition 5.22: The **worst case power consumption (WCPC)** of an embedded system is defined as the largest power consumption, computed as the maximum of the power consumption for all inputs and initial states.

The WCPC is relevant in the context of the dimensioning of the interconnect and the power supply. The WCEC is relevant in the context of the design of battery systems. We need to guarantee that the chosen battery system meets the WCEC requirements. A safe upper bound on the WCEC can be computed as follows:

$$\text{WCEC} \leq \int_0^{\text{WCET}} \text{WCPC} \, dt = \text{WCET} * \text{WCPC} \qquad (5.40)$$

Techniques for tighter WCEC estimation have been proposed, for example, by Jayaseelan et al. [259], Pallister et al. [421], and by Wägemann et al. [535].

5.5 Thermal Models

The quest for higher performances of embedded systems increased the chances of components becoming hot. Temperatures of the various components of embedded systems can have a serious impact on their usability, e.g. on sensor readouts. In the worst case, overheated components cause damages to other systems. For example, they may cause fire hazards. Hot components might also have other consequences, even in the absence of immediate failures. For example, the system life might be shortened, sometimes by large factors (see Black's equation on p. 271).

The thermal behavior of embedded systems is closely linked to the transformation of electrical energy into heat. Therefore, thermal models are usually linked to energy models. Thermal models are based on the laws of physics.

Consider a homogeneous plate made of a particular material and of area A and thickness L (see Fig. 5.20).

Fig. 5.20 Plate of thickness L

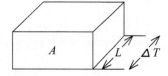

Suppose that there is a temperature difference of ΔT between the opposite sides. Then, the thermal power which gets transferred across the plate is equal to

$$P_{th} = \kappa \, \frac{\Delta T * A}{L} \tag{5.41}$$

where:

P_{th} : thermal power transferred

κ : thermal conductivity

A : area

ΔT : temperature difference

L : thickness

Definition 5.23: Due to Eq. (5.41), we can define **thermal conductivity** κ as the amount of the thermal power P_{th} transferred through a plate made of some material of area A and thickness L when the temperatures at the opposite side differ by one temperature unit (typically Kelvin).

κ depends on the material and environmental conditions. Values for some common materials for common conditions are included in Table 5.1. Refer to the cited sources for more information on the dependency on environmental conditions.

Table 5.1 Approximate thermal characteristics of materials for air, copper and silicon

Material	κ: Thermal conductivity (W/(K m))	c_p: Specific heat (J/(K g))	c_v: Volumetric heat capacity (J/(K m^3))
Air (25 C)	0.025 [555]	1.012 [552]	$1.21 * 10^3$ [552]
Copper	401 [555]	0.385 [543, 552]	$3.45 * 10^6$ [552]
Silicon ($\sim$ 26 C)	148 [142]	0.705 [142, 543]	$1.64 * 10^6$ [142][a]

[a]Calculated using Eq. (5.56)

Definition 5.24: Thermal conductance is defined as the amount of thermal energy which passes through a plate per unit of time if the temperatures at the two ends differ by one unit of temperature (typically Kelvin).

From Eq. (5.41), we have

$$\frac{P_{th}}{\Delta T} = \kappa * \frac{A}{L} \tag{5.42}$$

The reciprocal of this value is called thermal resistance R_{th}:

$$R_{th} = \frac{\Delta T}{P_{th}} = \frac{L}{\kappa * A} \tag{5.43}$$

Lemma 5.1: *Thermal resistances add up like electrical resistances. This allows us to map thermal modeling to electrical modeling.*

Example 5.3: Consider a microprocessor generating a thermal power P_{th}, the thermal resistance $R_{th,die}$ of the die (chip) and the thermal resistance $R_{th,fan}$ of the fan (see Fig. 5.21).

Fig. 5.21 Thermal model of microprocessor with fan

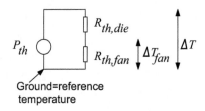

Then, we have

$$\Delta T = R_{th} * P_{th} \tag{5.44}$$

$$R_{th} = R_{th,die} + R_{th,fan} \tag{5.45}$$

Let us assume the following:

$$R_{th,die} = 0.4\,\text{W/K} \tag{5.46}$$

$$R_{th,fan} = 0.3\,\text{W/K} \tag{5.47}$$

$$P_{th} = 10\,W \tag{5.48}$$

Then, we compute:

$$\Delta T = 7\,\text{K} \tag{5.49}$$

$$\Delta T_{fan} = 3\,\text{K} \tag{5.50}$$

$$\nabla$$

Consumed power and thermal resistances play a key role in the computation of the thermal design power.

Definition 5.25 **([556]):** *"The **thermal design power (TDP)**, sometimes called thermal design point, is the maximum amount of heat generated by a computer chip or component (often the CPU or GPU) that the cooling system in a computer is designed to dissipate in typical operation. Rather than specifying CPU's real power dissipation, TDP serves as the nominal value for designing CPU cooling systems."*

We could try to compute the TDP from maximum temperatures and thermal resistances. In practice, however, published TDP values are actually not reflecting the WCPC. Hence, temperature sensors are required in order to obtain a safe operation.

So far, we have just considered the steady state. In general, transients and thermal capacitance (heat capacity) have to be considered.

Definition 5.26: The **thermal capacitance (heat capacity)** of some object is defined as the amount of thermal energy E_{th} which can be stored per difference ΔT in temperatures:

$$C_{th} = \frac{E_{th}}{\Delta T} \tag{5.51}$$

Primarily, C_{th} depends on the amount and type of matter contained in the object:

$$C_{th} = c_p * m \tag{5.52}$$

where c_p is the specific heat, and m the mass. We can also interpret Eq. (5.52) as the definition of the specific heat:

Definition 5.27: The **specific heat** c_p of some object made of some material of mass m is defined as

$$c_p = \frac{C_{th}}{m} \tag{5.53}$$

c_p depends on the type of matter used. c_p is temperature-dependent, but can be considered constant for small temperature ranges.

In our context, it is frequently more convenient to consider the heat capacity per volume instead of per unit of mass.

Definition 5.28: The **volumetric heat capacity** c_v is defined as

$$c_v = \frac{C_{th}}{\mathcal{V}} \tag{5.54}$$

where $\mathcal{V}$ is the volume of the object.

c_v and c_p are related by the mass density:

Definition 5.29: The **mass density** or **volume density** ρ is defined as

$$\rho = \frac{m}{\mathcal{V}} \tag{5.55}$$

Inserting $\mathcal{V} = m/\rho$ into the definition of c_v, we have

$$c_v = \frac{C_{th}}{\mathcal{V}} = \frac{C_{th} * \rho}{m} = c_p * \rho \tag{5.56}$$

This allows us to convert between tables published for c_p and c_v (see, e.g. Table 5.1).

Due to the correspondence to electrical circuits, we can also compute the transient behavior.

Example 5.4: We extend our microprocessor example as shown in Fig. 5.22 **(left)**.

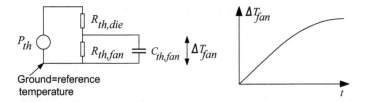

Fig. 5.22 Microprocessor with fan: **left**: thermal model; **right**: transient

The resulting transient for the temperature across the die and the fan are shown in Fig. 5.22 **(right)**. The system approaches the stable state like a network of resistors and capacitors.   ∇

Overall, it is feasible to model thermal behavior by using an equivalent electrical model. Equivalences are shown in Table 5.2. Well-known techniques for solving electrical network equations (see, for example, Chen et al. [98]) apply. However, there is no component corresponding to inductance on the thermal side.

Table 5.2 Equivalences between electrical and thermal models

Electrical model		Thermal model	
Current	I	Thermal flow, "power flow"	$P_{th} = \dot{Q}$
Total charge	$Q = \int I \, dt$	Thermal energy	$E_{th} = \int P_{th} \, dt$
Potential	ϕ	Temperature	T
Voltage = potential difference	$V = \Delta\phi$	Temperature difference	ΔT
Resistance[a]	$R = \rho_{el} \frac{L}{A}$	Thermal resistance	$R_{th} = \frac{1}{\kappa} \frac{L}{A}$
Ohm's law	$V = R * I$	Δ temperature at R_{th}	$\Delta T = R_{th} * P_{th}$
Capacitance	C	Thermal capacitance	C_{th}
Charge on capacitor	$Q = C * V$	Energy at capacitance	$E_{th} = C_{th} * \Delta T$
Capacitance of object[b]	$C = \rho_q V$	Capacitance of object	$C_{th} = c_v V$

[a] ρ_{el} is the specific electrical resistance or volume resistivity
[b] ρ_q is the volume charge density

This equivalence between thermal and electrical models is exploited in tools such as HotSpot [474]. Figure 5.23 shows a HotSpot model of a chip mounted on a heat spreader which in turn is mounted on a heat sink [475].

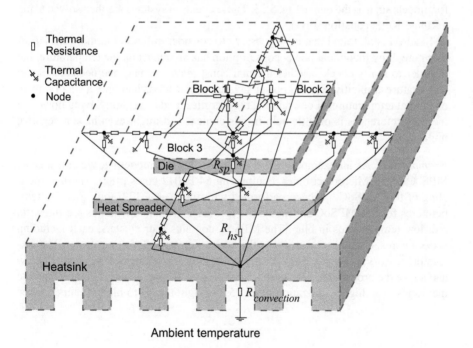

Fig. 5.23 HotSpot model of a chip mounted on a heat spreader and a heat sink

Skadron et al. [475] emphasize the fact that large temperature gradients can exist within a chip, a heat spreader or a heat sink. Hence, it is important not to assume a uniform temperature for these parts. In Fig. 5.23, the chip is assumed to comprise three micro-architectural components with each component forming one thermal zone. The heat spreader and the heat sink are modeled as five zones each: one beneath the chip and four on the sides. Each of the zones is shown as a node in the equivalent network in Fig. 5.23. The ambient temperature is assumed to be homogeneous. For each of the zones, there is one thermal capacitance. It is always considered to be connected to the ground. Furthermore, for each of the zones there is a pair of thermal resistors connecting adjacent zones. One of these is modeling the resistance between the node of the first zone and the boundary, the other the resistance between the boundary and the node of the second zone. This includes a pair of resistors connected to the zone beneath each of the zones. $R_{convection}$ is the thermal resistance to the environment. R_{hs} is thermal resistance between the heat spreader and the heat sink. R_{sp} is thermal resistance between the die (chip) and the heat spreader.

The heat source is actually not shown. In their experiments, Skadron et al. have used the Wattch (see p. 252) power simulator as heat source. Microarchitectural simulators such as SimpleScalar can be used to drive Wattch.

HotSpot contains mechanisms to create a system of partial differential equations for models such as the one in Fig. 5.23. These equation systems are then solved using a Runge-Kutta equation solver.

Skadron et al. found that it is necessary to consider different thermal zones. Furthermore, they found that power consumption has an impact on the temperature, but in order to really check whether thermal constraints are met, one needs to model temperature explicitly. Several power saving optimizations had only a small impact on crucial temperatures. For example, register files tend to get hot. Saving power on memory references is of little help in this context and might even have a negative impact.

Example 5.5: As an example of the results of thermal modeling, we consider an MPSoC of ST Microelectronics, comprising 64 P2012 cores [482]. Thermal modeling of this MPSoC has been performed with the 3D-ICE [22] tool. *Relative* temperatures for this MPSoC are shown in Fig. 5.24[5]. High temperatures are shown in red, low temperatures in blue. The MPSoC contains four clusters, each including sixteen cores. Each of the corners of the layout corresponds to one cluster. The sixteen processors are located at the center of the clusters. Memories are located below and above the processors. Simulation confirms that the processors are hotter than the memories. The higher utilization of Fig. 5.24 (**right**) leads to higher temperatures.

---

[5]Images are included with permission of David Atienza (EPFL). Images were obtained as part of the cooperation between EPFL and ST Microelectronics in the FP7 EU Project titled: "PRO3D: Programming for Future 3D Architectures with Many Cores".

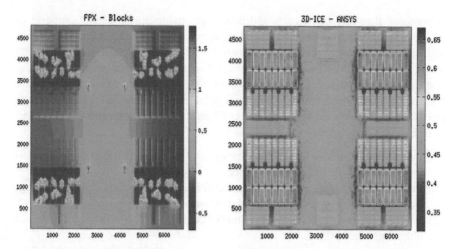

Fig. 5.24 Thermal simulation results for MPSoC: **left**: 50% utilization; **right**: 100% utilization

Detailed modeling of the layout avoided temperature overestimation.   ∇

Validation of thermal models requires precise temperature measurements [375].

5.6 Dependability and Risk Analysis

Next, we are going to look at dependability and risk analysis. This topic is linked to thermal modeling, since higher operating temperatures of devices imply a reduced lifetime. This will be described more quantitatively on p. 271 in the context of Black's equation.

5.6.1 Aspects of Dependability

Embedded and cyber-physical systems (like other products) can cause damages to properties and lives. The fact that such systems are potentially safety-critical was already included in Table 1.2 on p. 16. Hence, in general, we have to take this fact into account. It is not possible to reduce the risk of damages to zero. The best that we can do is to make the probability of damages small, hopefully orders of magnitude smaller than other risks. Allowed failures may be in the order of 1 failure per 10^9 h of operation or even significantly less for highly safety-critical systems like nuclear power plants. This may be several orders of magnitude less than the failure rates of chips. Hence, Kopetz [292] stressed that the system as a whole must be more dependable than any of its parts and that safety requirements cannot come in as an afterthought, but must be considered right from the beginning. Obviously, fault-tolerance mechanisms must be used. Due to the low acceptable failure rate, systems are not 100% testable. Instead, safety must be shown by a combination of testing and reasoning. Abstraction must be used to make the system explainable using a hierarchical set of behavioral models. Design faults and human faults must be taken into account. In order to address these challenges, Kopetz proposed the following twelve design principles:

1. Safety considerations may have to be used as **the** important part of the specification, driving the entire design process.
2. Precise specifications of design hypotheses must be made right at the beginning. These include expected failures and their probability.
3. Fault containment regions (FCRs) must be considered. Faults in one FCR should not affect other FCRs.
4. A consistent notion of time and state must be established. Otherwise, it will be impossible to differentiate between original and follow-up errors.
5. Well-defined interfaces must hide the internals of components.
6. It must be ensured that components fail independently.
7. Components should consider themselves to be correct unless two or more other components pretend the contrary to be true (principle of self-confidence).
8. Fault tolerance mechanisms must be designed such that they do not create any additional difficulty in explaining the behavior of the system. Fault tolerance mechanisms should be decoupled from the regular function.
9. The system must be designed for diagnosis. For example, it has to be possible to identify existing (but masked) errors.
10 The man-machine interface must be intuitive and forgiving. Safety should be maintained despite mistakes made by humans.
11. Every anomaly should be recorded. These anomalies may be unobservable at the regular interface level. This recording should involve internal effects, since otherwise they may be masked by fault-tolerance mechanisms.
12. Provide a never-give up strategy. Embedded systems may have to provide uninterrupted service. The generation of pop-up windows or going off line is unacceptable.

Dependability comprises a number of related issues, including safety, security, confidentiality and reliability.

Security of embedded and cyber-physical systems was not seen as a serious issue when these systems were not electronically accessible from the outside. This has changed for systems which can be accessed through communication channels and the two are now much more related, since security holes can cause physical malfunctions resulting in accidents.

A system can be dependable only if malfunctions due to external as well as internal reasons are very unlikely.

Definition 5.30: As system is **resilient** if internal or external changes of the assumptions made at design time will change the overall user experience only in a limited way.

A system which is **self-repairing** would provide some level of resiliency.

In the following subsections, we will address safety, security and reliability issues. Resiliency is beyond the scope of this book.

5.6.2 Safety Analysis

The evaluation of potentially safety-critical embedded systems should include an evaluation of the safety.

Typically, the minimum requirement for manufacturing safety-related products is to be ISO 9001 compliant. This standard defines requirements for quality management systems in general. Requirements as per this standard include the following principles [245]: customer focus, leadership, engagement of people, process approach, improvement, evidence-based decision making and relationship management. The first four principles are more or less self-explaining. The improvement principle requires work to proceed in plan, do, check and act (PDCA) cycles. The goal of planning includes establishing objectives and addressing risks and opportunities. The goal of the do phase is to implement the plan. This should be followed by checking the results and taking actions to improve if necessary.

For the design of safety-related systems, more specific guidelines have been developed and published as the IEC 61508 international standard [504]. Part 1 [224] of this standard defines standard techniques for technical systems in general. Part 2 [225] specifies *requirements for electrical/electronic/programmable electronic safety-related systems*. Software requirements are listed in part 3 [226]. Parts 4 to 6 contain less formal further recommendations. These standards assume that it is not feasible to design technical systems which always provide the expected service. Emphasis is placed on documented design procedures capable of tracing underlying reasons for incorrect decisions.

In standard IEC 61508, a distinction is made between four different levels of risks, called Safety Integrity Levels (SIL). For continuously operating devices, the standard specifies failure rates per hour of 10^{-5} to 10^{-6} for SIL-1, 10^{-6} to 10^{-7} for SIL-2, 10^{-7} to 10^{-8} for SIL-3, and 10^{-8} to 10^{-9} for SIL-4 [553]. SIL-4 is difficult to achieve and typically requires redundant execution. Problems arise from the current trend toward **mixed-criticality**, which means that subsystems of different SIL-levels are implemented, for example, on the same multi-core processor. Proper shielding of the different levels of criticality is difficult.

Standard IEC 61508 is expected to apply to several industries. There are specific extensions for specific industries. These consider, for example, the amount of time which is available for human interventions, the possibility of transitioning into a fail-safe mode and the impact of malfunctions. For example, there is very little time to react if something goes wrong in a car. However, cars can usually be stopped and parked in a "fail-safe" mode and a safe place (with the exception of some tunnels etc.). In contrast, there is usually some more time available in an airplane, but some safety-critical systems in an airplane cannot simply be turned off.

MISRA-C defines rules to be followed when using the C programming language for safety-critical systems [379].

ISO 26262 [243] is a standard more tailored for the automotive industry.

Standards IEC 62279 and CENELEC 50128 take the special situation for rail-based transportation into account [62].

For avionics, systems should comply with the Airworthiness Certification Specifications FAR-CS 25.1309 "Equipment, Systems and Installations", and AC-AMC 25.1309 "System design and analysis" [526]. This is extended for hardware by standard DO-254 and for software by standard DO-178B ("Software Considerations in Airborne Systems and Equipment Certification") [155, 450], in Europe also called ED-12B. DO-178C is a follow-up standard for DO-178B.

IEC 61511 [228] has been defined for applications in manufacturing and IEC 61513 [227] is a special standard for nuclear power plants.

5.6.3 Security Analysis

Security analysis needs to consider attacker models mentioned already in Sect. 3.8. This analysis needs to find out if attacks are feasible even without having physical access to the embedded system. If the system can be physically accessed, physical attacks and side channel attacks must be considered as well.

Furthermore, relationships between encryption and decryption protocols and achievable data rates must be analyzed, since it could easily happen that resource-constrained embedded devices do not provide the expected encryption and decryption rates.

5.6.4 Reliability Analysis

The design of dependable systems also requires an analysis of the reliability (the likelihood of initially correctly designed systems not to malfunction due to some internal fault). This task is expected to become more important and more difficult in the future, since **decreasing feature sizes of semiconductors will be resulting in a reduced reliability of semiconductor devices** (see, for example, http://variability. org). Transient as well as permanent faults are expected to become more frequent. Shrinking feature sizes will also cause an increased variability among device parameters. Therefore, dependability analysis and fault tolerant designs are becoming extremely important [173, 389]. Faults within semiconductors might lead to failures of the system. The terms **faults**, **failures** and the related terms **error** and **service** were defined by Laprie et al. [27, 309].

Definition 5.31: "The **service** delivered by a system (in its role as a **provider**) is its behavior as it is perceived by its user(s); ... The delivered service is a sequence of the provider's external states. ... **Correct service** is delivered when the service implements the system function".

Definition 5.32: A **service failure**, often abbreviated here to **failure**, is an event that occurs when the delivered service of a system deviates from the correct service. ... A service failure is a transition from correct service to incorrect service".

Definition 5.33: An **error** exists if one of the system's states is incorrect and may lead to its subsequent service failure.

Definition 5.34: "The adjudged or hypothesized cause of an error is called a **fault**. Faults can be internal or external of a system."

Some faults will not cause a system failure.

As an example, we might consider a transient *fault* flipping a bit in memory. After this bit flip, the memory cell will be in *error*. A *failure* will occur if the system service is affected by this error.

In line with these definitions, we will talk about *failure* rates when we consider systems that do not provide the expected system function. We will talk about *faults* whenever we consider the underlying **reasons** that might cause failures. There is a large number of possible reasons for faults, some of them resulting from reduced feature sizes of semiconductors. *Errors* will not be considered in the remaining part of this book.

As already mentioned, the computed upper bound of the rate of catastrophes has to be less than 10^{-9} per hour (SIL-4) [292] for many applications, corresponding to one case per 100,000 systems operating for 10,000 h. Reaching this level of dependability is only feasible if design evaluation also comprises the analysis of the reliability, the expected life-time and related objectives. Such an analysis is usually based on the probability of failures.

More precisely, we consider the probability densities of failures. Let x be the time until the first failure. x is a random variable. Let $f(x)$ be the probability density of this random variable.

As an example, we are frequently using the exponential probability density $f(x) = \lambda e^{-\lambda x}$. For this density function, failures are becoming less and less likely over time (after some time, it is likely that the system is not working any more and a system which is not working cannot fail). This density function is frequently used since it has a constant failure rate and, hence, describes in an appropriate way cases for which the failure rate is constant. We might even use this density function when the actual failure rate is unknown since a constant failure rate may be a good starting point. Moreover, this density function has nice mathematical properties. Figure 5.25 (**left**) shows this density function.

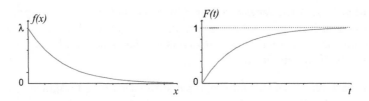

Fig. 5.25 Exponential distribution: **left**: density function; **right**: probability distribution

The probability distribution is frequently more interesting than the density. This distribution represents the probability of a system not working at time t. It can be obtained by integrating the density function until time t.

$$F(t) = Pr(x \leq t) \tag{5.57}$$

$$F(t) = \int_0^t f(x)\mathrm{d}x \tag{5.58}$$

For example, for the exponential distribution we obtain:

$$F(t) = \int_0^t \lambda e^{-\lambda x}\mathrm{d}x = -[e^{-\lambda x}]_0^t = 1 - e^{-\lambda t} \tag{5.59}$$

Figure 5.25 (**right**) contains the corresponding function. As time advances, this probability approaches 1. This means that, as time progresses, it becomes more likely that the system will have failed.

Definition 5.35: The **reliability** $R(t)$ of a system is the probability of the time until the first failure being larger than t:

$$R(t) = Pr(x > t), t \geq 0 \tag{5.60}$$

$$R(t) = \int_t^\infty f(x)\mathrm{d}x \tag{5.61}$$

$$F(t) + R(t) = \int_0^t f(x)\mathrm{d}x + \int_t^\infty f(x)\mathrm{d}x = 1 \tag{5.62}$$

$$R(t) = 1 - F(t) \tag{5.63}$$

$$f(x) = -\frac{\mathrm{d}R(t)}{\mathrm{d}t} \tag{5.64}$$

For the exponential distribution we have $R(t) = e^{-\lambda t}$ (see Fig. 5.26).

Fig. 5.26 Reliability for exponential distribution

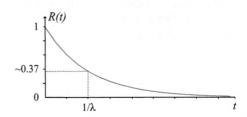

The probability for the system to be functional after time $t = 1/\lambda$ is about 37%.

Definition 5.36: The **failure rate** $\lambda(t)$ is the probability of a system failing between time t and time $t + \Delta t$.

$$\lambda(t) = \lim_{\Delta t \to 0} \frac{Pr(t < x \leq t + \Delta t | x > t)}{\Delta t} \qquad (5.65)$$

$Pr(t < x \leq t + \Delta t | x > t)$ is the conditional probability for the system failing within this time interval provided that it was working at time t. For conditional probabilities, there is the general equation $Pr(A|B) = Pr(AB)/Pr(B)$, where $Pr(AB)$ is the probability of A *and* B happening. $Pr(AB)$ is equal to $F(t + \Delta t) - F(t)$ in our case. $Pr(B)$ is the probability of the system working at time t, which is $R(t)$ in our notation. Therefore, Eq. (5.65) leads to:

$$\lambda(t) = \lim_{\Delta t \to 0} \frac{F(t + \Delta t) - F(t)}{\Delta t R(t)} = \frac{f(t)}{R(t)} \qquad (5.66)$$

For example, for the exponential distribution we obtain[6]:

$$\lambda(t) = \frac{f(t)}{R(t)} = \frac{\lambda e^{-\lambda t}}{e^{-\lambda t}} = \lambda \qquad (5.67)$$

Failure rates are frequently measured as multiples (or fractions) of 1 FIT, where "FIT" stands for *Failure unIT* and is also known as *Failures In Time*. 1 FIT corresponds to 1 failure per 10^9 h.

However, failure rates of real systems are frequently not constant. For many systems, we have a "bath tub curve"-like behavior (see Fig. 5.27).

Fig. 5.27 Bath tub curve-like failure rates

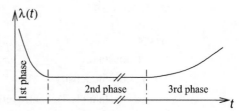

For this behavior, we are starting with an initially larger failure rate. This higher rate is a result of an imperfect production process or "infant mortality". The rate during the normal operating life is then essentially constant. At the end of the useful product life, the rate is then increasing again, due to wear-out.

---

[6]This result motivates denoting the failure rate and the constant of the exponential distribution with the same symbol.

Definition 5.37: The **Mean Time To Failure** (**MTTF**) is the average time until the next failure, provided that the system was initially working. This average can be computed as the expected value of random variable x:

$$\text{MTTF} = E\{x\} = \int_0^\infty x f(x) \mathrm{d}x \tag{5.68}$$

For example, for the exponential distribution we obtain:

$$\text{MTTF} = \int_0^\infty x \lambda \mathrm{e}^{-\lambda x} \mathrm{d}x \tag{5.69}$$

This integral can be computed using the product rule ($\int uv' = uv - \int u'v$ where in our case we have $u = x$ and $v' = \lambda \mathrm{e}^{-\lambda x}$). Therefore, Eq. (5.69) leads to the following equation:

$$\text{MTTF} = -[x \mathrm{e}^{-\lambda x}]_0^\infty + \int_0^\infty \mathrm{e}^{-\lambda x} \mathrm{d}x \tag{5.70}$$

$$= -\frac{1}{\lambda}[\mathrm{e}^{-\lambda x}]_0^\infty = -\frac{1}{\lambda}[0 - 1] = \frac{1}{\lambda} \tag{5.71}$$

This means that, for the exponential distribution the expected time until the next failure is the reciprocal value of the failure rate.

There is the following empirical relationship between MTTF and operating temperatures:

Lemma 5.2 (Black's Equation [52, 58]):

$$\text{MTTF} = \frac{A}{j_e^n} \mathrm{e}^{\frac{E_a}{kT}} \tag{5.72}$$

where

      A: constant

      j_e: current density

      n: constant (1..7), controversial, 2 according to Black

      E_a: activation energy (e.g. $\sim$0.6 eV)

      k: Boltzmann constant ($\sim$8.617 * $10^{-}5$ eV/K)

      T: temperature

Regardless of discussions about the correct value of n: this equation shows that the temperature has an exponential impact on the MTTF. Furthermore, current densities are also important: the larger the current densities, the shorter the lifetime of the product.

Definition 5.38: The **Mean Time To Repair (MTTR)** is the average time to repair a system, provided that the system is initially not working. This time is the expected value of the random variable denoting the time to repair.

Definition 5.39: The **Mean Time Between Failures (MTBF)** is the average time between two failures.

MTBF is the sum of MTTF and MTTR:

$$MTBF = MTTF + MTTR \tag{5.73}$$

Figure 5.28 shows a simplistic view of this equation: it is not reflecting the fact that we are dealing with probabilistic events and actual MTBF, MTTF, and MTTR values may vary randomly.

Fig. 5.28 Illustration of MTTF, MTTR and MTBF

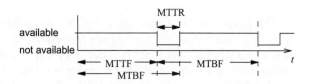

For many systems, repairs are not considered. Also, if they are considered, the MTTR should be much smaller than the MTTF. Therefore, the terms MTBF and MTTF are frequently mixed up. For example, the life-time of a hard disk may be quoted as a certain MTBF, even though it will never be repaired. Quoting this number as the MTTF would be more correct. Still, the MTTF provides only very rough information about dependability, especially if there are large variations in the failure rates over time.

Definition 5.40: The **availability** is the probability of a system being in an operational state.

The availability varies over time (just consider the bath tub curve!). Therefore, we can model availability by a time-dependent function $A(t)$. However, we are frequently only considering the availability A for large time intervals. Hence, we define

$$A = \lim_{t \to \infty} A(t) = \frac{MTTF}{MTBF} \tag{5.74}$$

For example, assume that we have a system which is repeatedly available for 999 days and then needs one day for repair. Such a system would have an availability of $A = 0.999$.

Allowed failure rates can be in the order of 1 FIT. This may be several orders of magnitude less than the failure rates of chips. This means that systems must be more reliable than their components! Obviously, the required level of reliability makes fault tolerance techniques a must!

Obtaining actual failure rates is difficult. Figure 5.29 shows one of the few published results [522].

Fig. 5.29 Failure rates of
TriQuint Gallium-Arsenide
devices (courtesy of
TriQuint, Inc., Hillsboro),
© TriQuint

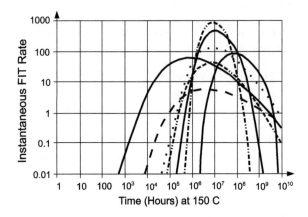

This figure contains failure rates for different Gallium-Arsenide (GaAs) devices with the hottest transistor operating at a temperature of 150 C. This example is used here to demonstrate that there exist devices for which the assumptions of constant failure rates or a bath tub-like behavior are oversimplifying[7]. As a result, citing a single MTTF number may be misleading. The actual distribution of failures over time should be used instead. In the particular case of this example, failure rates are less than 100 FIT for the first 20 years (175,300 hrs) of product life time, despite the high temperature. FIT numbers are actually very much temperature dependent and temperatures up to 275 C and known temperature dependences have been used at Triquint to compute failure rates for periods larger than the time available for testing. Triquint claims that their GaAs devices are more reliable than average silicon devices. Reports on FIT testing are also available for Xilinx FPGAs (see, for example, [571]).

5.6.5   Fault Tree Analysis, Failure Mode and Effect Analysis

It is frequently not possible to experimentally verify failure rates of complete systems. Requested failure rates are too small and failures may be unacceptable. We cannot fly 10^5 airplanes 10^4 hours each in an attempt to check if we reach a failure rate of less than 10^{-9} (SIL-4)! The only way out of this dilemma is to use a combination of checking failure rates of components and formally deriving from this guarantees for a reliable operation of the system. Design- and user-generated failures also must be taken into account. It is state of the art to use decision diagrams to compute the reliability of a system from that of its components [247].

Damages are resulting from **hazards** (chances for a failure). For each possible damage caused by a failure, there is a severity (the cost) and a probability. Risk can

---

[7]Therefore, the so-called log-normal distribution is sometimes considered.

be defined as the product of the two. Information concerning the damages resulting from component failures can be derived with at least two techniques [137, 434]:

- **Fault tree Analysis (FTA)**: FTA is a top-down method of analyzing risks. The analysis starts with a possible damage and then tries to come up with possible scenarios that lead to that damage. FTA is based on modeling a Boolean function reflecting the operational state of the system (operational or not operational). FTA typically includes symbols for AND- and OR-gates, representing conditions for possible damages. OR-gates are used if a single event could result in a hazard. AND-gates are used when several events or conditions are required for that hazard to exist. Figure 5.30 shows an example[8]. FTA is based on a **structural** model of the system, i.e. it reflects the partitioning of the system into components.

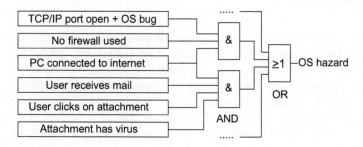

Fig. 5.30 Fault tree

The simple AND- and OR-gates cannot model all situations. For example, their modeling power is exceeded if shared resources of some limited amount (like energy or storage locations) exist. Markov models [69] may have to be used to cover such cases. Markov models are based on the notion of **states**, rather than on the structure of the system.

- **Failure mode and effect analysis (FMEA)**: FMEA starts at the components and tries to estimate their reliability. Using this information, the reliability of the system is computed from the reliability of its parts (corresponding to a bottom-up analysis). The first step is to create a table containing components, possible failures, probability of failures and consequences on the system behavior. Risks for the system as a whole are then computed from the table. Table 5.3 shows an example.

Table 5.3 FMEA table

Component	Failure	Consequences	Probability	Critical?
...	...	...	...	...
Processor	Metal migration	No service	10^{-7}/h	yes
...	...	...	...	...

---

[8]Consistent with the ANSI/IEEE standard 91, we use the symbols &, =1 and ≥1 to denote and-, xor-, and or-gates, respectively.

Tools supporting both approaches are available. Both approaches may be used in "safety cases". In such cases, an independent authority has to be convinced that certain technical equipment is indeed safe. One of the commonly requested properties of technical systems is that no single failing component should potentially cause a catastrophe.

The design of safe and dependable systems is a topic on its own. This book can only provide a few hints into this direction. There is an abundant amount of recent publications on the impact of reliability issues on system design. Examples include publications by Huang [214], Zhuo [585], and Pan [423]. For more information about dependability, consult books [175, 309, 325, 398, 489] on those areas.

5.7 Simulation

In this Chapter, we have so far placed an emphasis on design evaluation. Starting with this Section, we are now also considering **validation**. Simulation is a very common technique for evaluating and validating designs. Simulation consists of executing a design model on appropriate computing hardware, typically on general purpose digital computers. Obviously, this requires models to be executable. All the executable models and languages introduced in Chap. 2 can be used in simulations, and they can be used at various levels as described starting at p. 111. The level at which designs are simulated is always a compromise between simulation speed and accuracy. The faster the simulation, the less accuracy is available.

So far, we have used the term behavior in the sense of the functional behavior of systems (their input/output behavior). There are also simulations of some non-functional behaviors of designs, including the thermal behavior and the electromagnetic compatibility (EMC) with other electronic equipment. Due to the integration with physics, there is a large range of physical effects which may have to be included in the simulation model. As a result, it is impossible to cover all relevant approaches for simulating cyber-physical systems in this book. Law [311] provides an overview of approaches and topics in simulations on digital systems. A large amount of additional information on the simulation of systems (in particular of heterogeneous, cyber-physical systems) is available (see, for example, [121, 345, 420]). Some simulators specialize on specific application areas. Due to the large number of physical effects, it is impossible to provide a complete list of references.

For cyber-physical systems, simulations have serious limitations:

- Simulations are typically a lot slower than the actual design. Hence, if we interface the simulator with the actual environment, we can have quite a number of **violations of timing constraints**.
- Simulations in the physical environment may even be **dangerous** (who would want to drive a car with unstable control software?).
- For many applications, there may be huge amounts of data and it may be impossible to simulate enough data in the available time. Multimedia applications are notoriously known for this. For example, simulating the compression of some video stream takes an enormous amount of time.

- Most actual systems are too complex to allow simulating all possible cases (inputs). Hence, simulations can help us to find errors in our designs. They cannot guarantee absence of errors, since simulations cannot exhaustively be done for all possible combinations of inputs and internal states.

Due to these limitations, there is an increased emphasis on validation by formal verification (see p. 278). Nevertheless, sophisticated simulation techniques continue to play a key role for validation (see, for example, Braun et al.[68]). Academic solutions like gem5 (see http://gem5.org), SimpleScalar and OpenModelica as well as commercial solutions like the Synopsys® Virtualizer™ (see http://synopsys.com) are available. There are several tools for the simulation of networks (as required for the Internet of Things), including OMNET++ (see https://omnetpp.org/).

5.8 Rapid Prototyping and Emulation

Simulations are based on models, which are approximations of real systems. In general, there will be some difference between the real system and the model. We can reduce the gap by implementing some parts of our system under design (SUD) more precisely than in a simulator (for example, in a real, physical component).

Definition 5.41: Adopting a definition phrased by McGregor [218], we define **emulation** as the process of executing a model of the SUD where at least one component is **not** represented by simulation on some kind of host computer.

According to McGregor, *"Bridging the credibility gap is not the only reason for a growing interest in emulation – the above definition of an emulation model remains valid when turned around – an emulation model is one where part of the real system is replaced by a model. Using emulation models to test control systems under realistic conditions, by replacing the ... (real system) ... with a model, is proving to be of considerable interest to those responsible for commissioning, or the installation and start-up of automated systems of many kinds."*

In order to further improve credibility, we can continue replacing simulated components by real components. These components do not have to be the final components. They can be approximations of the real system itself, but should exceed the precision of simulations.

Note that it is now common to discuss the "emulation" of one computer on another computer by means of software. There is a lack of a precise definition of the use of the term in this context. However, it can be considered consistent with our definition, since the emulated computer is not just simulated. Rather, a speed faster than simulation speed is expected.

Definition 5.42: Fast prototyping is the process of executing a model of the SUD where **no** component is represented by simulation on some kind of host computer. Rather, all components are represented by realistic components. Some of these components should not yet be the finally used components (otherwise, this would be the real system).

There are many cases in which the designs should be tried out in realistic environments before final versions are manufactured. Control systems in cars are an excellent example for this. Such systems should be used by drivers in different environments before mass production is started. Accordingly, the automotive industry designs prototypes. These prototypes should essentially behave like the final systems, but they may be larger, more power consuming and have other properties which test drivers can accept. The term "prototype" can be associated with the entire system, comprising electrical and mechanical components. However, the distinction between rapid prototyping and emulation is also blurring. Rapid prototyping is by itself a wide area which cannot be comprehensively covered in this book.

Prototypes and emulators can be built, for example, using FPGAs. Racks containing FPGAs can be stored in the trunk while test drivers exercise the car. This approach is not limited to the automotive industry. There are several other fields in which prototypes are built from FPGAs. Commercially available **emulators** consist of a large number of FPGAs. They come with the required mapping tools which map specifications to these emulators. Using these emulators, experiments with systems which behave "almost" like the final systems can be run. However, catching errors by prototyping and emulation is already a problem for non-distributed systems. For distributed systems, the situation is even more difficult (see, for example, Tsai [523]).

5.9   Formal Verification

Formal verification[9] is concerned with formally proving a system correct, using the language of mathematics. First of all, a formal model is required to make formal verification applicable. This step can hardly be automated and may require some effort. Once the model is available, we can try to prove certain properties.

Formal verification techniques can be classified by the type of logic employed:

- **Propositional logic:** In this case, models consist of Boolean expressions. Tools are called **Boolean checkers**, **tautology checkers** or **equivalence checkers**. They can be used to verify that two representations of Boolean functions (or sets of Boolean functions) are equivalent. Since propositional logic is decidable, it is also decidable whether or not the two representations are equivalent (there will be no cases of doubt). For example, one representation might correspond to gates of an actual circuit and the other to its specification. Proving the equivalence then proves the effect of all design transformations (for example, optimizations for power or delay) to be correct. Boolean checkers can cope with designs which are too large to allow simulation-based exhaustive validation. The key reason for the power of Boolean checkers is the use of Binary Decision Diagrams (BDDs) [546]. The complexity of equivalence checks of Boolean functions represented with BDDs is linear in the number of BDD-nodes. In contrast, the equivalence check for

---

[9]This initial text on formal verification was based on a guest lecture given by Tiziana Margaria at TU Dortmund.

functions represented by sums of products is NP-hard. BDD-based equivalence checkers have therefore replaced simulators for this application and handle circuits with millions of transistors.

- **First order logic (FOL):** FOL adds ∃ and ∀ quantifiers to propositional logic. Some automation for verifying FOL models is feasible. However, since FOL is undecidable, there may be cases of doubt. Popular techniques include the **Hoare calculus**. Typically, operations on integers are also supported.
- **Higher order logic (HOL):** Higher order logic is based on lambda-calculus and allows functions to be manipulated like other objects [525]. For higher order logic, proofs can hardly ever be automated and typically must be done manually with some proof-support.

Propositional logic can be used to verify stateless logic networks, but cannot directly model finite state machines. For short input sequences, it may be sufficient to cut the feed-back loop in FSMs and to effectively deal with several copies of these FSMs, each copy representing the effect of one input pattern. However, this method does not work for longer input sequences. Such sequences can be handled with **model checking**.

For model checking, we have two inputs to the verification tool:

1. the model to be verified, and
2. properties to be verified.

States can be quantified with ∃ and ∀; numbers cannot. Verification tools can prove or disprove the properties. In the latter case, they can provide a counter-example. Model checking is easier to automate than FOL. It has been implemented for the first time in 1987, using BDDs. It was possible to locate several errors in the specification of the *future bus* protocol [104]. UPPAAL is a very popular tool for model checking[10].

This technique could be used, for example, to prove properties of the railway model of Fig. 2.52 (see p.78). It should be possible to convert the Petri net into a state chart and then confirm that the number of trains commuting between Cologne and Paris is indeed constant, confirming our discussion of Petri net place invariants on p. 77.

5.10 Problems

We suggest solving the following problems either at home or during a flipped classroom session:

5.1: Let us consider an example demonstrating the concept of Pareto-optimality. In this example, we study the results generated by task concurrency management (TCM) tools designed at the IMEC research center (*Interuniversitair Micro-Electronica Centrum*). TCM tools aim at establishing efficient mappings from applications to

---

[10]See http://www.uppaal.org for the academic and http://www.uppaal.com for the commercial version.

processors. Different multi-processor systems are evaluated and represented as sets of Pareto-optimal designs. Wong et al. [567] describe different options for the design of an MPEG-4-player. The authors assume that a combination of StrongARM-Processors and specialized accelerators should be used. Four designs meet the timing constraint of 30 ms (see Table 5.4).

Table 5.4 Processor configurations

Processor combination	1	2	3	4
Number of high speed processors	6	5	4	3
Number of low speed processors	0	3	5	7
Total number of processors	6	8	9	10

These different designs are shown in Fig. 5.31.

Fig. 5.31 Pareto points for multi-processor systems 2 and 3

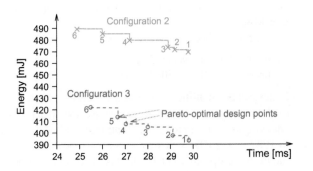

For combinations 1 and 4, the authors report that only one mapping of tasks to processors meets the timing constraints. For combinations 2 and 3, different time budgets lead to different task to processor mappings and different energy consumptions.

Which area in the objective space is dominated by at least one design of configuration 3? Is there any design belonging to configuration 2 which is not dominated by at least one design of configuration 3? Which area in the objective space dominates at least one design of configuration 3?

5.2: Which conditions must be met by computations of $WCET_{EST}$?

5.3: Let us consider cache states at a control flow join. Figure 5.32 shows abstract cache states before the join.

Fig. 5.32 Abstract cache
states

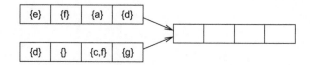

Now let us look at abstract cache states after the join. Which state would a *must*-analysis derive? Which state would a *may*-analysis derive?

5.4: Consider an incoming "bursty" event stream. The stream is periodic with a period of T. At the beginning of each period, two events arrive with a separation of d time units. Develop *arrival curves* for this stream! Resulting graphs should display times from 0 up to $3*T$.

5.5: Suppose that you are working with a processor having a maximum performance of b.

1. What do the *service curves* look like if the performance can deteriorate to b', due to cache conflicts?
2. How do the *service curves* change if some timer is interrupting the executed program every 100 ms and if servicing the interrupt takes 10 ms? Assume that there are no cache conflicts.
3. How do the *service curves* look like if you consider cache conflicts like in (1) **and** interrupts like in (2)?

Resulting graphs should display times from 0 up to 300 ms.

5.6: Suppose that we try to collect amber. However, there is the risk of also collecting white phosphorus. Suppose that we collect 50 objects. We keep all of them in water to avoid fire hazards. We classify 30 objects as amber and 20 as white phosphorus. However, two of the objects classified as amber are actually pieces of white phosphorus and 8 objects classified as white phosphorus are actually consisting of amber. Compute the precision, recall, accuracy and specificity for this classification!

5.7: Suppose that you try to compute the power consumption of your mobile phone using a shunt resistor. The following values are relevant for the computation of the power consumption at some time t: resistor: 0.47 Ohms, power supply voltage: 5.1 V, voltage across shunt: 0.23 V. What is the power consumption of your mobile at this time t?

5.8: Consider a copper plate of area $A = 1$ cm^2 and length 5 cm. How much thermal power is transferred if the difference between the temperatures at the two ends of the plate is 10 C?

5.9: Consider a hard disk drive for which we assume that half of the drives have failed after 5000 hours of operation. Let us assume that failures follow an exponential distribution. Compute the corresponding value of λ!

Chapter 6
Application Mapping

6.1 Definition of Scheduling Problems

6.1.1 Elaboration on the Design Problem

Once the specification has been completed, design activities can start. This is consistent with the simplified design information flow (see Fig. 6.1). Mapping applications to execution platforms is really a key activity. Therefore, we underline the importance of this Chapter.

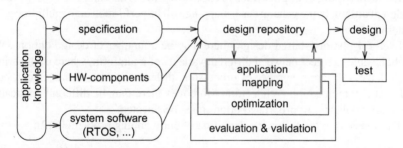

Fig. 6.1 Simplified design information flow

For embedded systems, we are frequently expecting that the system works with a certain combination of applications. For example, for a mobile phone, we expect being able to make a phone call while the Bluetooth stack is transmitting the audio signals to a head set and while we are looking up information in our "personal information manager" (PIM). At the same time, there may be a concurrent file transfer or even a video connection. We must make sure that these applications can be used together and that we are keeping the deadlines (no lost audio samples!). This is feasible through an analysis of the use cases.

© Springer International Publishing AG 2018
P. Marwedel, *Embedded System Design*, Embedded Systems,
DOI 10.1007/978-3-319-56045-8_6

It is a characteristic of embedded and cyber-physical systems that both hardware and software must be considered during their design. Therefore, this type of design is also called **hardware/software codesign**. The overall goal is to find the right combination of hardware and software resulting in the most efficient product meeting the specification. Therefore, embedded systems cannot be designed by a synthesis process taking only the behavioral specification into account. Rather, available components must be accounted for. There are also other reasons for this constraint: in order to cope with the increasing complexity of embedded systems and their stringent time-to-market requirements, reuse is essentially unavoidable. This led to the term **platform-based design**:

"*A platform is a family of architectures satisfying a set of constraints imposed to allow the reuse of hardware and software components. However, a hardware platform is not enough. Quick, reliable, derivative design requires using a platform application programming interface (API) to extend the platform toward application software. In general, a platform is an abstraction layer that covers many possible refinements to a lower level. Platform-based design is a meet-in-the-middle approach: in the top-down design flow, d esignern instance of the upper platform to an instance of the lower, and propagate design constraints*" [452]. The mapping is an iterative process in which performance evaluation tools guide the next assignment.

In this book, we focus on embedded system design based on available execution platforms. This reflects the fact that many modern systems are being built on top of some existing platform. Techniques other that the ones described in this book must be used when the execution platform needs to be designed as well. Due to our focus, the **mapping of applications to execution platforms** can be seen as the **main design problem**. In the general case, mapping will be performed onto multiprocessor systems.

Even for platform-based design, there may be a number of design options. We might be able to select between different variants of a platform, where each variant might have a different number of processors, different speeds of processors, or a different communication architecture. Moreover, there may be different applicable scheduling policies. Appropriate options must be selected.

This leads us to the following definition of our mapping problem [512]:

Given:

- a set of applications,
- use cases describing how the applications will be used,
- a set of possible candidate architectures:

  - (possibly heterogeneous) processors,
  - (possibly heterogeneous) communication architectures, and
  - possible scheduling policies.

Find:

- a mapping of applications to processors,
- appropriate scheduling techniques (if not fixed), and
- a target architecture (if not fixed).

Objectives:

- Keeping deadlines and/or maximizing performance, as well as
- minimizing cost, energy consumption, and possibly other objectives.

The exploration of possible architectural options is called **design space exploration** (DSE). As a special case, we may consider a completely fixed platform architecture.

Designing an AUTOSAR-based automotive system can be seen as an example: In AUTOSAR [26], we have a number of homogeneous execution units (called ECUs) and a number of software components. The question is: How do we map these software components to the ECUs such all real-time constraints are met? We would like to use the minimum number of ECUs.

For embedded systems, we can assume that the set of applications comprises a number of tasks which are released (are ready for execution) repeatedly. The executed code can be associated with tasks. For example, there may be the need to execute certain code once for every input sample. We denote each task by τ_i and sets of tasks by $\tau = \{\tau_1, \ldots, \tau_n\}$.

Definition 6.1: Each execution of a task is called a **job**. For each task τ_i, there is an associated set of jobs $J(\tau_i)$. Due to the repeated executions, the set of jobs of task τ_i is possibly not finite.

Definition 6.2: Tasks τ_i which are released once every T_i units of time are called **periodic tasks**, and T_i is called their **period**.

Definition 6.3: A task τ_i is called **sporadic** if there is a lower bound on the length of the interval between successive releases of this task. For each sporadic task τ_i, we call this interval length also T_i.

This minimum separation is important: without such a separation, arrival curves for any interval Δ could become unbounded. It would be impossible to find a schedule for a bounded set of resources.

Definition 6.4: Tasks which are neither periodic nor sporadic are called **aperiodic**.

For periodic and sporadic task systems, the concept of a hyper period is beneficial:

Definition 6.5: Let τ be a periodic or sporadic task system. Its **hyper period** is defined as the least common multiple of the periods of the individual tasks.

If tasks can be scheduled for one hyper period, they can be scheduled for all hyper periods, due to the repeating nature of the task structure.

6.1.2 Types of Scheduling Problems

The following notation is used in the remainder of this Chapter for jobs. Let $J = \{J_i\}$ be a set of jobs. Let (see Fig. 6.2):

- r_i be the release time of J_i (the time at which it becomes available for execution),
- C_i be the worst case execution time (WCET) of J_i,
- d_i be the (absolute) deadline of J_i,
- D_i be the **relative deadline**, that is, the time between a job J_i becoming available and the time until which the same job J_i has to finish execution.
- l_i be the **laxity** or **slack**, defined as

$$l_i = D_i - C_i \tag{6.1}$$

If $l_i = 0$, then J_i has to be started immediately after it it released.
- s_i be the actual starting time of J_i,
- f_i be the actual finishing time of J_i.
- In figures like Fig. 6.2, upward pointing vertical arrows indicate the release of jobs, and downward pointing arrows denote the deadline of jobs.

Fig. 6.2 Notation used for jobs

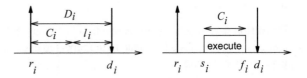

In the following, we will be using the triplet classification for scheduling problems which was presented by Pinedo [430], based on an notation introduced earlier by Graham, Lawler, Lenstra, and Kan [182]. According to the notation, scheduling problems can be classified by a triplet

$$(\alpha|\beta|\gamma). \tag{6.2}$$

The α Field

The α field describes the machine environment and consists of a single entry. Simple scheduling algorithms handle the case of single processors, whereas more complex algorithms also handle systems comprising multiple processors. In this book, we consider the following possible values of the α field:

- A value of 1 indicates a single processor.
- A value of Pm indicates m processors which can be used in parallel. Each job can be executed with the same speed on any of the m processors. In this case, processors are said to be **identical** (or homogeneous). The β field can be used to express constraints for the allocation of jobs to processors.
- A value of Qm denotes parallel processors with different performances. The performance is expressed as a scaling factor relative to the performance of the slowest processor. The scaling factor can be represented by a vector $(s_1, .., s_m)$, where component s_i is the scaling factor of processor π_i. In this case, processors are called **uniform**. The uniform processor model is very much simplified, we will hardly refer to it.

- A value of Rm indicates m processors with unrelated processing speeds. The execution time of the job or task i on processor k is $C_{i,k}$. Processors are called **heterogeneous**. Heterogeneous processors can be optimized for particular objectives, e.g., for high performance or a small energy consumption. Hence, heterogeneous processors are very important for embedded systems.

The β Field

The β field describes processing restrictions. This field may contain several components. In this book, we consider the following possible values of this field:

- An entry r_i denotes existing release times that are depending on the job i to be allocated.
- An entry $prmp$ indicates that preemptions are allowed. Non-preemptive scheduling is assumed if this entry is missing. Non-preemptive schedulers are based on the assumption that jobs are executed until they are done. As a result, the response time for external events[1] may be quite long if some jobs have a large execution time.

  Preemptive schedulers must be used if some jobs have long execution times or if the response time for external events is required to be short. However, preemption can result in unpredictable execution times of the preempted jobs. Therefore, restricting preemptions may be required in order to guarantee meeting the deadline of hard real-time jobs.

- Another possible entry would describe the type of timing constraints. We can distinguish between **soft and hard deadlines**.

Definition 6.6 (Kopetz [292]): *"A time constraint is called hard if not meeting that constraint could result in a catastrophe."*

All other time constraints are soft time constraints. Scheduling for soft deadlines is frequently based on extensions to standard operating systems. We will not discuss these systems further in this book. Therefore, the default assumption in this book is to have hard timing constraints.

- Entries *periodic* and *sporadic* may describe the type of task system considered.
- A value of $prec$ expresses the fact that precedence constraints exist. Precedences among the jobs require jobs to be executed according to certain partial orders. They may be caused by communication between jobs. For embedded systems, precedences are the rule rather than an exception.
- For sporadic and periodic task sets, we are frequently differentiating scheduling problems with respect to their deadlines:

  The case $D_i = T_i$, for all i, is called the case of **implicit-deadline tasks,** or **Liu-and-Layland (L & L) tasks**. This case is indicated by an entry $D_i = T_i$.

---

[1]This is the time from the occurrence of an external event until the completion of the reaction required for the event.

Task sets which must satisfy $\forall i \ : \ D_i \leq T_i$ are called **constrained-deadline tasks**. Tasks whose deadlines do not need to meet any constraints regarding their period are called **arbitrary-deadline tasks**. These cases can also be indicated by corresponding entries.

- We could use this field also to describe the type of scheduling employed. For example, we could use entries *fixed-job-prio* and *fixed-task-prio* for jobs and tasks with a fixed priority.

Furthermore, we could distinguish between static and dynamic scheduling. Dynamic schedulers take decisions at run-time. They are quite flexible, but generate overhead at run-time. Also, they are usually not aware of global contexts such as resource requirements or precedences between jobs. For embedded systems, such global contexts are typically available at design time and they should be exploited.

Static schedulers take their decisions at design time. They are based on planning the start times of jobs and generate tables of start times forwarded to a simple dispatcher. The dispatcher does not take any decisions, but is just in charge of starting jobs at the times indicated in the table. The dispatcher can be controlled by a timer, causing the dispatcher to analyze the table. Systems which are totally controlled by a timer are said to be **entirely time triggered** (TT systems). Such systems are explained in detail in the book by Kopetz [292]:

*"In an entirely time-triggered system, the temporal control structure of all tasks is established **a priori** by off-line support-tools. This temporal control structure is encoded in a **Task-Descriptor List (TDL)** that contains the cyclic schedule for all activities of the node[2] (Fig. 6.3). This schedule considers the required precedence and mutual exclusion relationships among the tasks such that an explicit coordination of the tasks by the operating system at run-time is not necessary. ... The dispatcher is activated by the synchronized clock tick. It looks at the TDL, and then performs the action that has been planned for this instant ...."*

Fig. 6.3 TDL in a time-triggered system

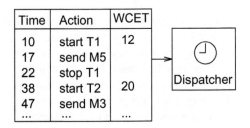

Time	Action	WCET
10	start T1	12
17	send M5	
22	stop T1	
38	start T2	20
47	send M3	
...	...	...

The main advantage of static scheduling is that it can be easily checked if timing constraints are met: *"For satisfying timing constraints in hard real-time systems, predictability of the system behavior is the most important concern; prerun-time*

---

[2]This term refers to a processor in this case.

scheduling is often the only practical means of providing predictability in a complex system" [576]. The main disadvantage is that the response to events may be quite poor.

- Multiprocessor scheduling algorithms can either be executed locally on one processor or can be distributed among a set of processors. Hence, we can also distinguish between **centralized and distributed scheduling**. This distinction could also be expressed in the β field.

The γ Field

The γ field describes the objective function. In this book, we consider the following possible values of this field:

- An entry of L_{max} means that the maximum lateness is to be minimized.

Definition 6.7: **Maximum lateness** is defined as the difference between the completion time and the deadline, maximized over all jobs. Maximum lateness is negative if all tasks complete before their deadline.

- An entry of MS_{max} denotes the case of minimizing the makespan (the time at which the last job finishes).

Definition 6.8: The **makespan** is defined as[3]

$$MS_{max} = max_i(f_i) \tag{6.3}$$

- In addition to the entries considered by Pinedo, other entries are relevant for embedded systems. For example, we might want to minimize the energy consumption or we might even consider trade-offs between several objectives.

A huge amount of scheduling algorithms is available and comprehensive coverage of existing algorithms is infeasible even if an entire book or course would be available. In a standard undergraduate curriculum, there is typically not enough headroom for a dedicated course on scheduling (but this may be different for courses for graduate students). Therefore, we provide only a brief introduction to scheduling in this book. Many scheduling problems are known to be very complex [40, 430]. In many cases, only approximately optimal mappings can be guaranteed. We will provide an overview of scheduling algorithms frequently considered in embedded systems. Table 6.1 comprises an overview of the techniques in this Chapter. From left to right, columns refer to the processor model, asynchronous arrival times, preemptiveness, precedences, periodic/sporadic tasks vs. aperiodic jobs, the deadline model (for periodic/sporadic tasks), job- vs. task-based priorities (for periodic/sporadic tasks), global vs. local scheduling (for multiprocessors), the objective, the subsection, and the name of algorithm(s). Algorithms like Earliest Deadline First are designed for non-periodic systems, but can be applied in periodic/sporadic systems as well. Note

---

[3]Pinedo denotes the makespan as C_{max}. We prefer to avoid confusion with execution times C_i.

Table 6.1 Scheduling techniques described or mentioned in this Chapter

α	β							γ	Section	Algorithm
Proc.	r_i	prmp	prec	periodic[a]	D_i	prio	glob	Objective		
1	-	-	-	-	-	-	-	L_{max}	6.2.1.1	Earliest due date
1	X	X	-	-	-	-	-	L_{max}	6.2.1.2	Earliest deadline first
1	X	X	-	-	-	-	-	L_{max}	6.2.1.3	Least laxity
1	X	-	-	-	-	-	-	L_{max}	6.2.1.4	(Theorem 6.3)
1	X	X	X	-	-	job	-	L_{max}	6.2.2.2	Latest deadline first
1	X	-	X	-	-	-	-	L_{max}	6.2.2.2	[484]
1	X	X	-	X	$= T_i$	task	-	$\leq D_i$	6.2.3.2	Rate monotonic
1	X	X	-	X	$\neq T_i$	task	-	$\leq D_i$	6.2.3.5	Deadline monotonic
Pm	-	-	-	-	-	-	X	$m = |\pi|$	6.3.1	Bin packing
Pm	-	-	-	-	-	-	X	$\sum b_i$	6.3.1	0/1 Multi-Knapsack
Pm	X	X	-	X	$= T_i$	-	-	$\leq D_i$	6.3.1	First fit decreasing

Table 6.1 (continued)

α	β							γ	Section	Algorithm
Proc.	r_i	prmp	prec	periodic[a]	D_i	prio	glob	Objective		
Pm	X	X	-	X	$=T_i$	job	X	$\leq D_i$	6.3.2	Pfair
Pm	X	X	-	-	-	job	X	$\leq D_i$	6.3.3	G-EDF, fpEDF, EDZL
Pm	X	X	-	X	$=T_i$	task	X	$\leq D_i$	6.3.4	G-RM, RM-US, RMZL
Pm	X	X	-	X	$\neq T_i$	task	X	$\leq D_i$	6.3.4.3	Density based
Pm	-	-	X	-	-	-	-	MS_{max}	6.4	ASAP, ALAP
Rm[b]	-	-	X	-	-	-	-	MS_{max}	6.4.3	List scheduling
Pm	-	-	X	-	-	-	-	MS_{max}	6.4.4	Integer linear progr.
Rm	-	-	X	-	-	-	-	MS_{max}	6.5.2	HEFT, CPOP
Rm	-	-	X	-	-	-	-	MS_{max}	6.5.3	e.g., [344]
Rm	X	X	X	-	-	-	(X)	various	6.5.4	DOL, HOPES, MAPS, ..

[a] Algorithms for aperiodic task sets can be applied to periodic/sporadic task sets
[b] List scheduling supports heterogeneous processors only in a limited way

that only the last three lines correspond to full support for heterogeneous processors, as can be seen in column one. Uniform processors will be mentioned only as a possible use of the 0/1 Multi-Knapsack model. If all jobs arrive at the same time (an entry of "-" for the second column), preemption is useless and, hence, the third column is not marked by an X. Entries for column D_i are relevant only for periodic/sporadic tasks. Regarding the objectives, we observe that lateness is the relevant objective in many cases. However, for periodic/sporadic scheduling, the key question is: Is there a schedule which meets the deadlines? Bin packing is designed to minimize the number of processors. For the HEFT and CPOP heuristics, the makespan is the relevant objective. Only the last line corresponds to a minimization of several objectives, either in the form of a single objective at a time or to real multi-objective optimization using Pareto optimality.

Scheduling is similar to performance evaluation in that it cannot be constrained to a single design step. Rather, scheduling algorithms may be required a number of times during the design of such systems. Very rough calculations may already be required while fixing the specification. Later, more detailed predictions of execution times may be required. After compilation, even more detailed knowledge exists about the execution times and, accordingly, more precise schedules can be made. Finally, it may be necessary to decide at run-time which task is to be executed next. In contrast, in time-triggered systems, RTOS scheduling may be limited to simple table look-ups for tasks to be executed.

In practice, it is very important to know whether or not a schedule exists for a given set of tasks and constraints. A set of tasks is said to be **schedulable** under a given set of constraints if a schedule exists for that set of tasks and constraints. For many applications, **schedulability tests** are important. Tests which always return precise results (called exact tests) are NP-hard in many situations [172]. Therefore, sufficient and necessary tests are used instead. For sufficient tests, sufficient conditions for guaranteeing a schedule are checked. There is a (hopefully small) probability of indicating that scheduling cannot be guaranteed even when a schedule exists. Necessary tests are based on checking necessary conditions. They can be used to show that no schedule exists. However, there may be cases in which necessary tests are passed and the schedule still does not exist.

6.2 Scheduling for Uniprocessors

Let us first consider the case of uniprocessor systems. According to the triplet notation, this corresponds to the case $(1|..|..)$. We are using some of the material from the book by Buttazzo [82] for this section. Refer to this book for additional references.

6.2.1 Independent Jobs

Furthermore, we are restricting our discussion initially to the even more special case of independent jobs executed on uniprocessors.

6.2.1.1 Earliest Due Date (EDD)-Algorithm

First of all, we are looking at the situation where all jobs arrive at the same time and we try to minimize lateness. If all jobs arrive at the same time, preemption is obviously useless. Therefore, according to the triplet notation, we are considering the case $(1| \ |L_{max})$.

A very simple rule for this case was found by Jackson in 1955 [251].

Theorem 6.1 (Jackson's Rule): *Given a set of n independent jobs with deadlines, any algorithm that executes the jobs in order of non-decreasing deadlines is optimal with respect to minimizing the maximum lateness.*

The algorithm following this rule is called the **Earliest Due Date** (EDD) algorithm. If the deadlines are known in advance, EDD can be implemented as a static scheduling algorithm. EDD requires all jobs to be sorted by their deadlines. Hence, its complexity is $O(n \log(n))$.

Proof (of the optimality of EDD): Let S be a schedule generated by any algorithm A. Suppose A does not lead to the same result as EDD. Then, there are jobs J_a and J_b such that the execution of J_b precedes the execution of J_a in J, even though the deadline of J_a is earlier than that of J_b ($d_a < d_b$). Now, consider a schedule S'. S' is generated from S by swapping the execution orders of J_a and J_b (see Fig. 6.4).

Fig. 6.4 Schedules S and S'

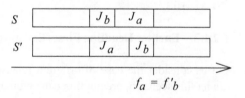

In schedule S, the deadline of J_a is earlier than that of J_b, but J_b is executed first. Hence, the maximum lateness among jobs J_a and J_b is that of J_a, or $L_{max}(a, b) = f_a - d_a$.

For schedule S', $L'_{max}(a, b) = max(L'_a, L'_b)$ is the maximum lateness among jobs J_a and J_b. L'_a is the maximum lateness of job J_a in schedule S'. L'_b is defined accordingly. There are two possible cases:

1. $L'_a > L'_b$: in this case, we have

$$L'_{max}(a, b) = f'_a - d_a$$

J_a terminates earlier in the new schedule. Therefore, we have

$$L'_{max}(a, b) = f'_a - d_a < f_a - d_a.$$

The right side of this inequality is the maximum lateness in schedule $\mathcal{S}$. Hence, the following holds:

$$L'_{max}(a, b) < L_{max}(a, b)$$

2. $L'_a \leq L'_b$:

In this case, we have:

$$L'_{max}(a, b) = f'_b - d_b = f_a - d_b \text{ (see Fig. 6.4).}$$

The deadline of J_a is earlier than the one of J_b. This leads to

$$L'_{max}(a, b) < f_a - d_a$$

Again, we have

$$L'_{max}(a, b) < L_{max}(a, b)$$

As a result, any schedule (which is not an EDD-schedule) can be turned into an EDD-schedule by a finite number of swaps. Maximum lateness can only decrease during these swaps. Therefore, EDD is optimal among for this class of scheduling problems.   □

6.2.1.2   Earliest Deadline First (EDF)-Algorithm

Let us consider the case of different release times for uniprocessor systems next. Under this scenario, preemption can potentially reduce maximum lateness. According to the triplet notation, this corresponds to the case $(1|r_i, prmp|L_{max})$.

The Earliest Deadline First (EDF) algorithm is optimal with respect to minimizing the maximum lateness. It is based on the following theorem [213]:

Theorem 6.2: *Given a set of n independent jobs with arbitrary arrival times, any algorithm that at any instant executes the job with the earliest absolute deadline among all the ready jobs is optimal with respect to minimizing the maximum lateness.*

EDF requires that each time a new ready job arrives, it is inserted into a queue of ready jobs, sorted by their deadlines. Hence, EDF is a dynamic scheduling algorithm. If a newly arrived job is inserted at the head of the queue, the currently executing job is **preempted**. If sorted lists are used for the queue, the complexity of EDF is $O(n^2)$. Bucket arrays could be used for reducing the execution time.

Example 6.1: Figure 6.5 shows a schedule derived with the EDF algorithm. At time 4, job J_2 has an earlier deadline. Therefore, it preempts J_1. At time 5, job J_3 arrives. Due to its later deadline it does not preempt J_2. The deadline of J_1 is lather than that of J_3 and, hence, it resumes only after J_3 has terminated. The deadline of J_1 is not included in the graph.

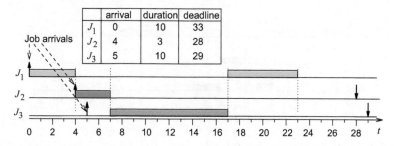

Fig. 6.5 EDF schedule

Priorities are obviously dynamic: they depend on which deadline is next. Since EDF uses dynamic priorities, it cannot be used with an operating system providing only fixed priorities. However, it has been shown that operating systems can be extended to simulate an EDF policy at the application level [127]. ∇

Proof (of Theorem 6.2): Let S be a schedule generated by some algorithm A, where A is different from EDF. Let S_{EDF} be a schedule generated by EDF. Now, we partition time into disjoint intervals of length 1^4. Each interval comprises times within the range $[t, t+1)$. Let $S(t)$ be the job which—according to schedule S—is executed during the interval $[t, t+1)$. Let $E(t)$ be the job which at time t has the earliest deadline among all jobs. Let $t_E(t)$ be the time $(\geq t)$ at which job $E(t)$ is starting its execution in schedule S.

S is not an EDF-schedule. Therefore, there must be a time t at which we are not executing the job having the earliest deadline. For t, we have $S(t) \neq E(t)$ (see Fig. 6.6).

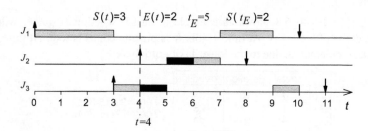

Fig. 6.6 Schedule S

Using the same arguments as for Jackson's Rule, we can show that swapping $S(t) \neq E(t)$ like in Fig. 6.7 does not increase maximum lateness.

Therefore, by a number of swaps, any non-EDF schedule can be turned into an EDF-schedule without increasing maximum lateness. This proves that EDF is optimal among all possible scheduling algorithms.

---

[4]This proof assumes a discrete time domain. It can be extended to a continuous time domain.

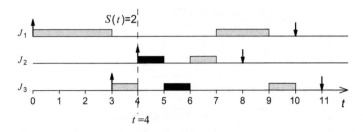

Fig. 6.7 Schedule after swapping jobs $\mathcal{S}(t)$ and $E(t)$

We can show that swapping will keep all deadlines, provided they were kept in schedule $\mathcal{S}$. According to the initial assumption, the maximum lateness in the schedule $\mathcal{S}$ is 0. Therefore, since EDF returns the optimal schedule for minimizing the maximum lateness, the maximum lateness of the EDF schedule is also 0. Hence, for this problem class, the EDF schedule is the optimal schedule to meet the deadlines.

$\square$

6.2.1.3   Least Laxity (LL) Algorithm

Focusing on laxity, we are now considering the case $(1|r_i, prmp, ..|..)$, with the goal of finding a schedule if one exists. Least Laxity (LL), Least Slack Time First (LST), and Minimum Laxity First (MLF) are three names for a laxity-based scheduling strategy [332]. According to LL scheduling, job priorities are a monotonically decreasing function of the laxity (see Eq. (6.1); the less laxity, the higher the priority). The laxity is dynamically changing and needs to be dynamically recomputed. Negative laxities provide an early warning for deadlines to be missed. LL scheduling is also preemptive. Preemptions are not restricted to times at which new jobs become available.

Example 6.2: Figure 6.8 shows an example of an LL schedule, together with the computations of the laxity. At time 4, job J_1 is preempted, as before. At time 5, J_2 is now also preempted, due to the lower laxity of job J_3.

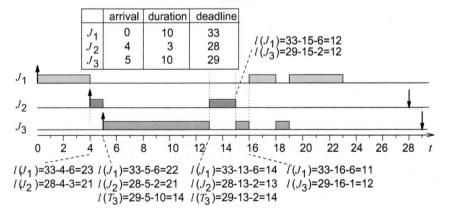

Fig. 6.8 Least laxity schedule

∇

It can be shown (this is left as an exercise in [332]) that LL is also an optimal scheduling policy for uniprocessor systems with meeting deadlines as the objective. This means that it will find a schedule if one exists. Due to its dynamic priorities, it cannot be used with a standard OS providing only fixed priorities. Furthermore, LL scheduling—in contrast to EDF scheduling—requires the knowledge of the execution time and typically generates many context switches. Its use is therefore restricted to special situations where its properties are attractive. Also, laxity can play a role in multiprocessor scheduling, as will be shown in Sects. 6.3.3.2 and 6.3.4.2.

6.2.1.4 Scheduling Without Preemption

Let us now consider the case of not allowing preemptions, denoted as $(1|r_i|L_{max})$.

Theorem 6.3: *If preemption is not allowed, optimal schedules must leave the processor idle at certain times in order to finish jobs with early deadlines arriving late.*

Proof: Let us assume that an optimal non-preemptive scheduler (not having knowledge about the future) never leaves the processor idle. This scheduler must schedule the example of Fig. 6.9 optimally (it must find a schedule if one exists). For the example of Fig. 6.9, we assume we are given two jobs. Let τ_1 be a periodic task with $C_1 = 2$, $T_1 = 4$, $D_1 = 4$, and $r_1 = 0$. Let τ_2 be a sporadic task with $C_2 = 1$, $D_2 = 1$, $T_2 = 4$, and $r_2 = 1$, i.e., sporadically becoming available at times $4 * n + 1$.

Fig. 6.9 Scheduler needs to leave processor idle

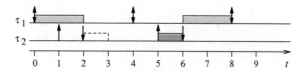

Under the above assumptions, our scheduler has to start the execution of job τ_1 at time 0, since it is supposed not to leave any idle time. Since the scheduler is non-preemptive, it cannot start τ_2 when it becomes available at time 1. Hence, τ_2 misses its deadline. If the scheduler had left the processor idle (as shown in Fig. 6.9 at time 4), a legal schedule would have been found. Hence, the scheduler is not optimal. This is a contradiction to the assumptions that optimal schedulers not leaving the processor idle at certain times exist. □

We conclude: in order to avoid missed deadlines, the scheduler needs knowledge about the future. Such algorithms are called **clairvoyant**. An algorithm leaving the processor idle in the presence of executable tasks is not **work conserving**:

Definition 6.9: A scheduling algorithm is **work-conserving** if it does not allow there to be a time at which a processor is idle and there is an executable task [116].

If no knowledge about the arrival times is available a priori, then no online algorithm can decide whether or not to keep the processor idle.

If arrival times are known a priori, the scheduling problem becomes NP-hard in general and branch and bound techniques are typically used for generating schedules.

6.2.2   Scheduling with Precedence Constraints

Next, let us consider precedence constraints, according to the triplet notation denoted as $(1|r_i, prmp, prec|L_{max})$.

6.2.2.1   Task Graphs

Precedence constraints are expressed by directed acyclic graphs (DAGs) $G = (\tau, E)$. The set τ represents the vertices (or nodes) of the DAG, and $E \subseteq \tau \times \tau$ its edges.

Example 6.3: Figure 6.10 shows an example. Edges express the fact that source nodes (the first components of the tuples represented an edge) must be executed

Fig. 6.10   Task DAG

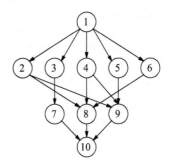

before their sink nodes (the second components of the tuples represented an edge). Vertex labels denote task numbers, edge label represent communication volumes.   ∇

There may be several reasons for describing applications as DAGs:

1. On the one hand, each vertex might correspond to an instance of a task and edges would then represent dependencies between tasks.
2. On the other hand, the availability of multiprocessors leads to the idea of splitting tasks into subtasks and executing these subtasks in an overlapping manner on different processors. Each vertex could then correspond to a subtask. Automatic partitioning of tasks into subtasks such that parallel processors can be efficiently exploited is called **automatic parallelization**. Automatic parallelization is even more difficult than automatic scheduling for a given number of subtasks.

Both cases of creating DAGs can be used in combination: we can have dependencies among tasks and tasks can be split into subtasks. In the following, we assume that the DAG represents any of the situations just described and we will call the DAGs **task graphs**. For scheduling, it is not relevant how the DAG was actually generated.

Example 6.4: A legal schedule for a simpler task graph including message transmission is shown in Fig. 6.11. Task τ_3 can be executed only after task τ_1 and τ_2 have completed and sent messages to τ_3.

Fig. 6.11 Precedence graph and schedule

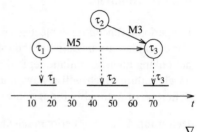

∇

6.2.2.2 Latest Deadline First (LDF) Algorithm

An optimal algorithm for minimizing the maximum lateness for the case of simultaneous arrival times of dependent tasks or jobs was presented by Lawler [312]. The algorithm is called **Latest Deadline First** (LDF). LDF reads the task graph. Among all tasks with no successors, it picks the one with the latest deadline and puts it into a queue. It then repeats this process, always selecting the task with the latest deadline among tasks whose successors have all been selected and inserting it into the queue. At run-time, the tasks are executed in an order **opposite** to the order in which tasks have been entered into the queue. LDF is non-preemptive and is optimal for uniprocessors.

Example 6.5: Consider the case of Fig. 6.11. LDF would first store τ_3 in a queue, since it has no successor. As a result, successors of τ_1 and τ_2 have all been selected already. Which of the two is stored in the queue first depends on their deadline. The node having the later deadline is stored in the queue first. At run-time, the queue is processed in reverse order, starting, for example, with τ_1. ∇

The case of asynchronous arrival times can be handled with a modified EDF algorithm. The key idea is to transform the problem from a given set of dependent jobs into a set of independent jobs with different timing parameters [99]. This algorithm is again optimal for uniprocessor systems.

If preemption is not allowed, the heuristic algorithm developed by Stankovic and Ramamritham [484] can be used.

6.2.3 Periodic Scheduling Without Precedence Constraints

Next, we will consider the periodic case. We will consider mostly tasks instead of jobs, since most properties for periodic systems can be derived for tasks. We will

restrict ourselves to a description of the case in which tasks are independent, described
as $(1|r_i, prmp, periodic|...)$ in the triplet notation.

6.2.3.1 Notation

For periodic scheduling, objectives relevant for aperiodic scheduling are less useful.
For example, minimization of the total length of the schedule is not an issue if we
are talking about an infinite repetition of jobs. The best that we can do is to design
an algorithm which will always find a schedule if one exists. This motivates the
definition of optimality for periodic schedules.

Definition 6.10: For periodic scheduling, a scheduler is defined to be **optimal** iff it
will find a feasible schedule if one exists.

Definition 6.11: For periodic **and sporadic** task systems $\tau = \{\tau_1, .., \tau_n\}$, we define
task utilization as

$$u_i = \frac{C_i}{T_i} \tag{6.4}$$

This means that, for sporadic task systems, we are using the same definition as for
periodic systems, even though T_i just denotes the minimum separation of jobs.

Definition 6.12: For a task system $\tau = \{\tau_1...\tau_n\}$ with utilization u_i of task τ_i, we
define the maximum and the total utilization by:

$$U_{max} = \max_i (u_i) \tag{6.5}$$

$$U_{sum} = \sum_i u_i \tag{6.6}$$

6.2.3.2 Rate Monotonic Scheduling

Rate monotonic (RM) scheduling [331] is probably the most well-known scheduling
algorithm for independent periodic tasks. Rate monotonic scheduling is based on the
following assumptions (**"RM assumptions"**):

1. All tasks that have hard deadlines are periodic.
2. All tasks are independent.
3. $D_i = T_i$, for all tasks.
4. C_i is constant and is known for all tasks. Self-suspension (voluntarily relinquish-
   ing the execution) is not allowed.
5. The time required for context switching is negligible.
6. For a single processor and for n tasks, the following equation holds for the accu-
   mulated utilization U_{sum}:

$$U_{sum} = \sum_{i=1}^{n} \frac{C_i}{T_i} \leq n(2^{1/n} - 1) \tag{6.7}$$

Figure 6.12 shows the right-hand side of Eq. (6.7).

Fig. 6.12 Right-hand side of Eq. (6.7)

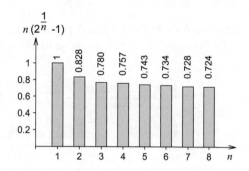

The right-hand side is about 0.7 for large n:

$$\lim_{n \to \infty} n * (2^{1/n} - 1) = log_e(2) = ln(2) \ (\sim 0.7) \tag{6.8}$$

Then, according to the policy for rate monotonic scheduling, **the priority of tasks is a monotonically decreasing function of their period**. In other words, tasks with a short period will get a high priority and tasks with a long period will be assigned a low priority. RM scheduling is a **preemptive scheduling policy** with **fixed priorities**.

Example 6.6: Figure 6.13 shows an example of a schedule generated with RM scheduling. Task τ_2 is preempted several times. Double-headed arrows indicate the arrival time of a job as well as the deadline of the previous job. Tasks τ_1 to τ_3 have a period of 2, 6, and 6, respectively. Execution times are, 0.5, 2, and 1.75. Task τ_1 has the shortest period and, hence, the highest rate and priority. Each time task τ_1 becomes available, its jobs preempt the currently active task. Task τ_2 has the same period as task τ_3, and neither of them preempts the other.

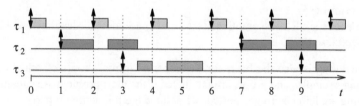

Fig. 6.13 Example of a schedule generated with RM scheduling

∇

Equation (6.7) requires that some of the computing power of the processor is not used in order to make sure that all requests are honored in time. What is the reason for this bound on the utilization? The key reason is that RM scheduling, due to its static priorities, will possibly preempt a task which is close to its deadline in favor of some higher priority task with a much later deadline. The task having a lower priority can then miss its deadline.

Example 6.7: In Fig. 6.14, task parameters are: $T_1 = 5$, $C_1 = 3$, $T_2 = 8$, and $C_2 = 3$. Not enough idle time is available to guarantee schedulability for RM scheduling.

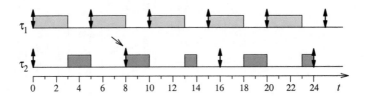

Fig. 6.14 RM schedule does not meet deadline at time 8

In this case, we have $U_{sum} = \frac{3}{5} + \frac{3}{8} = \frac{39}{40} = 0.975$. $2 * (2^{\frac{1}{2}} - 1)$ is about 0.828. Hence, schedulability is not guaranteed for RM scheduling and, in fact, the deadline is missed at time 8. We assume that the missing computations are not scheduled in the next period. ∇

Such missed deadlines cannot happen if the utilization of the processor is very low and, obviously, they can happen when the utilization is high, as in Fig. 6.14. If the condition of equation (6.7) is met, the utilization is guaranteed to be low enough to prevent problems like that of Fig. 6.14. Equation (6.7) is a **sufficient** condition. This means: we might still find a schedule if the condition is not met. Other sufficient conditions exist [57].

RM scheduling has the following important advantages:

- We can show that it is an optimal fixed-priority preemptive scheduling algorithm for uniprocessor systems [57].
- It is based on **static** priorities, enabling its application in an operating system with fixed priorities.
- If the above six RM-assumptions (see p. 318) are met, all deadlines will be met (see Buttazzo [82]).

RM scheduling is also the basis for a number of formal proofs of schedulability. Designing examples and proofs are facilitated if the most problematic situations for scheduling are known. To get started, we assume the following property:

Property 6.1: We assume that every job completes before the next job of the same task is released.

Definition 6.13: A critical instant for a task τ_i is defined to be an instant t at which a release of that task will have the largest response time.

Theorem 6.4 **(Critical instant theorem):** *For fixed priority scheduling, the response time for execution on a uniprocessor system is maximized for each task τ_i if τ_i is released at the same time as all tasks having a higher priority.*

Proof: Here, we present the original proof by Liu and Layland [331], using the wording of these authors (except for making the notation consistent with ours): "Let $\tau = \{\tau_1, ..., \tau_n\}$ denote a set of priority-ordered tasks with τ_n being the task with the lowest priority. Consider a particular request for τ_n that occurs at t_1. Suppose that between t_1 and $t_1 + T_n$, the time at which the subsequent request of τ_n occurs, requests for task τ_i, $i < n$, occur at t_2, $t_2 + T_i$, $t_2 + 2T_i$, ..., $t_2 + kT_i$, as illustrated in Fig. 6.15. Clearly, the preemption of τ_n by τ_i will cause a certain amount of delay in the completion of the request for τ_n that occurred at t_1, unless the request for τ_n

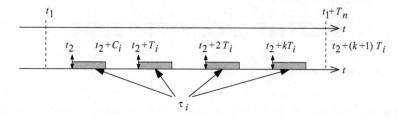

Fig. 6.15 Delaying task τ_n by some τ_i of higher priority

is completed before t_2. Moreover, from Fig. 6.15 we see immediately that advancing the request time t_2 will not speed up the completion of τ_n. The completion time is either unchanged or delayed by such an advancement. Consequently, the delay in the completion of τ_n is largest when t_2 coincides with t_1. Repeating the argument for all τ_i, $i = 2, ..., m - 1$, we prove the theorem". □

Implicitly, we have used Property 6.1 in the proof. If we consider the general case (i.e., the situation in which the assumption of Property 6.1 does not hold, see e.g., Baker [32]), Theorem 6.4 remains valid, but the proof becomes more complex, as shown by Devillers et al. [124] and Bril [71][5].

The critical instant theorem is of great help when scheduling uniprocessor systems. In general, the critical instant theorem does not hold for multiprocessor systems, which makes proofs much harder. So, the validity of this theorem should really be appreciated!

Let us look at other properties of RM scheduling now. The idle time or *spare capacity* of the processor is not always required.

---

[5]I owe this hint to J.J. Chen of TU Dortmund.

Theorem 6.5: *Let τ be a system of periodic tasks. If the period of all tasks is a multiple of the period of the task having the next higher priority, τ can be scheduled with RM-scheduling if*

$$U_{sum} \leq 1 \qquad\qquad (6.9)$$

Example 6.8: This requirement is met if tasks in a TV set must be executed at rates of 25, 50, and 100 Hz (or 30, 60 and 120 Hz). $\qquad\qquad\qquad\qquad\qquad \nabla$

Proof (of theorem 6.5): Let tasks be sorted by priorities, such that $\forall i : T_i \leq T_{i+1}$. Consider some task τ_i and the task with the next lower priority, task τ_{i+1} (see Fig. 6.16). Note that the second deadline of τ_{i+1} matches the fourth deadline of

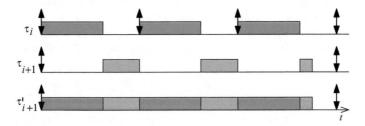

Fig. 6.16 Folding of tasks of adjacent priorities

τ_i neatly. Therefore, we can fold the execution times of task τ_{i+1} into the execution times of τ_i and create a new task τ'_{i+1}, containing the execution times of the two original tasks. This folding is feasible if the total execution time of the two tasks does not exceed the period of τ_{i+1}. The process can be repeated in the same way with the next lower priority task. Overall, folding is feasible as long as the overall utilization does not exceed 1. $\qquad\qquad\qquad\qquad\qquad\qquad\qquad\qquad\qquad\qquad\qquad \Box$

Equations (6.7) or (6.9) provide easy means to check conditions for schedulability.

Due to the critical instant theorem, the proof of optimality of RM scheduling needs to consider only the case in which tasks are released concurrently with all other tasks of higher priority.

6.2.3.3 Earliest Deadline First Scheduling

EDF can also be applied to periodic task sets.

The hyper period for the example of Fig. 6.14 is 40. Obviously, it is sufficient to solve the scheduling problem for a single hyper period. This schedule can then be repeated for the other hyper periods. It follows from the optimality of EDF for non-periodic schedules that EDF is also optimal for a single hyper period and therefore also for the entire scheduling problem. No additional constraints must be met to guarantee optimality. This implies that EDF is optimal also for the case of $U_{sum} = 1$.

Example 6.9: No deadline is missed if the example of Fig. 6.14 is scheduled with EDF (see Fig. 6.17). At time 5, the behavior is different from that of RM scheduling: due to the earlier deadline of τ_2, it is not preempted.

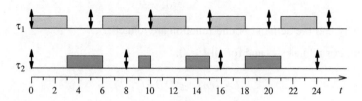

Fig. 6.17 EDF generated schedule for the example of 6.14

▽

6.2.3.4 Explicit-Deadline Tasks

Now, we move toward the consideration of tasks whose deadline is not the same as the period. Such tasks are called **explicit-deadline tasks**. Each task τ_i in such a system is characterized a triple (C_i, D_i, T_i), where D_i is the relative deadline with $D_i \leq T_i$. The case $D_i \leq T_i$ is called the **constrained deadline** case. The **arbitrary deadline** case is characterized by the absence of such constraint. Obviously, the class of explicit-deadline tasks is more general than the class of implicit-deadline tasks and each implicit-deadline task is also an explicit-deadline task.

Utilization is of limited value for the characterization of computational demands of explicit-deadline tasks. To some extent, density plays the role which utilization played to far. Density is defined as

$$dens_i = \frac{C_i}{\min(D_i, T_i)} \tag{6.10}$$

$$dens_{sum}(\tau) = \sum_{\tau_i \in \tau} dens_i \tag{6.11}$$

$$dens_{max}(\tau) = \max_{\tau_i \in \tau}(dens_i) \tag{6.12}$$

Density values characterize computational requirements. A tighter bound is provided by the so-called demand-bound function (DBF):

Definition 6.14: *For any sporadic task τ_i and any real number $t \geq 0$, the **demand bound function** $DBF(\tau_i, t)$ is the largest cumulative execution requirement of all jobs that can be generated by τ_i to have both their release times and their deadlines within a contiguous interval of length t.*

The overall execution requirements of task τ_i over an interval $[t_0, t_0 + t)$ are maximized if one of its jobs arrives at the start of the interval—i.e., at time instant t_0—and its subsequent jobs arrive as rapidly as permitted, i.e., at instants $t_0 + T_i, t_0 + 2T_i, t_0 + 3T_i, \ldots$ This observation leads to Eq. (6.13) [40, 42]:

$$DBF(\tau_i, t) = \max\left(0, \left(\left\lfloor \frac{t - D_i}{T_i} \right\rfloor + 1\right) * C_i\right) \qquad (6.13)$$

Density and the demand bound function are related:

Lemma: For all tasks τ_i and for all $t \geq 0$:

$$t * dens_i \geq DBF(\tau_i, t) \qquad (6.14)$$

Proof: Let us compare the graphs depicting density and DBF as a function of time. Figure 6.18 shows both functions.

Fig. 6.18 Comparison of density and DBF

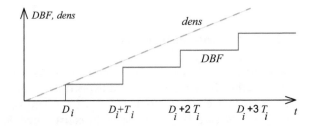

The left-hand side of Eq. (6.14) is visualized as the straight line with slope $dens_i$. DBF is a step function with steps of height C_i. Whenever a task must be executed, the step function increases by C_i. The first step is at $t = D_i$. By definition of the density, this step does not exceed the straight line. The next steps will be at $t = D_i + T_i$, $t = D_i + 2T_i$, $t = D_i + 3T_i$, and so on, since these are the intervals of time after which the demand increases by C_i. Again, these steps will not exceed the straight line. $\qquad\square$

EDF can be easily extended to handle the case when deadlines are different from the periods. For RM scheduling, the extension is called Deadline Monotonic Scheduling.

6.2.3.5 Deadline Monotonic Scheduling

Explicit-deadline tasks can be dealt with in **deadline monotonic (DM) scheduling**. For DM scheduling, static task priorities are based on non-increasing deadlines: for any two tasks τ_i and $\tau_{i'}$, the priority of τ_i will be higher than that of $\tau_{i'}$ if $D_i < D_{i'}$.

For constrained-deadline tasks, Eq. (6.7) can be generalized into Eq. (6.15) which is sufficient, but not necessary [82].

$$\sum_{i=1}^{n} \frac{C_i}{D_i} \leq n(2^{1/n} - 1) \qquad (6.15)$$

6.2.4 *Periodic Scheduling with Precedence Constraints*

Scheduling-dependent tasks are more difficult than scheduling-independent tasks, in particular in the non-preemptive case ($(1|r_i, prec, periodic|L_{max})$ in the triplet notation). The problem of deciding whether or not a non-preemptive schedule exists for a given set of dependent tasks and a given deadline is NP-complete [172]. In order to reduce the scheduling effort, different strategies are used:

- adding additional resources such that scheduling becomes easier, and
- partitioning of scheduling into static and dynamic parts. With this approach, as many decisions as possible are taken at design time and only a minimum of decisions is left for run-time.

6.2.5 *Sporadic Events*

In the case of sporadic events, we could connect sporadic events to interrupts and execute them immediately if their interrupt priority is the highest in the system. However, quite unpredictable timing behavior would result for all the other tasks. Therefore, special **sporadic task servers** are used which execute at regular intervals and check for ready sporadic tasks. This way, sporadic tasks are essentially turned into periodic tasks, thereby improving the predictability of the whole system.

6.3 Scheduling for Independent Jobs on Identical Multiprocessors

Next, we are going to consider multiprocessors, due to their widespread use in the form of multi-cores in contemporary embedded systems. A large number of issues has to be considered during the transition from uniprocessors to multiprocessors. Initially, we assume having m identical processors (or "cores"). Furthermore, we assume dealing with a task system $\tau = \{\tau_1, ..., \tau_n\}$ where each task i is characterized by its worst case execution time (WCET) C_i and—in case of periodic or sporadic tasks—its period T_i which is considered to also define the deadline unless otherwise noted. Whenever the periodic or sporadic nature of tasks is not relevant, we may also consider a set of jobs with explicit deadlines d_i instead.

For multiprocessors, it is not sufficient to decide when to execute tasks or their jobs. Rather, we must decide when to execute jobs and **where** to execute them. Thus, a one-dimensional problem becomes a two-dimensional problem.

For m identical processors, obvious necessary conditions for schedulability are:

$$\forall i : u_i \leq 1 \tag{6.16}$$

$$U_{sum} \leq m \tag{6.17}$$

6.3.1 Partitioned Scheduling of Implicit-Deadline Tasks

Our presentation in the next sections is based predominantly on a book written by
Baruah et al. [40] and complemented by material from other sources like a survey
paper by Davis et al. [116] and slides by I. Puaut [436, 437]. Baruah et al. focus on
sporadic task systems. This is partly motivated by the fact that for such systems—
in contrast to periodic task systems—no global time synchronization is required
for releasing jobs. Rather, it is sufficient to maintain a time base which ensures
that the minimum intervals T_i are kept. Also, sporadic tasks systems are considered
for complexity reasons. We start by considering sporadic implicit-deadline tasks
on identical multiprocessors. In the triplet notation, this corresponds to the case
$(Pm|D_i = T_i, sporadic|..)$.

Furthermore, we are initially restricting ourselves to the case of partitioned
scheduling. This means that each task is allocated to a particular processor. Task
migration is not allowed. Partitioned scheduling for synchronous arrival times can
be done by bin packing, defined in a notation adjusted for real-time scheduling as
follows:

Definition 6.15 (Souza [479], Chap. 10): Let $\tau = \{1, ..., n\}$ be at set of items,
where each item $i \in \tau$ has a size $c_i \in (0, 1]$. Let $\pi = \{1, ...m\}$ be a set of bins with
capacity one. The problem of finding an assignment $a : \tau \to \pi$ such that the number
of non-empty bins $m \leq n$ is minimal and such that allocated sizes do not exceed the
bin capacity is called the **bin packing problem**.

Bin packing is known to be NP-hard [172]. Hence, optimal algorithms such as
the one proposed by Korf [294] need large run-times. Formalization of the scheduling problem as a bin packing problem aims at the minimization of the number of
processors m.

For a given number m of processors, it is more appropriate to model scheduling for
synchronous arrival times as a Knapsack problem, more precisely as a 0/1 Multiple-
Knapsack problem. This problem can be defined as follows, again using a notation
adjusted for real-time scheduling:

Definition 6.16 (Martello [351]): Let $\tau = \{1, ..., n\}$ be a set of n items, each with
a size c_i and a benefit b_i. Let π be a set of m knapsacks, each with a capacity κ_k,
with $(m \leq n)$. Suppose that we can partially allocate a subset of items to knapsacks
such that

$$a : \tau \to \pi \text{ such that } \forall k : \sum_{i, a:i \to k} c_i \leq \kappa_k. \tag{6.18}$$

The problem of selecting disjoint subsets of items so that the total profit $\sum_i b_i$ for
items in knapsacks is maximized is called **the 0/1 Multiple Knapsack Problem
(MKP)**.

Given an algorithm for the 0/1 Multiple-Knapsack problem, we can allocate jobs to m processors. For identical processors, capacities would all be equal. For uniform processors, we can use capacities to take processor speeds into account. The 0/1 Multiple-Knapsack problem is NP-hard as well. Note that we would possibly not schedule all tasks.

Due to the complexity of scheduling for synchronous arrival times, there is no hope for efficient optimal algorithms for the general problem and in practice, heuristics are used. Common heuristics are considering tasks and processors in a certain sequence. Heuristics differ by the sequence they use. Lopez et al. [338] have compared several heuristics. They restrict themselves to so-called reasonable allocation algorithms, defined as follows:

Definition 6.17: A **reasonable allocation (RA) algorithm** is defined as one that fails to allocate a task to a multiprocessor platform only when the task does not fit into any processor upon the platform.

Definition 6.18: A **reasonable allocation decreasing (RAD) algorithm** is defined as an RA algorithm considering tasks in a non-increasing order of utilization.

The algorithms studied by Lopez et al. are obtained by combining all possible combinations of two characteristics:

1. The order in which tasks are considered: tasks can be considered in decreasing order of utilization (denoted by **D**), increasing order of utilization (denoted by **I**), and in arbitrary order (denoted by an empty character).
2. The search strategy for processor allocation. We consider processors to be ordered in some way. Then, the **first fit** strategy (**FF**) will allocate the first processor on which it fits. The worst fit strategy (**WF**) will allocate the processor with the largest remaining capacity. The best fit strategy (**BF**) will allocate the processor with the minimum remaining capacity on which it fits.

There are a total of nine combinations. All combinations can be implemented efficiently. For example, algorithm **FFD** can be detailed as follows:

```
Sort task set according to non-increasing utilizations uᵢ = Cᵢ / Tᵢ;
/* Assume task set is renumbered according to the sorting;*/
for (mt=0; mt ≤ m; mt++) K[mt] =1;              /* initialize capacity */
for (i=1; i≤n; i++) {                            /* for each task */
   for (mt=1; (uᵢ >K[mt]) and (mt≤m); mt++);  /* sufficient capacity? */
      if (mt > m) mt=0 ;                        /* no solution, use index 0 */
   a[i]=mt;                        /* return processor allocation in array */
   K[mt]=K[mt]-uᵢ;                              /* update remaining capacity */
}
```

The heuristic algorithm is certainly not optimal. There may be the question: How far are we off the optimum? Many publications discuss upper bounds on the number of additional processors needed, if compared to the minimum number of processors needed for optimal bin packing. The paper by Dosa [131] is an example of this. For real-time systems, a different question is relevant: For a given number of processors, is there any bound on the overall utilization up to which schedulability is guaranteed? One utilization bound was proved by Lopez et al. [338]:

Theorem 6.6: *Any reasonable allocation algorithm has a utilization bound no smaller than*

$$U_{B1}(U_{max}) = m - (m - 1)U_{max} \tag{6.19}$$

Proof: When a task with utilization u_i cannot be allocated, every processor must have tasks allocated to it with a per processor utilization exceeding $(1 - u_i)$. The overall utilization over all allocated tasks and including τ_i must then exceed

$$m(1 - u_i) + u_i = m - (m - 1)u_i \tag{6.20}$$
$$\geq m - (m - 1)U_{max} \tag{6.21}$$

This condition must be met for allocation not to be feasible.   □

Furthermore, define β as

$$\beta = \left\lfloor \frac{1}{U_{max}} \right\rfloor \tag{6.22}$$

β is a lower bound on the number of tasks of our task set which we can run on a single processor. Let us assume that EDF is used for local scheduling on each processor. Lopez et al. also showed the following theorem:

Theorem 6.7: *No allocation algorithm can have a utilization bound larger than*

$$U_{B2}(\beta) = \frac{\beta m + 1}{\beta + 1} \tag{6.23}$$

Proof: See Lopez et al. [338].

Lopez et al. also proved that **WF** and **WFI** have Eq. (6.19) as their lower bound, and the remaining algorithms have Eq. (6.23) as their lower bound. Whenever U_{max} approaches 1, the bound in Eq. (6.19) also approaches 1:

$$U_{B1}(1) = 1 \tag{6.24}$$

When U_{max} gets close to 1, β becomes 1 and U_{B2} becomes:

$$U_{B2}(1) = \frac{m + 1}{2} \tag{6.25}$$

The bound in Eq. (6.25) allows us to use multiple processors in a much more efficient way compared to the bound in Eq. (6.24). Hence, with respect to these bounds, **WF** and **WFI** are inferior to the other seven algorithms. Experimentally, it has been shown that **FFD** seems to be superior to **FF** or **FFI** and **BFD** seems to be superior to **BF** and **BFI** [40]. There is also some theoretical evidence which supports this observation [40].

The sketched nine algorithms are relatively simple algorithms. We refrain from presenting more elaborate algorithms for the same problem since the problem considered is too much simplified to apply to realistic applications:

- The scheduling problem, as it has been addressed in this section, is a very much restricted one. There are no precedences, no preemption, and only identical processors.
- Partitioned scheduling may lead to unused processor resources even in situations where jobs are available. This means that partitioned scheduling is not work-conserving. Therefore, optimality is not guaranteed.

Hence, the information in this section provides fundamental knowledge, but practical applications require more sophisticated approaches, like the ones to be presented in the following sections.

6.3.2 Global Dynamic-Priority Scheduling for Implicit Deadlines

Having unused processors in the presence of available jobs can be avoided with global scheduling. For global scheduling, the allocation of processors to tasks or jobs is dynamic. This gives us more flexibility, especially in the presence of changing workloads or changing processor availabilities. In the absence of execution constraints, upper bounds on the utilization like the ones in Eqs. (6.19) and (6.23) are replaced by

$$U_{sum} \leq m \tag{6.26}$$

However, this better utilization bound and flexibility comes at the price of a certain overhead for scheduling decisions, preemptions, and job migrations.

The key idea of proportional fair (pfair) scheduling [41] is to execute each task at a rate corresponding to its utilization[6]. For example, if $u_i = 0.5$ for a set of tasks, then each task should be executed approximately half of the time, regardless of the number of processors. For pfair scheduling, we assume that time is quantized and enumerated with integers. Also, C_i and T_i parameters are assumed to be represented by integers.

---

[6]The presentation of pfair scheduling is based on slides by I. Puaut [437].

Definition 6.19: The **lag** of a task τ_i at time t with respect to schedule $\mathcal{S}$, denoted as $lag(\mathcal{S}, \tau_i, t)$, is the difference between the number of slots that a task has received and the number of slots that it should have received:

$$lag(\mathcal{S}, \tau_i, t) = u_i * t - \sum_{u=0}^{t-1} alloc(\mathcal{S}, \tau_i, u) \qquad (6.27)$$

The first term is the target execution time of task τ_i, and the second is the time during which this task has been executed in schedule $\mathcal{S}$. A schedule is said to be a pfair schedule if the lag remains in the interval $(-1, +1)$.

Example 6.10: Figure 6.19 shows the function of actually executed time as a function of real time. The amount of executed time should not reach the two dashed lines.

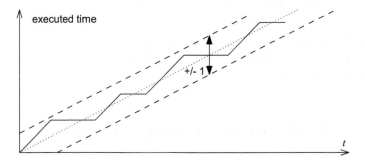

Fig. 6.19 Execution time as a function of real time

∇

For pfair scheduling, we divide each task τ_i into subtasks $\tau_i^{\,j}$, where j enumerates the execution intervals. For each subtask, we define a pseudo release time and a pseudo deadline:

$$r(\tau_i^{\,j}) = \left\lfloor \frac{j-1}{u_i} \right\rfloor \qquad (6.28)$$

$$d(\tau_i^{\,j}) = \left\lceil \frac{j}{u_i} \right\rceil \qquad (6.29)$$

Example 6.11: Consider a task τ_i with $C_i = 8$, $T_i = 11$. Possible intervals for the number of allocated execution slots for each j are shown in Fig. 6.20.

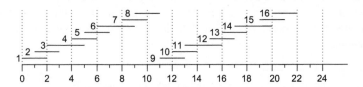

Fig. 6.20 Intervals for allocated execution time

For example:

$$r(\tau_i^6) = \left\lfloor \frac{6-1}{8/11} \right\rfloor = \left\lfloor \frac{55}{8} \right\rfloor = 6$$

$$d(\tau_i^6) = \left\lceil \frac{6}{8/11} \right\rceil = \left\lceil \frac{66}{8} \right\rceil = 9$$

Hence, the sixth subtask of task τ_i must be executed in time interval (6:9). ▽

One particular approach for allocation of a correct number of execution slots is presented in the book by Baruah et al. [40]. In general, there are variations of this scheme: we can apply EDF to pseudo deadlines or we can modify EDF by defining rules which are applied in case of ties. It is feasible to obtain schedulability for full processor utilization, i.e., for $U_{sum} \leq m$.

Pfair scheduling potentially suffers from a large number of migrations between processors. Also, due to the integer (over-)approximation of execution times, it is not work conserving. Variants have been proposed which reduce the overhead for job migrations. Also, the overall complexity can be reduced with some variants.

Pfair scheduling finds many applications in operating systems, for example for scheduling virtual machines.

6.3.3 Global Fixed-Job-Priority Scheduling for Implicit Deadlines

6.3.3.1 G-EDF Scheduling

We can also try to solve the two-dimensional problem with extensions of uniprocessor scheduling algorithms. For example, we could use Global EDF (G-EDF). G-EDF, just like EDF, defines job priorities based on the closeness of the next deadlines. If m processors are available, those m jobs having the highest priorities among all available jobs are executed. Obviously, such **priorities are job-dependent**, and not just task-dependent. In a global scheduling strategy, we would like to keep preemptions and task migrations to a minimum. For G-EDF, these numbers depend on how we allocate tasks/jobs to a particular processor [181].

Lemma 6.1: *G-EDF is not optimal.*

Proof: The proof is by counter-example, adopted from Cho et al. [102]. Suppose $m = 2$ and $C_1 = 3$, $D_1 = 4$, $C_2 = 2$, $D_2 = 3$, $C_3 = 2$, and $D_3 = 3$. As shown in Fig. 6.21 (**left**), G-EDF schedules J_2 and J_3 first, due to their earlier deadline. J_1 misses its deadline. However, a schedule is feasible, as shown in Fig. 6.21 (**right**).

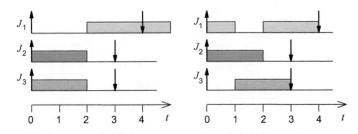

Fig. 6.21 Left: G-EDF violates deadline at $t = 4$; **right**: feasible schedule   ☐

Obviously, the problem for G-EDF results from not being able to use the second processor for $t > 2$.

In general, G-EDF may suffer anomalies like the so-called **Dhall effect** [125]: periodic task sets for which one task has a utilization close to one cannot be scheduled with G-EDF.

Example 6.12: To demonstrate the effect, let us consider the case of $n = m + 1$ and

$$\forall i \in [1..m] : T_i = 1, C_i = 2\varepsilon, u_i = 2\varepsilon \tag{6.30}$$

$$T_{m+1} = 1 + \epsilon, C_{m+1} = 1, u_{m+1} = \tfrac{1}{1+\varepsilon} \tag{6.31}$$

A corresponding schedule is shown in Fig. 6.22. Initially, only tasks $\tau_1, ..\tau_m$ are executed. The execution of task τ_{m+1} starts only after the first m tasks have completed

Fig. 6.22 Dhall effect

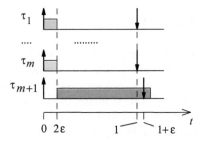

their execution and it will miss its deadline. The presence of a single task τ_{m+1} with a high utilization is sufficient to cause a deadline miss at $t = 1 + \varepsilon$. This happens even though the utilization of the other tasks is very small. In fact, the utilization of tasks $\tau_1, ..\tau_m$ can be arbitrarily small and we will still have a deadline miss.   ▽

This motivates using variants of algorithms which assign high priorities to tasks with a high utilization, regardless of their deadline or period.

Algorithm fpEDF is such an algorithm. We assume that we are given an implicit-deadline sporadic task system $\tau = \{\tau_1, ...\tau_n\}$ and that tasks are ordered by non-increasing utilizations u_i. Our goal is to schedule these tasks on m identical processors while avoiding the Dhall effect. Algorithm fpEDF works as follows [40]:

```
for  (i=1; i ≤ m − 1; i++){
  if (u_i >0.5) τ_i's jobs obtain highest priority (ties broken arbitrarily)
  else break;
}                        /* Remaining jobs get priorities according to EDF. */
```

This means that the $m - 1$ tasks of largest utilization will obtain the highest priority if their utilization exceeds a value of 0.5.

Theorem 6.8: *Algorithm $fpEDF$ has a utilization bound no smaller than $\frac{m+1}{2}$.*

This is the best bound which we can expect unless some additional information is known, as is evident from the following theorem.

Theorem 6.9: *No m-processor fixed-job-scheduling algorithm has a schedulable utilization greater than $\frac{m+1}{2}$.*

The proofs of both theorems can be found in [40]. As in the case of partitioned scheduling, stronger bounds are feasible if the largest utilization is known.

A similar idea is used in scheduling algorithm $EDF(k)$: for $EDF(k)$, k tasks of highest utilization obtain the highest priority, breaking ties arbitrarily. All other tasks are scheduled according to EDF.

Theorem 6.10: *Let τ be an implicit-deadline sporadic task system. $EDF(k)$ will schedule τ on m unit-speed processors, where*

$$m = (k - 1) + \left\lceil \frac{U(\tau^{(k+1)})}{1 - u_k} \right\rceil \tag{6.32}$$

and $U(\tau^{(k+1)})$ is the utilization for the task set with the first k tasks removed.

The proof of this theorem can again be found in [40].

6.3.3.2 EDZL Scheduling

Obviously, G-EDF can miss deadlines for task sets that are schedulable. We can improve G-EDF by adding a consideration of laxity: the EDZL algorithm applies G-EDF as long as the laxity of jobs is greater than zero (see [40], Chap. 20). However, whenever the laxity of a job becomes zero, the job gets the highest priority among all jobs, even including currently executing jobs.

Example 6.13: Consider the example in Fig. 6.23, adopted from Puaut [436].

In this example, parameters are as follows: $n = 3, m = 2, T_1 = T_2 = T_3 = 3, C_1 = C_2 = C_3 = 2$. For this example G-EDF misses the deadlines for τ_3 at times $t = 3n$ for $n = 1, 2, 3..,$ as can be seen in Fig. 6.23 (**left**). However, EDZL keeps the deadlines as can be seen in Fig. 6.23 (**right**). The detailed behavior depends somewhat on the processor allocation used by EDZL.

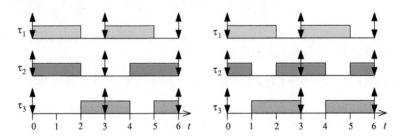

Fig. 6.23 G-EDF: **left**: missed deadlines; **right**: ZL improvement

∇

EDZL is strictly superior to EDF, as shown by Choi et al. [456]. Informally, this can be shown as follows[7]: Suppose that S is a schedule from EDF, and S' is a schedule from EDZL for the same input task set. If a job at time t is scheduled in EDZL but not in EDF, then the job misses the deadline in EDF but not in EDZL. If both schedule the job, then the schedule remains the same. That is, the first moment when S differs from S' has the following results:

- either EDZL remains feasible, but EDF becomes infeasible
- or both EDZL and EDF are infeasible.

Therefore, EDZL is superior to EDF.

Piao et al. [429] proved the following utilization bound for EDZL

$$U_{sum} \leq \frac{m + 1}{2} \tag{6.33}$$

6.3.4   Global Fixed-Task-Priority Scheduling for Implicit Deadlines

6.3.4.1   Global Rate Monotonic Scheduling

In a similar way, we can extend rate monotonic scheduling to global rate monotonic scheduling (G-RM). For G-RM, there is an anomaly concerning relaxed schedules:

---

[7]I owe this informal explanation to J.J. Chen, TU Dortmund.

Lemma 6.2: *For G-RM, there may be situations in which schedules exist for a certain task system, but deadlines are violated if periods are extended.*

Proof: We prove the existence of such situations by means of an example, adopted from Puaut [436]. Consider the case $m = 2, n = 3, T_1 = 3, C_1 = 2, T_2 = 4, C_2 = 2, T_3 = 12, C_3 = 7$. Figure 6.24 shows a schedule generated by G-RM.

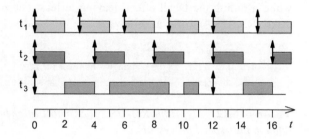

Fig. 6.24 Schedule generated by G-RM

If we extend the period of τ_1 to $T_1 = 4$, τ_3 will miss its deadline (see Fig. 6.25). This counter-intuitive result makes the design of proofs and examples much more complex, compared to the uniprocessor case. □

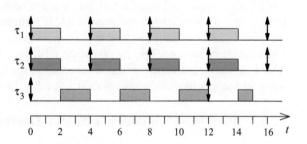

Fig. 6.25 Schedule with a missed deadline at $t = 12$ generated by G-RM

The critical instant theorem for uniprocessors (see p. 316) is also not valid for multi-core systems.

The following utilization bound has been shown for G-RM[8]:

Theorem 6.11: *Any implicit-deadline periodic or sporadic task system τ satisfying*

$$U_{sum} \leq \frac{m}{2}(1 - U_{max}(\tau)) + U_{max}(\tau) \tag{6.34}$$

is successfully scheduled by G-RM on m unit-speed processors [53].

---

[8] A tighter bound has been shown by Chen et al. [95].

G-RM also suffers from the Dhall effect: Note that U_{sum} in equation (6.34) approaches zero as U_{max} approaches one. Also, like G-EDF, the algorithm cannot fully exploit the presence of multiple processors.

Therefore, algorithm RM-US(ξ) with threshold ξ has been proposed, where US stands for utilization threshold. Given an implicit-deadline sporadic task system $\tau = \{\tau_1, \ldots \tau_n\}$ and tasks ordered by nonincreasing utilizations u_i, the goal is to schedule these up to $(m - 1)$ high utilization tasks on $m - 1$ identical processors while avoiding the Dhall effect, leaving at least one processor for the remaining tasks.. RM-US(ξ) works as follows:

```
for (i=1; i≤ m − 1; i++) {
  if (u_i > ξ) τ_i is assigned highest priority
  else break; }    /* remaining tasks are allocated according to G-RM */
```

Theorem 6.12: *Algorithm RM-US(ξ) has a utilization bound no smaller than $\frac{m^2}{(3m-2)}$ upon m unit-speed processors.*

The proof was published by Andersson et al. [14]. For $3m \gg 2$, this bound approaches $\frac{m}{3}$. A tighter bound was shown by Chen et al. [95].

6.3.4.2 RMZL Scheduling

G-RM might miss deadlines for task sets that are schedulable and we can consider improvements. One such improvement is RMZL scheduling. For RMZL scheduling, we use (G-) RM scheduling as long as the current laxity is larger than zero. However, when the laxity becomes zero for one of the jobs, we raise its priority to the highest. RMZL scheduling is superior to RM scheduling, since schedules are changed only when RM scheduling could have missed a deadline [40].

6.3.4.3 Partitioned Scheduling for Explicit Deadlines

Partitioned scheduling for explicit-deadline task systems can be done similar to partitioned scheduling for implicit-deadline task systems by replacing sorting by utilization with sorting by density. However, this approach is not recommended, since density can be unbounded in certain cases. Baruah et al. present a better approach for partitioned scheduling [40].

6.4 Dependent Jobs on Homogeneous Multiprocessors

In general, results presented in the previous section constitute fundamental basic knowledge, but the restriction to independent tasks and identical processors inhibits their application for many design problems. Next, we will be dropping these restrictions. First of all, we will be dropping the restriction to independent tasks. We will be focusing on some simple algorithms which have been used in the design automation community. For example, as-soon-as-possible (ASAP), as-late-as-possible (ALAP), list (LS), and force-directed scheduling (FDS) are very popular for automated synthesis from algorithmic design descriptions, so-called high-level synthesis (HLS) (see Coussy [112] for an overview).

6.4.1 As-Soon-As-Possible Scheduling

Considering precedence constraints, as-soon-as-possible (ASAP) scheduling tries to schedule each task as soon as feasible. ASAP scheduling, as used in HLS considers a mapping of tasks to integer start times ≥ 0. Preemptions are not allowed. Allocation to specific processors has to be performed after ASAP scheduling has computed start times. Therefore, ASAP scheduling provides a mapping only to start times: $S : \tau \rightarrow \mathbb{N}_0$.

For this algorithm, we assume that the execution times of all tasks are known and that they are independent of the processor executing the tasks. Hence, we are assuming that processors are homogeneous. The algorithm does not consider any constraints on the number of processors and assumes that the number of processors needed for the resulting schedule is available. The ASAP algorithm works as follows:

```
for (t=0; there are unscheduled tasks; t++)   {
    τ'={all tasks for which all predecessors finished};
    set start time of all tasks in τ' to t;      }
```

Example 6.14: Let us assume that the task graph of Fig. 6.26 (**left**) is given. Each node labeled i represents a task τ_i. Furthermore, let us assume that execution times correspond to those listed in Fig. 6.26 (**right**).

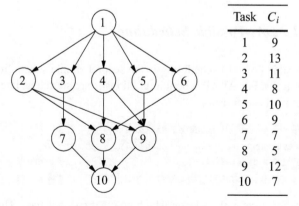

Task	C_i
1	9
2	13
3	11
4	8
5	10
6	9
7	7
8	5
9	12
10	7

Fig. 6.26 **Left**: task graph; **right**: execution times of tasks

Then, ASAP scheduling will generate the schedule shown in Fig. 6.27.

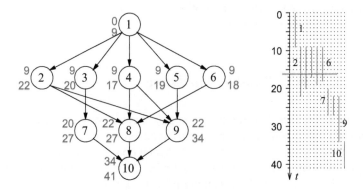

Fig. 6.27 **Left**: ASAP scheduled task graph; **right**: time line

Numbers in blue denote start times, and numbers in green denote finish times. Tasks τ_2 to τ_6 all start immediately after task τ_1 has finished, since they do not depend on any other task. Also, tasks τ_7 to τ_9 start as soon as the last of their predecessors has finished and the same holds for task τ_{10}. The red line in Fig. 6.27 (**right**) shows that a maximum of five processors is needed, since ASAP scheduling does not consider any constraints on the number of processors. ∇

ASAP scheduling minimizes the makespan, since all tasks are scheduled as early as possible. The presented algorithm could be extended to also cover real numbers as execution times. We may consider ASAP scheduling to be of linear complexity, provided that we use a clever technique for computing τ'. The algorithm can also be applied to personal life, corresponding to a situation where each person is eager to perform available work as early as possible.

6.4.2 As-Late-As-Possible Scheduling

As-late-as-possible (ALAP) scheduling is the second simple scheduling algorithm for dependent tasks. For ALAP scheduling, all tasks are started as late as possible. The algorithm works as follows:

```
for (t=0; there are unscheduled tasks; t--)                                        {
      τ'={all tasks on which no unscheduled task depends};
      set start time of all tasks in τ' to (t - their execution time); }
Shift all times such that the first tasks start at t=0.
```

The algorithm starts with tasks on which no other task depends. These tasks are assumed to finish at time 0. Their start time is then computed from their execution time. The loop then iterates backwards over time steps. Whenever we reach a time step, at which a task should finish the latest, its start time is computed and the task is scheduled. After finishing the loop, all times are shifted toward positive times such that the first task starts at time 0. We could also consider ALAP scheduling as a case of ASAP scheduling starting at the "other" end of the graph.

Example 6.15: For the task graph in Fig. 6.26, ALAP scheduling would generate the result shown in Fig. 6.28. The color coding is the same as for the ASAP example.

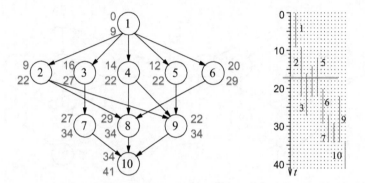

Fig. 6.28 **Left**: ALAP scheduled task graph; **right**: time line

Note that each task finishes as late as possible. In particular, tasks τ_7 to τ_9 finish only at time 34. Tasks τ_4 to τ_6 finish later than for the ASAP schedule. Tasks τ_1, τ_2, τ_9, and τ_{10} are scheduled as in the ASAP schedule, since these tasks determine the makespan. Tasks which determine the makespan are said to be on the **critical path**. Four processors are needed, as indicated by the red line. This reduction of the number of processors needed happens by coincidence. It is not a general property of ALAP scheduling. ∇

This scheduling strategy can also be applied to personal life. It corresponds to a situation where each person (is lazy and) finishes tasks as late as possible.

6.4.3 List Scheduling

With list scheduling (LS), we try to maintain the low complexity of ASAP and ALAP scheduling while making the algorithm aware of available processors. Processors may be of different types, but we do still assume that there is a one-to-one mapping between tasks and processor types. Hence, processors may be heterogeneous, but the crucial mapping from tasks to processor types is not generated by list scheduling.

We assume that we have a set L of processor types. List scheduling respects upper bounds B_l on the number of processors for each type $l \in L$.

List scheduling requires the availability of a priority function reflecting the urgency of scheduling a certain task τ_i. The following urgency metrics are in use [505]:

- **Mobility**: mobility is defined as the difference between the start times for the ASAP and ALAP schedule. Figure 6.29 (**left**) shows the mobility for our running example in red. Obviously, scheduling is *urgent* for the four tasks on the critical path for which mobility is zero.

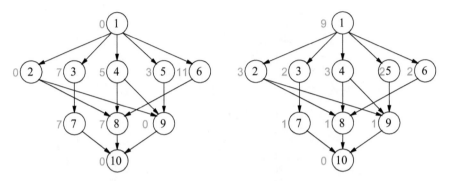

Fig. 6.29 Running example: **left**: mobility; **right**: number of successors

- **The number of nodes below task** τ_i in the tree (see Fig. 6.29 (**right**)).
- **Path length**: the path length for a task τ_i is defined as the length of the path from starting at τ_i to finishing the entire graph G. Path length is typically weighted by the execution time associated with the nodes, assuming that this information is known. In Fig. 6.30 (**left**), path length has been added.

List scheduling requires the knowledge of the task graph $G = (\tau, E)$ to be scheduled, a mapping from each node of the graph to the corresponding resource type $l \in L$, an upper bound B_l for each l, a priority function (as just explained) and the execution time for each task $\tau_i \in \tau$. List scheduling then fits nodes of maximum priority into each of the time steps such that the constraints are not violated [505]:

```
for (t=0; there are unscheduled tasks; t++)           /* loop over times */
    for (l ∈ L) {                              /* loop over resource types */
        τ*_{t,l} = set of tasks of type l still executing at time t;
        τ**_{t,l} = set of tasks of type l ready to start execution at time t;
        Compute set τ'_t ⊆ τ**_{t,l} of maximum priority such that
        |τ'_t| + |τ*_{t,l}| ≤ B_l .
        Set start times of all τ_i ∈ τ'_t to t: s_i = t;      }
```

Example 6.16: Figure 6.30 shows the result of list scheduling as applied to our example in Fig. 6.26, using path length as priority. We assume that all processors are of

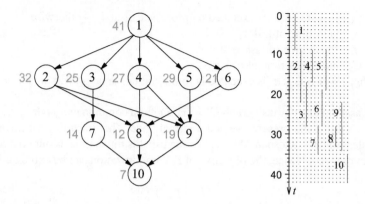

Fig. 6.30 **Left**: task graph with path length; **right**: time line for path length based list scheduling

the same type and that we allow no more than three processors ($B_1 = 3$). At time 9, tasks τ_2, τ_4, and τ_5 have the longest path length and hence the highest priority. τ_4 finishes at time 17, and τ_3 as well as τ_6 have the longest path length of the remaining tasks. We assume that we schedule τ_3. At time 19, τ_5 finishes and τ_6 can be started. At time 28, τ_3 and τ_6 finish, freeing processors for τ_7 and τ_8. τ_7 finishes at time 35, enabling dependent task τ_{10} to start and to finish at time 42, only slightly later than in the ASAP and ALAP schedules, despite using only three processors.   ∇

LS—such as ASAP and ALAP scheduling—does not allocate tasks to processors, but there is also no need for doing this for the restricted resource model. LS can also be extended to real numbers as execution times. The algorithm typically generates good results and is easy to adapt to various scenarios. These two features make LS a very popular scheduling algorithm for tasks with precedences.

Force-directed scheduling (FDS) is another heuristic scheduling algorithm for dependent tasks. FDS aims at an efficient use of processors. It tries to balance the number of processors that may be needed at any particular time. For the details, see Paulin et al. [426].

6.4.4 Optimal Scheduling with Integer Linear Programming

Next, we will be describing an approach for mapping tasks to multiple processors for which decisions are taken on a more global view of the design problem. It is based on integer linear programming (ILP) (see Appendix A). In this way, constraints and optimization goals are made explicit. We are adopting material from a publication of Coscun et al. [111] in our presentation.

ILP models consist of a linear cost function and a set of linear constraints. We will use the following variables in these two parts of the model:

$x_{i,k} := 1$ if task τ_i is executed on processor π_k and $=0$ otherwise
s_i : start time of task τ_i
f_i : finish time of task τ_i
C_i : execution time of task τ_i
$b_{i,j} := 1$ if task τ_i is executed before τ_j on the same processor, else $= 0$

Let us assume that our task graph $G = (\tau, E)$ has a common exit node τ_{exit}. If such node is not initially present, we add a virtual node. The finish time of this node is equivalent to the makespan MS_{max}. We can use this finish time as our cost function to be minimized. Hence, the objective of ILP minimization can be expressed as:

$$Min(f_{\tau_{exit}}) \tag{6.35}$$

Now, let us look at the constraints. First of all, we require that each task is executed on some processor:

$$\forall \tau_i \in \tau : \sum_{k \in \{1..m\}} x_{i,k} = 1 \tag{6.36}$$

Second, the different times are related by the following equations:

$$\forall \tau_i \in \tau : f_i = s_i + C_i \tag{6.37}$$

Third, in order to respect precedence relations, the following equations can be used:

$$\forall (\tau_i, \tau_j) \in E : s_j - f_i \geq 0 \tag{6.38}$$

Fourth, in a single core, we have to execute in a sequence as determined by variable $b_{i,j}$:

$$\forall (\tau_i, \tau_j) : f_i \leq s_j \text{ if } b_{i,j} = 1 \tag{6.39}$$

Furthermore, we have to take into account that each processor can execute only a single task at a time. This can be expressed in the following way:

$$\forall (\tau_i, \tau_j) : b_{i,j} + b_{j,i} = 1 \text{ if } \exists \pi_k : x_{i,k} = x_{j,k} = 1 \tag{6.40}$$

Equations (6.39) and (6.40) can be turned into the linear form required for ILP [111].

The resulting ILP model can be fed into some available ILP solver. ILP models, such as the one presented, have the advantage of precisely modeling the design problem and the objectives. They enable optimizations from a global viewpoint, using mathematical optimization techniques and stepping away from imperative programming.

The ILP problem is NP-hard. Therefore, run-times of ILP solvers can become large. However, there has been significant progress in the design of ILP solvers in recent years. Hence, moderately large problems can be solved in acceptable times. However, due to the complexity of ILP, these approaches do not scale to really large designs and run-times may be unacceptable. Nevertheless, these models can still be used for exact optimization of moderately large design problems and serve as a good starting point for heuristics for very large design problems.

6.5 Dependent Jobs on Heterogeneous Multiprocessors

6.5.1 Problem Description

Next to dropping the restriction to independent tasks, we would like to drop the restriction to homogeneous processors. We assume that the processing speeds of the different processors of our execution platform $\pi = \{\pi_1, ..., \pi_m\}$ are unrelated. According to Pinedo's triplet notation, we are now considering the case $(R_m | r_i, prec, ... | ...)$. This allows us to also model platforms comprising a mixture of execution units, including FPGAs and GPUs.

The theory of the resulting scheduling problems has not been studied comprehensively. As a result, Baruah et al. [40] state (in Chap. 22): *although unrelated multiprocessors are becoming increasingly more important in real-time systems implementation, the resulting scheduling theoretic study of such systems is, relatively speaking, still in its infancy.* Some first results are presented in the book by Baruah, but we resort to presenting methods published in the design automation community. These methods are able to handle realistic design tasks, sacrificing proofs of optimality.

6.5.2 Static Scheduling with Local Heuristics

In the following, we will describe the Heterogeneous-Earliest-Finish-Time (HEFT) and Critical-Path-on-a-Processor (CPOP) algorithms for static scheduling of tasks in a task graph $G = (\tau, E)$ onto a heterogeneous multiprocessor system $\pi = \{\pi_1, ..., \pi_m\}$ [521]. These two algorithms are standard examples of fast algorithms. In a way, they extend ASAP and ALAP scheduling for heterogeneous processors.

This is the notation we need:

- We assume that the task graph has a common entry node τ_{entry}. If such a common node is not initially present, we will add an artificial node having zero execution time and communication bandwidth requirements.
- We assume that the task graph has a common exit node τ_{exit}. If such a common node is not initially present, we will add an artificial node having zero execution time and communication bandwidth requirements.
- Matrix $C = (c_{i,k})$ denotes the execution time of task τ_i on processor π_k.

- Matrix $B = (b_{k,l})$ denotes the communication bandwidth for communication from processor π_k to processor π_l.
- Matrix $data = (data_{i,j})$ represents the amount of data which must be transmitted from task τ_i to task τ_j.
- Vector $\kappa = (\kappa_k)$ contains the communication startup costs on processor π_k.
- Matrix $H = (h_{i,j,k,l})$ describes the communication cost from task τ_i to task τ_j under the assumption that τ_i is mapped to processor π_k and task τ_j is mapped to processor π_l[9].
  In general, in our notation, we will be using index i for the source of precedences and index k for the processor it is allocated to. Index j denotes the sink of precedences and index l the processor it is allocated to.
- For a mapping to processors π_k and π_l, $h_{i,j,k,l}$ represents the communication cost from task τ_i to task τ_j:

$$h_{i,j,k,l} = \kappa_k + \frac{data_{i,j}}{b_{k,l}} \text{ if } k \neq l \tag{6.41}$$

$$= 0 \text{ if } k = l \tag{6.42}$$

- The average communication cost is defined as

$$\overline{h_{i,j}} = \overline{\kappa} + \frac{data_{i,j}}{\overline{B}} \tag{6.43}$$

where $\overline{\kappa}$ is the average communication startup time and $\overline{B}$ is the average communication bandwidth.

- Given a partial schedule, $s_e(\tau_i, \pi_k)$ is the earliest start time for task τ_i on processor π_k. Obviously, $s_e(\tau_{entry}, \pi_k)$ is zero, for any k.
  We define $f_e(\tau_i, \pi_k)$ as the earliest finishing time for task τ_i on heterogeneous processor π_k. $f_e(\tau_{entry}, \pi_k)$ is equal to $c_{entry,k}$.
  Once the decision to schedule task τ_i on processor π_k has been taken, the actual start time $s(\tau_i, \pi_k)$ and the actual finish time $f(\tau_i, \pi_k)$ can be computed.
  $s_e(\tau_j, \pi_l)$ and $f_e(\tau_j, \pi_l)$ can be computed from a partial schedule iteratively as follows:

$$s_e(\tau_j, \pi_l) = max \left\{ avail(l), max_{\tau_i \in pred(\tau_j)}(f(\tau_i) + h_{i,j,k,l}) \right\} \tag{6.44}$$

$$f_e(\tau_j, \pi_l) = c_{j,l} + s_e(\tau_j, \pi_l) \tag{6.45}$$

where $pred(\tau_j)$ is the set of immediate predecessor tasks of task τ_j, k is the processor task τ_i is mapped to in the partial schedule, and $avail(l)$ is the time that processor π_l completed the execution of its last task. The max expression in the inner term is the time when all data needed by τ_j has arrived at processor π_l.

- For HEFT and CPOP, we assume that the $makespan$ is to be minimized. The $makespan$ is computed from the actual finish time of the exit node:

---

[9]Indexes k and l are not explicit in the original paper.

$$makespan = f(\tau_{exit}) \tag{6.46}$$

- The average execution time $\overline{c_i}$ is the execution time $c_{i,k}$ averaged over all processors.
- The **upward rank** $rank_u(\tau_i)$ of a task τ_i is the length of the critical path from the exit node up to and including node τ_i:

$$rank_u(\tau_{exit}) = \overline{c_{exit}} \tag{6.47}$$

$$rank_u(\tau_i) = \overline{c_i} + \max_{\tau_j \in succ(\tau_i)} (\overline{h_{i,j}} + rank_u(\tau_j)) \tag{6.48}$$

- The **downward rank** $rank_d(\tau_j)$ of a task τ_j is the length of the critical path from the start node up to and **excluding** node τ_j:

$$rank_d(\tau_{entry}) = 0 \tag{6.49}$$

$$rank_d(\tau_j) = \max_{\tau_i \in pred(\tau_j)} (rank_d(\tau_i) + \overline{c_i} + \overline{h_{i,j}}) \tag{6.50}$$

The HEFT algorithm is shown below:

```
Set the computation and communication costs to mean values;
Compute ranku(τi)∀τi (upward traversal starting at τexit);
Sort tasks in non-increasing order of ranku values;
while there are unscheduled tasks in the list do {
 select the first task τi in the list for scheduling;
 for each processor πk ∈ π {
    compute fe(τi,πk) using an insertion-based scheduling policy;
 }
 assign task τi to processor πk minimizing fe(τi,πk);
}
```

In this context, "insertion-based policy" means that the algorithm searches for a sufficiently large gap among already scheduled tasks such that an allocation into this gap would respect precedence constraints.

Example 6.17: Suppose that execution times are given by the table in Fig. 6.31 (**left**). Note that for each task, the execution times in Fig. 6.26 (**right**) have been selected as the minimum time among the three processors.

Figure 6.31 (**center**) shows the schedule obtained by HEFT for the DAG shown in Fig. 6.26 (**left**). Precedences have been correctly taken into account. We cannot expect to generate the same short schedule as for ASAP or ALAP scheduling as these policies ignore resource constraints.

Task	π_1	π_2	π_3
1	14	16	9
2	13	19	18
3	11	13	19
4	13	8	17
5	12	13	10
6	13	16	9
7	7	15	11
8	5	11	14
9	18	12	20
10	21	7	16

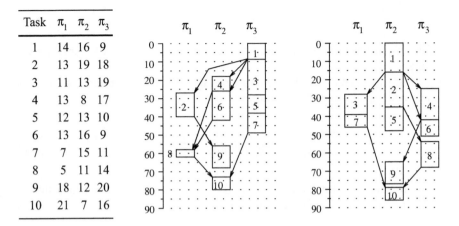

Fig. 6.31 **Left**: execution times; **center**: results for HEFT; **right**: results for CPOP

$\triangledown$

The CPOP algorithm focuses on the critical path in the DAG and uses different task priorities and different processor allocation strategies. The CPOP algorithm works as follows:

```
Set the computation and communication costs to mean values;
Compute ∀i : rank_u(τ_i) and rank_d(τ_i);
Compute ∀i : priority(τ_i) = rank_d(τ_i) + rank_u(τ_i);
|CP| = priority(τ_entry); /* length of the critical path */
SET_CP = {τ_entry}, where SET_CP ist the set of tasks on the critical path;
τ_i = τ_entry;
while τ_i is not the exit task {
  Select τ_j ∈ succ(τ_i), where priority(τ_j) == |CP|.
  SET_CP = SET_CP ∪ {τ_j};
  τ_i = τ_j };
Select processor π_CP minimizing execution time on the critical path;
Initialize the priority queue with the entry task;
while there is an unscheduled task in the priority queue {
  Select the highest priority task τ_i from the priority queue;
  if τ_i ∈ SET_CP {assign task τ_i on π_CP }
  else {assign task τ_i to the processor which minimizes f_e(τ_i, π_k)};
  Update priority queue with successors of τ_i if they become ready;
}
```

Example 6.18: Figure 6.31 (**right**) shows the scheduling result for algorithm CPOP. $\triangledown$

The HEFT and CPOP algorithms are fast and relatively simple algorithms. Obviously, these algorithms make use of several approximations (e.g., average communication costs) and heuristics. They were selected for this book to demonstrate some key issues of scheduling algorithms for heterogeneous scheduling algorithms. However, it is possible to improve over the results of these two algorithms. For example, Kim et al. [282] present more complex algorithms generating better results. A mapping for KPNs aiming at makespan minimization has been published by Castrillon et al. [87].

6.5.3 Static Scheduling with Integer Linear Programming

Integer linear programming can also be applied to the case of heterogeneous processors. One approach has been published by Maculan et al. [344]. Most importantly, processor-dependent execution times are taken care of. However, the presented equations require some refinement before they can be fed into an ILP solver and applications have not been included. Also, it is possible to adapt techniques published in the context of high-level synthesis [45, 301].

In most of the publications, optimizations aim at optimizing a single objective. In general, more objectives should be considered. For example, Fard et al. [154] present an algorithm taking four different objectives into account.

6.5.4 Static Scheduling with Evolutionary Algorithms

Integer programming-based approaches potentially suffer from long execution times. In many cases, the use of evolutionary algorithms allows a better optimization while still keeping execution times reasonably short. We will demonstrate this by means of the distributed Operation Layer (DOL) tools from ETH Zürich. The DOL tools from ETH Zürich [514] incorporate

- **Automatic selection of computation templates**: Processor types can be completely heterogeneous. Standard processors, micro-controllers, DSP processors, FPGAs etc. are all possible options.
- **Automatic selection of communication techniques**: Various interconnection schemes such as central buses, hierarchical buses, and rings are feasible.
- **Automatic selection of scheduling and arbitration**: DOL design space exploration tools automatically choose between rate monotonic scheduling, EDF, TDMA-based schemes, and priority-based schemes.

The input to DOL consists of a set of tasks together with use cases. The output describes the execution platform, the mapping of tasks to processors together with task schedules. This output is expected to meet constraints (such as memory size and timing constraints) and to minimize objectives (such as size and energy). Applications are represented by so-called problem graphs, which in essence are special task graphs. Figure 6.32 shows a simple DOL-problem graph. This graph models computations (see nodes 1, 2, 3, 4) and communication (see nodes 5, 6, 7).

Fig. 6.32 DOL problem
graph

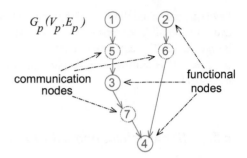

In addition, possible execution platforms are represented by so-called **architecture graphs**. Figure 6.33 shows a simple hardware platform together with its architecture graph. Again, communication is modeled explicitly.

Fig. 6.33 DOL architecture graph

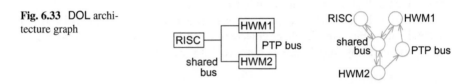

The problem graph and the architecture graph are connected in the **specification graph**. Figure 6.34 shows a DOL specification graph. Specification graphs consist of a problem graph and a architecture graph. Edges between the two subgraphs represent feasible implementations. For example, computation 1 can be implemented only on the RISC processor, computation 3 on the RISC processor or on HWM1.

Fig. 6.34 DOL specification
graph

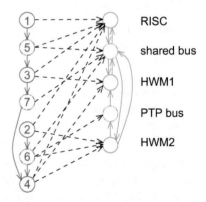

Communication 5 can be implemented on the shared bus or locally on the processor if computations 1 and 3 are both mapped to the processor.

Implementations are represented by a triple:

- **An allocation** A: A is a subset of the architecture graph, representing hardware components allocated (selected) for a particular design.
- **A binding** b: a selected subset of the edges between specification and architecture identifies a relation between the two. Selected edges are called **bindings**.
- **A schedule** S: S assigns start times to each node τ_i in the problem graph.

Example 6.19: Figure 6.35 shows how the specification of Fig. 6.34 can be turned into an implementation. HWM2 and the PTP bus are not used and not included in

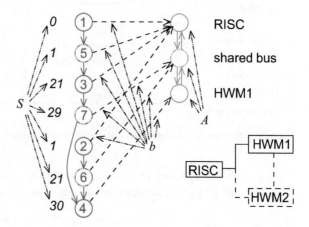

Fig. 6.35 DOL implementation

the set A. A subset b of the edges have been selected for mapping. Nodes 1, 2, 3, 5 have indeed all been mapped to the RISC processor, turning communication 5 into local communication. Node 4 is mapped to HWM1 and communicates via shared bus. Schedule S specifies that computation 1 starts at time 0, communication 5 and computation 2 start at time 1, computation 3 and communication 6 start at time 21, communication 7 starts at time 29, and finally, computation 4 starts at time 30. ∇

In DOL, implementations are generated with evolutionary algorithms. With such algorithms, solutions are represented as strings in chromosomes of "individuals" [29, 30, 107]. Using evolutionary algorithms, new sets of solutions can be derived from existing sets of solutions. The derivation is based on evolutionary operators such as mutation, selection, and recombination. The selection of new sets of solutions is based on **fitness values**. Evolutionary algorithms are capable of solving complex optimization problems not tractable by other types of algorithms. Finding appropriate ways of encoding solutions in chromosomes is not easy. On the one hand, the decoding should not require too much run-time. On the other hand, we must deal with the situation after the evolutionary transformations. These transformations could generate infeasible solutions, except for some carefully designed encodings.

In DOL, chromosomes encode allocations and bindings. In order to evaluate the fitness of a certain solution, allocations and bindings must be decoded from the individuals (see Fig. 6.36). In DOL, schedules are not encoded in the chromosomes. Rather, they are derived from the allocation and binding. This way overloading evolutionary algorithms with scheduling decisions is avoided. Once the schedule has been computed, the fitness of solutions can be evaluated.

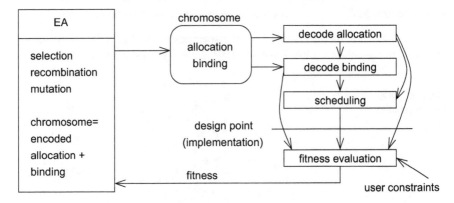

Fig. 6.36 Decoding of solutions from chromosomes of individuals

The overall architecture of DOL is shown in Fig. 6.37. Initially, the task graph, use cases, and available resources are defined. This can be done with a specialized

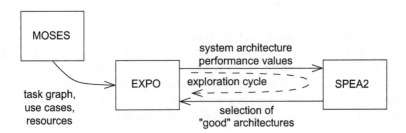

Fig. 6.37 DOL tool

editor called MOSES. This initial information is evaluated in the evaluation framework EXPO. Performance values computed by EXPO are then sent to SPEA2, an evolutionary algorithm-based optimization framework. SPEA2 selects good candidate architectures. These are sent back to EXPO for an evaluation. Evaluation results are then communicated again to SPEA2 for another round of evolutionary optimizations. This kind of ping-pong game between EXPO and SPEA2 continues until good solutions have been found. The selection of solutions is based on the principle of Pareto optimality. A set of Pareto-optimal designs is returned to the designer, who can then analyze the trade-off between the different objectives.

Example 6.20: Figure 6.38 shows the resulting visualization of the Pareto front. Trade-offs between the performance for two applications and the cost of the application-specific platform can be seen.

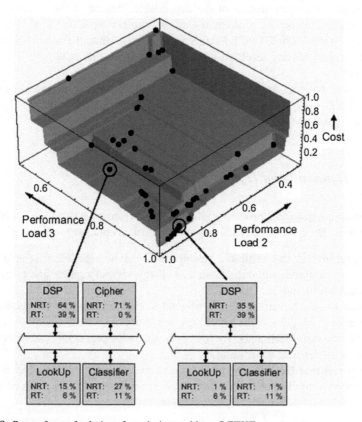

Fig. 6.38 Pareto front of solutions for a design problem, © ETHZ

∇

Evolutionary algorithms have become a standard technique for more advanced scheduling problems, beyond the problems solved by HEFT or CPOP.

The functionality of the SystemCodesigner [272] is somewhat similar to that of DOL. However, it differs in the way specifications are described (they can be represented in SystemC) and in the way the optimizations are performed. The mapping of applications is modeled as an ILP model. A first solution is generated using an ILP optimizer. This solution is then improved by switching to evolutionary algorithms[10].

Daedalus [401] incorporates automatic parallelization. For this purpose, sequential applications are mapped to Kahn process networks. Design space exploration is then performed using Kahn process networks as an intermediate representation.

---

[10]A more recent version uses a satisfiability (SAT) solver for the same purpose.

Other approaches start from a given task graph and map to a fixed architecture. For example, Ruggiero maps applications to Cell processors [451]. The HOPES-system is able to map to various processors [188], using models of computation supported by the Ptolemy tools. Some tools take additional objectives into account. For example, Xu considers the optimization of the dependable lifetime of the resulting system [577]. Simunic incorporates thermal analysis into her work and tries to avoid too hot areas on the MPSoC [467]. Further work includes that of Popovici et al. [432]. This work uses several levels of modeling, employing Simulink and SystemC as languages.

Auto-parallelizing approaches for fixed architectures include the Mnemee tool set [346] and work at the University of Edinburgh [163]. MAPS tools [90] combine automatic parallelization with a limited DSE.

6.5.5 Dynamic and Hybrid Scheduling

For dynamic scheduling, processor allocation is performed at run-time rather than at design time. Dynamic scheduling has a number of advantages [468]:

- **Adaptability to the available resources**: Dynamic scheduling is able to take changing resource availabilities such as energy, memory space, and communication bandwidth into account.
- **Ability to enable unforeseeable upgrades**: changing application requirements are easier to integrate when scheduling is dynamic.
- **Resilience to defects**: defective resources like failed processors can be taken into account by dynamic scheduling.
- **Use of non-real-time platforms**: dynamic scheduling is the standard for non-real-time computing. Hence, techniques for non-real-time computing can be applied, which helps to reduce development efforts.

However, there are also disadvantages:

- **Lacking real-time guarantees**: in a fully dynamically scheduled system, it is difficult if not impossible to give real-time guarantees.
- **Run-time overhead**: dynamic scheduling requires run-time for taking scheduling decisions. Therefore, complex scheduling techniques must be avoided.
- **Limited knowledge**: at run-time, there is typically limited knowledge concerning the task system and its parameters.

There are two approaches for dynamic scheduling: **on-the-fly mapping** and hybrid **mapping using previously analyzed (DSE) results**.

Singh et al. [468] provide an overview of 25 different approaches for on-the-fly mapping. This type of mapping is closest to mapping in non-real-time systems.

Hybrid mapping techniques using previously analyzed (DSE) results try to avoid some of the disadvantages listed above by making results from design-time analysis available at run-time. For example, we could precompute schedules for likely run-time scenarios and then select at run-time the schedule for the current scenario. Singh et al. distinguish between multiple mappings precomputed for a single application, multiple mappings precomputed for a multiple applications, and reliability-aware analysis[11]. The authors provide an overview of 21 different approaches for performing design-time analysis and run-time mapping in a sequence.

6.6 Problems

We suggest solving the following problems either at home or during a flipped classroom session:

6.1: Suppose that we have a set of 4 jobs. Release times r_i, deadlines D_i, and execution times C_i are as follows:

- J_1: $r_1 = 10$, $D_1 = 18$, $C_1 = 4$
- J_2: $r_2 = 0$, $D_2 = 28$, $C_2 = 12$
- J_3: $r_3 = 6$, $D_3 = 17$, $C_3 = 3$
- J_4: $r_4 = 3$, $D_4 = 13$, $C_4 = 6$

Generate a graphical representation of schedules for this task set, using *Earliest Deadline First* (EDF) and *Least Laxity* (LL) scheduling algorithms! For LL scheduling, indicate laxities for all tasks at all context switch times. Will any task miss its deadline?

6.2: Suppose that we have a task set of six tasks τ_1 to τ_6. Their execution times and their deadlines are as follows:

- τ_1: $D_1 = 15$, $C_1 = 3$
- τ_2: $D_2 = 13$, $C_2 = 5$
- τ_3: $D_3 = 14$, $C_3 = 4$
- τ_4: $D_4 = 16$, $C_4 = 2$
- τ_5: $D_3 = 20$, $C_3 = 4$
- τ_6: $D_4 = 22$, $C_4 = 3$

Precedences are as shown in Fig. 6.39. Tasks τ_1 and τ_2 are available immediately.

---

[11]We merge Singh's hybrid mappings with these three classes.

Fig. 6.39 Precedences

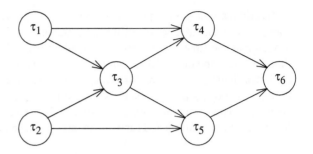

Generate a graphical representation of schedules for this task set, using the *Latest Deadline First* (LDF) algorithm!

6.3: Suppose that we have a system comprising two tasks. Task 1 has a period of 5 and execution time of 2. The second task has a period of 7 and an execution time of 4. Let the deadlines be equal to the periods. Assume that we are using *Rate Monotonic Scheduling* (RMS). Could any of the two tasks miss its deadline, due to a too high processor utilization? Compute this utilization and compare it to a bound which would guarantee schedulability! Generate a graphical representation of the resulting schedule! Suppose that tasks will always run to their completion, even if they missed their deadline.

6.4: Consider the same task set as in the previous assignment. Use *Earliest Deadline First* (EDF) for scheduling. Can any of the tasks miss its deadline? If not, why not? Generate a graphical representation of the resulting schedule! Suppose that tasks will always run to their completion.

Chapter 7
Optimization

Embedded systems have to be efficient. In particular, this applies to sensor networks embedded in the Internet of Things. In order to achieve this goal, many optimizations have been developed. Only a small subset of those can be mentioned in this book. In this chapter, we will present a selected set of such optimizations. As indicated in our design flow, these optimizations complement the tools mapping applications to the final systems, as described in Chap. 6 and as shown in Fig. 7.1.

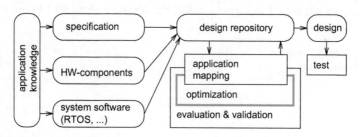

Fig. 7.1 Context of the current chapter

Mapping tools may be optimizing, and optimization techniques may involve scheduling. Hence, the scopes of the current and of Chap. 6 are partially overlapping. The focus of Chap. 6 is on fundamental knowledge for mapping to platforms, while the current chapter deals mostly with improvements over basic techniques and is similar to the character of an elective.

This chapter is structured as follows: First of all, we will present some high-level optimization techniques, which could precede compilation of source code or could be integrated into it. We will then describe concurrency management for tasks. Section 7.3 comprises advanced compilation techniques. The final Sect. 7.4 introduces power and thermal management techniques.

© Springer International Publishing AG 2018
P. Marwedel, *Embedded System Design*, Embedded Systems,
DOI 10.1007/978-3-319-56045-8_7

7.1   High-Level Optimizations

In the next section, we will be considering optimizations which can be applied to the
source code of embedded software, before compilation or during early compilation
phases. Detecting regular structures such as array access patterns may be easier at
the source code level than at the machine code level. Also, optimization effects can
usually be expressed by rewriting the source program, i.e., the modified code can
be expressed in the source language. This helps in understanding the effect of such
transformations. We do also consider cases, in which it may be necessary to annotate
the source code with compiler directives and hints. Such code transformations are
called **high-level optimizations**. They have the potential to improve the efficiency
of embedded software.

7.1.1   Simple Loop Transformations

There is a number of loop transformations that can be applied to source code. The
following is a list of standard loop transformations:

- **Loop permutation**: Consider a two-dimensional array. According to the C standard [276], two-dimensional arrays are laid out in memory as shown in Fig. 7.2.
  Adjacent index values of the second index are mapped to a contiguous block of
  locations in memory. This layout is called **row-major order** [388]. For row-major

Fig. 7.2 Memory layout for
two-dimensional array
p[j][k] in C

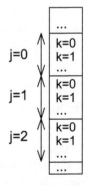

layout, it is usually beneficial to organize loops such that the last index corresponds
to the innermost loop.

Note that the layout for arrays is different for FORTRAN: Adjacent values of
the first index are mapped to a contiguous block of locations in memory (**column-**

major order). Switching between publications describing optimizations for FOR-
TRAN and for C can, therefore, be confusing.

Example 7.1: The following is a loop permutation:

```
for (k=0; k<m; k++)                    for (j=0; j<n; j++)
  for (j=0; j<n; j++)        ⇔           for (k=0; k<m; k++)
    p[j][k]= ...                           p[j][k]= ...
```

Such permutations may have a positive effect on the reuse of array elements in the
cache, since the next iteration of the innermost loop will access an adjacent location
in memory.   ▽

Caches are normally organized such that adjacent locations can be accessed significantly faster than locations that are further away from the previously accessed
location. In this way, caches are exploiting **spatial locality**.

Definition 7.1: Consider memory references to memory addresses a and b. Suppose
that we assume an access a. We observe **spatial locality** if – under this condition
– the probability of also accessing b increases for small differences of addresses
a and b.

- **Loop unrolling**: Loop unrolling is a standard transformation creating several
  instances of the loop body.

Example 7.2: In this example, we unroll the loop:

```
for (j=0; j<n; j++)                    for (j=0; j<n; j+=2)
  p[j]= ... ;               ⇔            {p[j]= ... ;
                                          p[j+1]= ...}
```

In this particular case, the loop is unrolled once.   ▽

The number of copies of the loop is called the **unrolling factor**. Unrolling factors
larger than two are possible. Unrolling reduces the loop overhead (less branches per
execution of the original loop body), and therefore typically improves the speed. As
an extreme case, loops can be completely unrolled, removing control overhead and
branches altogether. Unrolling typically enables a number of following transformations and may, therefore, be beneficial even in cases where just unrolling the program
does not give any advantages. However, unrolling increases code size. Unrolling is
normally restricted to loops with a constant number of iterations.

- **Loop fusion, loop fission**: There may be cases in which two separate loops can
  be merged, and there may be cases in which a single loop is split into two.

Example 7.3: Consider the two versions of the following code:

```
for (j=0; j<n; j++)                    for (j=0; j<n; j++)
  p[j]= ... ;                            {p[j]= ... ;
for (j=0; j<n; j++)        ⇔            p[j]=p[j]+ ...}
  p[j]=p[j]+ ...
```

The left version may be advantageous if the target processor provides a zero-overhead loop instruction which can only be used for small loops. Also, the left version may provide good candidates for unrolling, due to the simple loops. The right version might lead to an improved cache behavior (due to the improved locality of references to array p) and also increases the potential for parallel computations within the loop body. As with many other transformations, it is difficult to know which of the transformations lead to the best code.   ∇

7.1.2   Loop Tiling/Blocking

Since small memories are faster than large memories (see p. 166), the use of memory hierarchies may be beneficial. Possible "small" memories include caches and scratch-pad memories. A significant reuse factor for the information in those memories is required. Otherwise, the memory hierarchy cannot be exploited.

Example 7.4: Reuse effects can be demonstrated by an analysis of the following example. Let us consider matrix multiplication for arrays of size $N \times N$:

```
for (i=0; i<N; i++)
  for(j=0; j<N; j++)    {
      r=0;
      for (k=0; j<N; k++)
          r+=X[i][k]*Y[k][j];
      Z[i][j]=r;
  }
```

Let us consider access patterns for this code. Scalar variable r represents Z[i,j] in all iterations of the innermost loop. This is supposed to help the compiler to allocate this element temporarily to a register. We assume that array elements are allocated in row-major order (as it is standard for C). This means that array elements with adjacent row (right most) index values are stored in adjacent memory locations. Accordingly, adjacent locations of X are fetched during the iterations of the innermost loop. This property is beneficial if the memory system uses prefetching (whenever a word is loaded into the cache, loading of the next word is started as well). Figure 7.3 shows access patterns for this code. However, accesses to Y do not exhibit spatial locality.

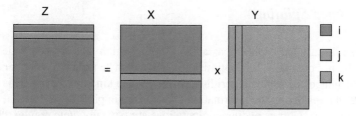

Fig. 7.3 Access pattern for unblocked matrix multiplication

If the cache is not large enough to hold a full cache row, every access to Y will be a cache miss. Hence, there will be N^3 references to elements of Y in main memory.

Research on scientific computing led to the design of **blocked** or **tiled algorithms** [306, 578], which improve the **locality of references**. The following is a tiled version of the above algorithm[1] for a block size parameter B:

```
for (ii=0; kk<N; ii+=B)
  for (jj=0; jj<N; jj+=B)
    for (kk=0; kk<N; kk+=B)
      for (i=ii; i<min(ii+B-1,N); ii++)
        for (j=jj; j<min(jj+B-1,N); jj++) {
          r=0;
          for (k=kk; k<min(kk+B-1,N); k++)
            r+= X[i][k]*Y[k][j];
          Z[i][j]=r;
        }
```

Now, the two innermost loops are constrained to traverse a block of size N^2 for array Y. Suppose that a block of size N^2 fits into the cache. Then, the first execution of the innermost loop will load this block into the cache. During the second execution of the innermost loop, these elements will be reused. Overall, there will be B−1 reuses of elements of Y. Hence, the number of accesses to main memory for elements of this array will be reduced to $N^3/(B-1)$.   ∇

Optimizing the reuse factor has been an area of comprehensive research. Initial research focused on the performance improvements that can be obtained by tiling. Performance improvements for matrix multiplication by a factor between 3 and 4.3 were reported by Lam [306]. Improvements increase with an increasing gap between processor and memory speeds. Tiling can also reduce the energy consumption of memory systems [103].

---

[1]This code was adopted from http://www.netlib.org/utk/papers/autoblock/node2.html.

7.1.3   Loop Splitting

Next, we discuss loop splitting as another optimization that can be applied before compilation. Potentially, this optimization could also be added to compilers.

Many image processing algorithms perform some kind of filtering. This filtering consists of considering the information about a certain pixel as well as that of some of its neighbors. Corresponding computations are typically quite regular. However, if the considered pixel is close to the boundary of the image, not all neighboring pixels exist and the computations must be modified. In a straightforward description of the filtering algorithm, these modifications may result in tests being performed in the innermost loop of the algorithm. A more efficient version of the algorithm can be generated by splitting the loops such that one loop body handles the regular cases and a second loop body handles the exceptions. Figure 7.4 is a graphical representation of this transformation. Margin checking is required for the yellow areas.

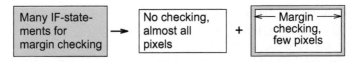

Fig. 7.4 Splitting image processing into regular and special cases

Performing this loop splitting manually is very difficult and error-prone. Falk et al. have published an algorithm [152] which also works for larger dimensions automatically. It is based on a sophisticated analysis of accesses to array elements in loops. Optimized solutions are generated using genetic algorithms.

Example 7.5: The following code shows a loop nest from the MPEG-4 standard performing motion estimation:

```
for (z=0; z<20; z++)
 for (x=0; x<36; x++) {x1=4*x;
  for (y=0; y<49; y++) {y1=4*y;
   for (k=0; k<9; k++) {x2=x1+k-4;
    for (l=0; l<9; l++) {y2=y1+l-4;
     for (i=0; i<4; i++) {x3=x1+i; x4=x2+i;
      for (j=0; j<4; j++) {y3=y1+j; y4=y2+j;
       if (x3<0 || 35<x3 || y3<0 || 48<y3)
        then_block_1; else else_block_1;
       if (x4<0 || 35<x4 || y4<0 || 48<y4)
        then_block_2; else else_block_2;
}}}}}}
```

Using Falk's algorithm, this loop nest is transformed into the following one:

```
for (z=0; z<20; z++)
 for (x=0; x<36; x++) {x1=4*x;
  for (y=0; y<49; y++)
   if (x>=10 || y>=14)
     for (; y<49; y++)
       for (k=0; k<9; k++)
         for (l=0; l<9; l++ )
           for (i=0; i<4; i++)
             for (j=0; j<4; j++) {
               then_block_1; then_block_2}
   else {y1=4*y;
     for (k=0; k<9; k++) {x2=x1+k-4;
       for (l=0; l<9; l++) {y2=y1+l-4;
         for (i=0; i<4; i++) {x3=x1+i; x4=x2+i;
           for (j=0; j<4; j++) {y3=y1+j; y4=y2+j;
             if ( 0 || 35 <x3 || 0 || 48 < y3)     /* x3<0, y3<0 never true */
             then_block_1; else else_block_1;
             if (x4 < 0|| 35 < x4 || y4 < 0 || 48 < y4)
             then_block_2; else else_block_2; }}}}}}
```

Instead of complex tests in the innermost loop, we have a splitting if-statement after the third for-loop statement. Regular cases are handled in the then-part of this statement. The else-part handles the relatively small number of remaining cases. ∇

Run-times can be reduced by loop splitting for various applications and architectures. Resulting relative run-times are shown in Fig. 7.5. For the motion estimation

Fig. 7.5 Results for loop splitting

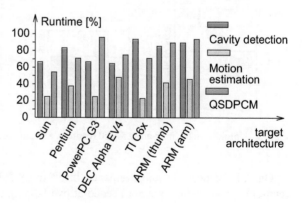

rithm, cycle counts can be reduced by up to about 75% (to 25% of the original value). Substantial savings (larger than for the simple transformations mentioned earlier) are possible. In order to exploit this potential, Falk's algorithm can be implemented, e.g., as a compiler pre-pass tool.

7.1.4 Array Folding

Some embedded applications, especially in the multimedia domain, include large arrays. Since memory space in embedded systems is limited, options for reducing the storage requirements of arrays should be explored. Figure 7.6 represents the addresses used by five arrays as a function of time. At any particular time, only a subset of array elements is needed. The maximum number of elements needed is called the **address reference window** [118]. In Fig. 7.6, this maximum is indicated by a double-headed arrow.

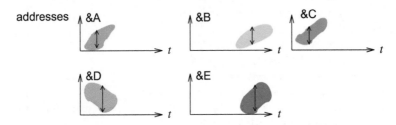

Fig. 7.6 Reference patterns for arrays

A classical memory allocation for arrays is shown in Fig. 7.7 (**left**). Each array allocated the maximum of the space it requires during the entire execution time (if we consider global arrays).

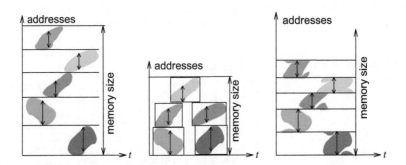

Fig. 7.7 Unfolded (**left**), inter-array folded (**center**), and intra-array folded (**right**) arrays

One of the possible improvements, **inter-array folding**, is shown in Fig. 7.7 (**center**). Arrays which are not needed at overlapping time intervals can share the same memory space. A second improvement, intra-array folding [117], is shown in Fig. 7.7 (**right**). It takes advantage of the limited sets of components needed **within** an array. Storage can be saved at the expense of more complex address computations. The two kinds of foldings can also be combined.

Other forms of high-level transformations have been analyzed by Chung, Benini, and De Micheli [103, 501]. There are many additional contributions in this domain in the compiler community.

7.1.5 Floating-Point to Fixed-Point Conversion

Floating-point to fixed-point conversion is a commonly used optimization technique. This conversion is motivated by the fact that many signal processing standards (such as MPEG-2 or MPEG-4) are specified in the form of C programs using floating-point data types. It is left to the designer to find an efficient implementation of these standards.

For many signal processing applications, it is possible to replace floating-point numbers with fixed-point numbers (see p. 150). The benefits may be significant. For example, a reduction of the cycle count by 75% and of the energy consumption by 76% has been reported for an MPEG-2 video compression algorithm [216]. However, some loss of precision is normally incurred. More precisely, there is a trade-off between the cost of the implementation and the quality of the algorithm (evaluated e.g., in terms of quality metrics, see Sect. 5.3 on p. 245). For small word lengths, the quality may be seriously affected. Consequently, the quality loss has to be analyzed. This replacement was initially performed manually. However, it is a very tedious and error-prone process.

Therefore, researchers have tried to support this replacement with tools. One of such tools is FRIDGE (fixed-point programming design environment) [271, 560]. The functionality of FRIDGE has been made available commercially as part of the Synopsys System Studio tool suite [495].

In FRIDGE, the design process starts with an algorithm described in C, including floating-point numbers. This algorithm is then converted to an algorithm described in **fixed-C**. Fixed-C extends C by two fixed-point data types, using the type definition features of C++. Fixed-C is a subset of C++ and provides two data types fixed and Fixed. Fixed-point data types can be declared very much like other variables.

Example 7.6: The following declaration declares a scalar variable, a pointer, and an array to be fixed-point data types.

```
fixed a,*b,c[8];
```

These declarations are valid for their entire scope which is possibly the whole program.   ▽

Providing parameters of fixed-point data types can be delayed until assignment time.

Example 7.7

```
a=fixed(5,4,s,wt,*b)
```

The word-length parameter of a is set to 5 bits, the fractional word-length to 4 bits, sign to present (s), overflow handling to wraparound (w), and the rounding mode to truncation (t).   ▽

The parameters for variables that are read in an assignment are determined by the assignment(s) to those variables. The data type Fixed is similar to fixed, except that a consistency check between parameters used in the declaration and those used in the assignment is performed. For every assignment to a variable, parameters (including the word-length) can be different. This parameter information can be added to the original C-program before the application is simulated. Simulation provides value ranges for all assignments. Based on that information, FRIDGE adds parameter information to all assignments. FRIDGE also infers parameter information from the context. For example, the maximum value of additions is considered to be the sum of the arguments. Added parameter information can be either based on simulations or on worst case considerations. Being based on simulations, FRIDGE does not necessarily assume the worst case values, in contrast to a formal analysis. The resulting C++ program is simulated again to check for the quality loss. SystemC can be used for simulating fixed-point data types.

An analysis of the trade-offs between the additional noise introduced and the word-length needed was proposed by Shi and Brodersen [461] and also by Menard et al. [372]. The topic continues to attract researchers [321].

7.2 Task Level Concurrency Management

As mentioned on p. 35, the task graphs' **granularity** is one of their most important properties. Even for hierarchical task graphs, it may be useful to change the granularity of the nodes. The partitioning of specifications into tasks or processes does not necessarily aim at the maximum implementation efficiency. Rather, during the specification phase, a clear separation of concerns and a clean software model are more important than caring about the implementation too much. For example, a clear separation of concerns includes a clear separation of the implementation of abstract data types from their use. Also, we might be using several tasks in a pipelined fashion in our specification, while merging some of them might reduce context switching overhead. Hence, there will not necessarily be a one-to-one correspondence between the tasks in the specification and those in the implementation. This means that a regrouping of tasks may be advisable. Such a regrouping is indeed feasible by merging and splitting of tasks.

Merging of task graphs can be performed, whenever some task τ_i is the immediate predecessor of some other task τ_j and if τ_j does not have any other immediate predecessor (see Fig. 7.8 with $\tau_i = \tau_3$ and $\tau_j = \tau_4$). This transformation can lead to a reduced overhead of context switches if the node is implemented in software, and it can lead to a larger potential for optimizations in general.

Fig. 7.8 Merging of tasks

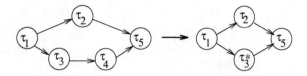

On the other hand, splitting of tasks may be advantageous for the following reasons:

Tasks may be holding resources (like large amounts of memory), while they are waiting for some input. In order to maximize the use of these resources, it may be best to constrain the use of these resources to the time intervals during which these resources are actually needed. In Fig. 7.9, we are assuming that task τ_2 requires some input somewhere in its code.

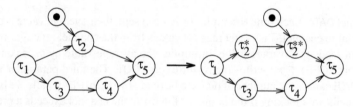

Fig. 7.9 Splitting of tasks

In the initial version, the execution of task τ_2 can only start if this input is available. We can split the node into τ_2^* and τ_2^{**} such that the input is only required for the execution of τ_2^{**}. Now, τ_2^* can start earlier, resulting in more scheduling freedom. This improved scheduling freedom might improve resource utilization and could even enable meeting some deadline. It may also have an impact on the memory required for data storage, since τ_2^* could release some of its memory before terminating and this memory could be used by other tasks, while τ_2^{**} is waiting for input.

One might argue that the tasks should release resources like large amounts of memory anyway before waiting for input. However, the readability of the original specification could suffer from caring about implementation issues in an early design phase.

Quite complex transformations of the specifications can be performed with a Petri net-based technique described by Cortadella et al. [110]. Their technique starts with a specification consisting of a set of tasks described in a language called *FlowC*. FlowC extends C with process headers and inter-task communication specified in the form of READ- and WRITE-function calls.

Example 7.8: Figure 7.10 shows an input specification using FlowC. The example uses input ports IN and COEF, as well as an output port OUT. Point-to-point interprocess communication between processes is realized through a unidirectional buffered channel DATA. Task GetData reads data from the environment and sends it

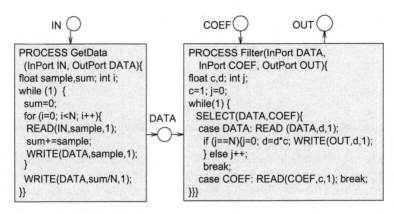

Fig. 7.10 System specification

to channel DATA. Each time N samples have been sent, their average value is also sent via the same channel. Task Filter reads N values from the channel (and ignores them) and then reads the average value, multiplies the average value by c. (c can be read from port COEF). Filter writes the result to port OUT. The third parameter in READ and WRITE calls is the number of items to be read or written. READ calls are blocking, WRITE calls are blocking if the number of items in the channel exceeds a predefined threshold. The SELECT statement has the same semantics as the statement with the same name in Ada (see p. 107): Execution of this task is suspended until input arrives from one of the ports. This example meets all criteria for splitting tasks that were mentioned in the context of Fig. 7.9. Both tasks will be waiting for input while occupying resources. Efficiency could be improved by restructuring these tasks. However, the simple splitting of Fig. 7.9 is not sufficient. The technique proposed by Cortadella et al. is more comprehensive: FlowC-programs are first translated into (extended) Petri nets. Petri nets for each of the tasks are then merged into a single Petri net. Using results from Petri net theory, new tasks are then generated. Figure 7.11 shows a possible new task structure. In this new task structure, there is one task which performs all initializations: In addition, there is one task for each of the input ports. An efficient implementation would raise interrupts each time new input is received for a port. There should be a unique interrupt per port. The tasks could then be started directly by those interrupts, and there would be no need to invoke the operating system for that. Communication can be implemented as a single shared global variable (assuming a shared address space). The operating system overhead would be small, if required at all.

The code for task tau_in shown in Fig. 7.11 is the one that is generated by the Petri net-based inter-task optimization of the task structure. It should be further optimized by intra-task optimizations, since the test performed for the first if-statement is always false (j is equal to i-1 in this case, and i and j are reset to 0 whenever i becomes equal to N). For the third if-statement, the test is always true, since this point of control is only reached if i is equal to N and i is equal to j whenever label L0 is reached. Also,

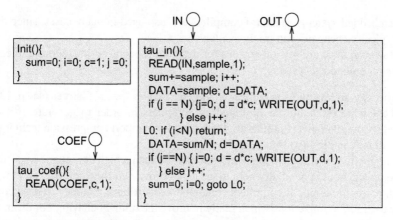

Fig. 7.11 Generated software tasks

the number of variables can be reduced. The following is an optimized version of tau_in:

```
tau_in () {
    READ(IN,sample,1);
    sum+=sample;   i++;
    DATA=sample;   d=DATA;
L0:  if (i<N) return;
    DATA=sum/N;  d=DATA;
    d=d*c;  WRITE(OUT,d,1);
    sum=0;  i=0;
    return;
}
```

The optimized version of tau_in could be generated by a clever compiler. Hardly any of today's compilers will generate this version, but the example shows the type of transformations required for generating "good" task structures.   ∇

For more details about the task generation refer to Cortadella et al. [110]. Similar optimizations are described in the book by Thoen [515] and in a publication by Meijer et al. [371].

7.3 Compilers for Embedded Systems

7.3.1 Introduction

Obviously, optimizations and compilers are available for the processors used in PCs and servers. Compiler generation for commonly used processors is well understood.

For embedded systems, standard compilers are also used in many cases, since they are typically cheap or even freely available.

However, there are several reasons for designing special optimizations and compilers for embedded systems:

- Processor architectures in embedded systems exhibit special features (see p. 139). These features should be exploited by compilers in order to generate efficient code. Compilation techniques might also have to support compression techniques described on pp. 144–146.
- A high efficiency of the code is more important than a high compilation speed.
- Compilers could potentially help to meet and prove real-time constraints. First of all, it would be nice if compilers contained explicit timing models. These could be used for optimizations which really improve the timing behavior. For example, it may be beneficial to freeze certain cache lines in order to prevent frequently executed code from being evicted and reloaded several times.
- Compilers may help to reduce the energy consumption of embedded systems. Compilers performing energy optimizations should be available.
- For embedded systems, there is a larger variety of instruction sets. Hence, there are more processors for which compilers should be available. Sometimes, there is even the request to support the optimization of instruction sets with **retargetable** compilers. For such compilers, the instruction set can be specified as an input to a compiler generation system. Such systems can be used for experimentally modifying instruction sets and then observing the resulting changes for the generated machine code. This is one particular case of **design space exploration** and is supported, for example, by Tensilica tools [83].

Some approaches for retargetable compilers are described in a book on this topic [361]. Optimizations can be found in books by Leupers et al. [323, 324]. In this section, we will present examples of compilation techniques for embedded processors.

7.3.2 Energy-Aware Compilation

Many embedded systems are mobile systems which must run on batteries. While computational demands on mobile systems are increasing, battery technology is expected to improve only slowly [248]. Hence, the availability of energy is a serious bottleneck for new applications.

Saving energy can be done at various levels, including the fabrication process technology, the device technology, circuit design, the operating system, and the application algorithms. Adequate translation from algorithms to machine code can also help. High-level optimization techniques such as those presented on pp. 337–343 can also help to reduce the energy consumption. In this section, we will look at compiler optimizations which can reduce the energy consumption (frequently called low-*power* optimizations). **Energy models** are very essential ingredients of all energy optimizations. Energy models were presented in Chap. 5. Using models like those, the following compiler optimizations have been used for reducing the energy consumption:

- **Energy-aware scheduling**: The order of instructions can be changed as long as the meaning of the program does not change. The order can be changed such that the number of transitions on the instruction bus is minimized. This optimization can be performed on the output generated by a compiler and, therefore, does not require any change to the compiler.
- **Energy-aware instruction selection**: Typically, there are different instruction sequences for implementing the same source code. In a standard compiler, the number of instructions or the number of cycles is used as a criterion (cost function) for selecting a good sequence. This criterion can be replaced by the energy consumed by that sequence. Steinke and others found that energy-aware instruction selection reduces the energy consumption by some percent [485].
- **Replacing the cost function** is also possible for other standard compiler optimizations, such as register pipelining and loop invariant code motion. Possible improvements are also in the order of a few percent.
- **Exploitation of the memory hierarchy**: As already explained on p. 164, smaller memories provide faster access and consume less energy per access. Therefore, a significant amount of energy can be saved if memory hierarchies are exploited. Of all the compiler optimizations analyzed by Steinke [486, 488], the energy savings enabled by memory hierarchies are the largest. It is, therefore, beneficial to use small scratchpad memories (SPMs, see p. 169) in addition to large background memories. All accesses to the corresponding address range will then require less energy and are faster than accesses to the larger memory. The compiler should be responsible for allocating variables and instructions to the scratchpad. This approach does, however, require that frequently accessed variables and code sequences are identified and mapped to that address range.

7.3.3 Memory-Architecture Aware Compilation

7.3.3.1 Compilation Techniques for Scratchpads

The advantages of using SPMs have been clearly demonstrated [36]. Therefore, exploiting SPMs is the most prominent case of memory hierarchy exploitation. Available compilers are usually capable of mapping memory objects to certain address ranges in the memory. Toward this end, the source code typically has to be annotated.

Example 7.9: For ARM® tools, memory segments can be introduced in the source code by using pragmas like

```
# pragma arm section rwdata = "foo", rodata = "bar"
```

Variables declared after this pragma would be mapped to read-write segment
"foo" and constants would be mapped to read-only segment "bar". Linker commands can then map these segments to particular address ranges, including those
belonging to the SPM.   ▽

This is the approach taken in compilers for ARM processors [21]. This is not a
very comfortable approach and it would be nice if compilers could perform such
a mapping automatically for frequently accessed objects. Therefore, optimization
algorithms have been designed. A survey has been presented at the ARTIST summer
school [359]. Available SPM optimizations can be classified into two categories:

- **Non-overlaying** (or "static") memory allocation strategies: For these strategies,
  memory objects will stay in the SPM, while the corresponding application is
  executed.
- **Overlaying** (or "dynamic") memory allocation strategies: For these strategies,
  memory objects are moved in and out of the SPM at run-time. This is a kind of
  "compiler-controlled paging", except the migration of objects happens between
  the SPM and some slower memory and does not involve any disks.

7.3.3.2 Non-overlaying Allocation

For non-overlaying allocation, we can start by considering the allocation of functions
and global variables to the SPM. For this purpose, each function and each global
variable can be modeled as a memory object. Let

- S be the size of the SPM,
- sf_i and sv_i be the sizes of function i and variable i, respectively,
- g be the energy consumption saved per access to the SPM (i.e., the difference
  between the energy required per access to the slow main memory and the one
  required per access to the SPM),
- nf_i and nv_i be the number of accesses to function i and variable i, respectively,
  and
- xf_i and xv_i be defined as

$$xf_i = \begin{cases} 1 & \text{if function } i \text{ is mapped to the SPM} \\ 0 & \text{otherwise} \end{cases} \tag{7.1}$$

$$xv_i = \begin{cases} 1 & \text{if variable } i \text{ is mapped to the SPM} \\ 0 & \text{otherwise} \end{cases} \tag{7.2}$$

Then, the goal is to maximize the gain

$$G = g \left(\sum_i nf_i * xf_i + \sum_i nv_i * xv_i \right) \tag{7.3}$$

while respecting the size constraint

$$\sum_i sf_i * xf_i + \sum_i sv_i * xv_i \leq S \qquad (7.4)$$

The problem is known as a (simple) **knapsack** problem (see p. 309 for the more general case). Standard knapsack algorithms can be used for selecting the objects to be allocated to the SPM. However, Eqs. (7.3) and (7.4) also have the form of an integer linear programming (ILP) problem (see Appendix A) and ILP solvers can be used as well. g is a constant factor in the objective function and is not needed for the solution of the ILP problem. The corresponding optimization can be implemented as a pre-pass optimization (see Fig. 7.12).

Fig. 7.12 Pre-pass optimization

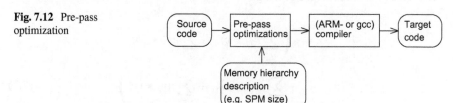

The optimization impacts addresses of functions and global variables. Compilers typically allow a manual specification of these addresses in the source code. Hence, no change to the compiler itself is required. The advantage of such a pre-pass optimization is that it can be used with compilers for many different target processors. There is no need to modify a large number of target-specific compilers.

The knapsack model can be extended into various directions:

• **Allocation of basic blocks**: The approach just described only allows the allocation of entire functions or variables to the SPM. As a result, a major fraction of the SPM may remain empty if functions and variables are large. Therefore, we try to reduce the granularity of the objects which are allocated to the SPM. The natural choice is to consider **basic blocks** as memory objects. In addition, we do also consider sets of adjacent basic blocks, where adjacency is defined as being adjacent in the control flow graph [485]. We call such sets of adjacent blocks as **multi-blocks**. Figure 7.13 shows a control flow graph and the set of considered multi-blocks.

Fig. 7.13 Basic blocks and multi-blocks

The ILP model can be extended accordingly. Let:

- sb_i and sm_i be the sizes of basic blocks i and multi-blocks i, respectively,
- nb_i and nm_i be the number of accesses to basic block i and multi-blocks i, respectively, and
- xb_i and xm_i be defined as

$$xb_i = \begin{cases} 1 & \text{if basic block } i \text{ is mapped to the SPM} \\ 0 & \text{otherwise} \end{cases} \tag{7.5}$$

$$xm_i = \begin{cases} 1 & \text{if multi-block } i \text{ is mapped to the SPM} \\ 0 & \text{otherwise} \end{cases} \tag{7.6}$$

Then, the goal is to maximize the gain

$$G = g\left(\sum_i nf_i * xf_i + \sum_i nb_i * xb_i \right.$$
$$\left. + \sum_i nm_i * xm_i + \sum_i nv_i * xv_i \right) \tag{7.7}$$

while respecting the constraints

$$\sum_i sf_i * xf_i + \sum_i sb_i * xb_i + \sum_i sm_i * xm_i + \sum_i sv_i * xv_i \leq S \tag{7.8}$$

$$\forall \text{ basic blocks } i : xb_i + xf_{fct(i)} + \sum_{i' \in multiblock(i)} xm_{i'} \leq 1 \tag{7.9}$$

where $fct(i)$ is the function containing basic block i

and $multiblock(i)$ is the set of multi-blocks containing basic block i

The constraint (7.9) ensures that a basic block is mapped to the SPM only once, instead of potentially being mapped as a member of the enclosing function and a member of a multi-block.

Experiments using this model were performed by Steinke et al. [488]. For some benchmark applications, energy reductions of up to about 80% were found, even though the size of the SPM was just a small fraction of the total code size of the application. Results for the bubble sort program are shown in Fig. 7.14.

Obviously, larger SPMs lead to a reduced energy consumption in the main memory (see white boxes). The energy required in the CPU is also reduced, since less wait cycles are required. The SPM needs only small amounts of energy (see the tiny blue boxes). Supply voltages have been assumed to be constant, even though a faster execution could have allowed us to scale down frequencies and voltages, leading to an even larger energy reduction.

- **Partitioned memories** [547]: Small memories are faster and require less energy per access. Therefore, it makes sense to partition memories into several smaller

Fig. 7.14 Energy reduction
by compiler-based mapping
to a SPM

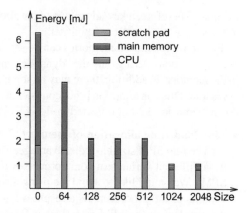

memories. The ILP model can be extended easily to also model several memories.
We do not distinguish between the various types of memory objects (functions,
basic blocks, variables, etc.) in this case. An index i represents any memory object.
Let

- S_j be the size of the memory j,
- s_i be the size of object i (as before),
- e_j be the energy consumption per access to memory j,
- n_i the number of accesses to object i (as before), and
- $x_{i,j}$ be defined as

$$x_{i,j} = \begin{cases} 1 & \text{if object } i \text{ is mapped to memory } j \\ 0 & \text{otherwise} \end{cases} \qquad (7.10)$$

Instead of maximizing the energy saving, we are now minimizing the overall energy
consumption. Hence, the goal is now to minimize

$$C = \sum_j e_j \sum_i x_{i,j} * n_i \qquad (7.11)$$

while respecting the constraints

$$\forall j : \sum_i s_i * x_{i,j} \le S_j \qquad (7.12)$$

$$\forall i : \sum_j x_{i,j} = 1 \qquad (7.13)$$

Partitioned memories are advantageous especially for varying memory requirements.
Storage locations accessed frequently are called the **working set** of an application.
Applications with a small working set could use a very small fast memory, whereas
applications requiring a larger working set could be allocated to a somewhat larger

memory. Therefore, a key advantage of partitioned memories is their ability to adapt to the size of the current working set.

Furthermore, unused memories can be shut down to save additional energy. However, we are considering only the "dynamic" energy consumption caused by accesses to the memory. In addition, there may be some energy consumption even if the memory is idle. This consumption is not considered here. Therefore, savings from shutting down memories are not reflected in Eqs. (7.11) and (7.12).

- **Link/load-time allocation of memory** [399]: Optimizing code at compile time for a certain SPM size has a disadvantage: The code might perform badly if we run it on different variants of some processor if these variants have differently sized SPMs. We would like to avoid requiring different executable files for the different variants of the processor. As a result, we are interested in executables which are independent of the SPM size. This is feasible if we perform the optimization at link-time. The proposed approach computes the ratio of the number of accesses divided by the size of a variable at compile time and stores this value together with other information about variables in the executable. At load time, the OS is queried for the size of the SPM. Then, the code is patched such that as many profitable variables as possible are allocated to the SPM.

7.3.3.3 Overlaying Allocation

Large applications may have multiple hot spots (multiple areas of code containing compute-intensive loops). Non-overlaying approaches fail to provide the best possible results in this context. For such applications, the SPM should be exploited for each of the hot spots. This requires an automatic migration between the layers in the memory hierarchy. For overlaying algorithms, memory objects are migrated between different levels of the hierarchy[2]. This migration can be either explicitly programmed in the application or inserted automatically. Overlaying algorithms are beneficial for applications with multiple hot spots, for which the code or data can be evicting each other. For overlaying algorithms, we are typically assuming that all applications are known at design time such that memory allocation can be considered at this time. Algorithms by Verma [531] and by Udayakumararan et al. [524] are early examples of such algorithms.

Verma's algorithm starts with the CFG of the application to be optimized. For edges of the graph, Verma considers potentially freeing the SPM for locally used memory objects.

Example 7.10: In Fig. 7.15, we are considering control blocks B1 to B10 and control flow branching at B2. We assume that array A is defined, modified, and used along the left path. T3 is only used in the right part of the branch. We consider potentially freeing the SPM so that T3 can be locally allocated to the SPM. This requires spill and load operations in potentially inserted blocks B9 and B10 (thin and dotted lines:

---

[2]Some of the material in this subsection has also been included in a separate book by the same author and publisher [360].

Fig. 7.15 Potential spill
code

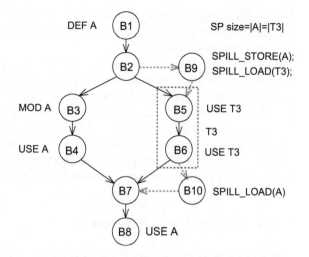

potential inserts). Cost and benefit of these spill operations are then incorporated into a global ILP. Solving the ILP yields an optimal set of memory copy operations. ∇

For a set of benchmarks, the average reductions in energy consumption and execution time, compared to the non-overlaying case, are 34 and 18%, respectively. Blocks of code are handled as if they were arrays of data.

Udayakumararan's algorithm is similar, but it evaluates memory objects according to their number of memory accesses divided by their size. This metric is then used to heuristically guide the optimization process. This approach can also take heap objects into account.

Large arrays are difficult to allocate to SPM. In fact, even a single array can be too large to fit into an SPM. The splitting strategy of Verma [153] is restricted to a single array splitting. Loop tiling is a more general technique, which can be applied either manually or automatically [327]. Furthermore, array indexes can be analyzed in detail such that frequently accessed array components can be kept in the SPM [340].

Our explanations have so far mainly addressed code and global data. *Stack* and *heap data* require special attention. In both cases, two trivial solutions may be feasible: In some cases, we might prefer not to allocate code or heap data to the SPM at all. In other cases, we could run stack [5] and heap size analysis [211] to check whether stack or heap fit completely into the SPM and, if they do, allocate them to the SPM.

For the heap, Dominguez et al. [129] proposed to analyze the liveness of heap objects. Whenever some heap object is potentially needed, code is generated to ensure that the object will be in the SPM. Objects will always be at the same address, so that the problem of dangling references to heap objects in the SPM is avoided. McIllroy et al. [366] propose a dynamic memory allocator taking characteristics of SPM into account. Bai et al. [31] suggest that the programmer should enclose accesses to global pointers by two functions $p2s$ and $s2p$. These functions provide conversions between global and local (SPM) addresses and also ensure a proper copying of memory contents.

For stack variables, Udayakumararan et al. [524] proposed to use two stacks, one for calls to short functions with their stack being in main memory and one for calls to computationally expensive functions whose stack area is in the SPM. Kannan et al. [269] suggested to keep the top stack frames in the SPM in a circular fashion. During function calls, a check for a sufficient amount of space for the required stack frame is made. If the space is not available, old stack frames are copied to a reserved area in main memory. During returns from function calls, these frames can be copied back. Various optimizations aim at minimizing the necessary checks.

7.3.3.4 Multiple Threads/Processes

The above approaches are still limited to handling a single process or thread. For multiple threads, moving objects into and out of the SPM at context switch time have to be considered. Verma [532] proposed three different approaches:

1. For the first approach, only a single process owns space in the SPM at any given time. At each context switch, the information of the preempted process in the occupied space is saved and the information for the process to be executed is restored. This approach is called the **saving/restoring approach**. This approach does not work well with large SPMs, since the copying would consume a significant amount of time and energy.
2. For the second approach, the space in the SPM is partitioned into areas for the various processes. The size of the partitions is determined in a special optimization. The SPM is filled during initialization. No further compiler-controlled copying is required. Therefore, this approach is called the **non-saving approach**. This approach makes sense only for SPMs large enough to contain areas for several processes.
3. The third approach is a **hybrid** approach: The SPM is split into an area jointly used by processes and a second area, in which processes obtain some exclusively allocated space. The size of the two areas is determined in an optimization.

In more dynamic cases, the set of applications may vary during the use of the system. For such cases, dynamic memory managers are appropriate. Pyka [438] published an algorithm based on an SPM manager which is part of the operating system.

This additional level of indirection can be avoided if a memory management unit (see Appendix C) is available. Egger et al. [143] developed a technique exploiting MMUs: At compile time, sections of code are classified as either benefiting or not benefiting from an allocation to the SPM. The code benefiting is stored in a certain area in the virtual address space. Initially, this area is not mapped to physical memory. Therefore, a page fault occurs when the code is accessed for the very first time. Page fault handling then invokes the SPM manager (SPMM) and the SPMM allocates (and deallocates) space in the SPM, always updating the virtual-to-real addresses translation tables as needed. The approach is designed to handle code and is capable of supporting a dynamically changing set of applications. Unfortunately, the size of current SPM corresponds to just a few entries in today's page tables, resulting in a coarse-grained SPM allocation.

7.3.3.5 Supporting Different Architectures and Objectives

We have so far considered different allocation types. Another dimension in SPM allocation is the architectural dimension. Implicitly, we have so far considered single core systems with a single-memory hierarchy layer and a single SPM. Other architectures exist as well. For example, there may be hybrid systems containing both caches and SPM. We can try to reduce cache misses by selectively allocating SPM space in case of cache conflicts [94, 268, 584]. Also, we can have different memory technologies, like flash memory or other types of nonvolatile RAM [538]. For flash memory, load balancing is important. Also, there might be multiple levels of memories.

SPM can possibly be shared across cores. Also, there may be multiple memory hierarchy levels, some of which can be shared. Liu et al. [333] present an ILP-based approach for this.

Still another dimension in SPM allocation is the objective function. So far, we have focused on energy or run-time minimization. Other objectives can be considered as well. Implicitly, we have modeled the average case energy consumption. We could have modeled the worst case energy consumption (WCEC) instead. The WCEC is an objective considered, for example, by Liu [333]. Reliability and endurance are relevant for the design of reliable applications, in particular, in the presence of aging [541]. It may also be necessary to avoid overheating of memories.

7.3.4 Reconciling Compilers and Timing Analysis

Almost all compilers which are available today do not include a timing model. Therefore, the development of real-time software typically has to follow an iterative approach: Software is compiled by a compiler which is unaware of any timing information. The resulting code is then analyzed using a timing analyzer such as aiT [4]. If the timing constraints are not met, some of the inputs to the compiler run must be changed and the procedure has to be repeated. We call this "trial-and-error"-based development of real-time software. This approach suffers from several problems. First of all, the number of required design iterations is initially unknown. Furthermore, the compiler used in this approach is "optimizing", but a precise evaluation of objectives apart from the code size is usually impossible. Hence, compiler writers can only hope that their "optimizations" have a positive impact of the quality of the code in terms of relevant objectives. Due to the complex timing behavior of modern processors, this hope is hardly supported by evidence. Finally, the "trial-and-error"-

based development of real-time software requires the designer to find appropriate modifications of the input to the compiler such that the real-time constraints will eventually be met.

This "trial-and-error"-based approach can be avoided if timing analysis is integrated into the compiler. This is the aim of the development of the worst case execution time aware compiler WCC at TU Dortmund. The development of WCC started with an integration of the timing analyzer aiT into an experimental compiler for the TriCore architecture. Figure 7.16 shows the resulting overall structure.

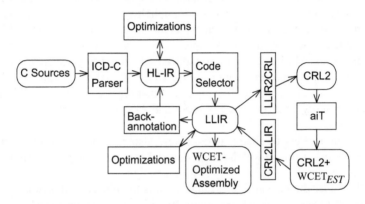

Fig. 7.16 Worst case execution time aware compiler WCC

WCC uses the ICD-C compiler infrastructure [222] to read and parse C source code. The source is then converted into a "high-level intermediate representation" (HL-IR). The HL-IR is an abstract representation of the source code. Various optimizations can be applied to the HL-IR. The optimized HL-IR is passed to the code selector. The code selector maps source code operations to machine instructions. WCC supports the Infineon TriCore and the ARM7 architectures. TriCore instructions are represented in the low-level intermediate representation LLIR. In order to estimate the $WCET_{EST}$, the LLIR is converted into the CRL2 representation used by aiT (using the converter LLIR2CRL). aiT is then able to generate $WCET_{EST}$ for the given machine code. This information is converted back into the LLIR representation (using the converter CRL2LLIR). WCC uses this information to consider $WCET_{EST}$ as the objective function during optimizations. This can be done straightforward for optimizations at the LLIR level. However, many optimizations are performed at the HL-IR level. $WCET_{EST}$-directed optimizations at this level require using back annotation from the LLIR level to the HL-IR level. ICD-C includes this back annotation.

WCC has been used to study the impact of optimizing for a reduced $WCET_{EST}$ in the compiler. The numerous results include a study of the impact of this objective for register allocation [151]. Results shown in Fig. 7.17 indicate a dramatic impact.

$WCET_{EST}$ can be reduced down to 68.8% of the original $WCET_{EST}$ on the average by just using WCET-aware register allocation in WCC. The largest reduction yields a $WCET_{EST}$ of only 24.1% of the original $WCET_{EST}$. The combined effect

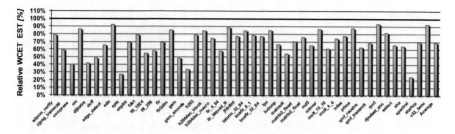

Fig. 7.17 Reduction of WCET$_{EST}$ by WCET-aware register allocation

of several such optimizations has been analyzed by Lokuciejewski et al. [337]. For the considered benchmarks, Lokuciejewski found a reduction of down to 57.1% of the original WCET$_{EST}$.

7.4 Power and Thermal Management

7.4.1 Dynamic Voltage and Frequency Scaling (DVFS)

Some embedded processors support dynamic power management (see p. 143) and dynamic voltage scaling (see p. 141). An additional optimization step can be used to exploit these features. Typically, such an optimization step follows code generation by the compiler. Optimizations at this step require a global view of all tasks of the system, including their dependencies and slack times.

Example 7.11: The potential of dynamic voltage scheduling is demonstrated in this following example [242]. We assume that we have a processor which runs at three different voltages, 2.5, 4.0, and 5.0 V. Assuming an energy consumption of 40 nJ per cycle at 5.0 V, Eq. (3.14) can be used to compute the energy consumption at the other voltages (see Table 7.1, where 25 nJ is a rounded value).

Table 7.1 Characteristics of processor with DVFS

V_{dd} [V]	5.0	4.0	2.5
Energy per cycle [nJ]	40	25	10
f_{max} [MHz]	50	40	25
Cycle time [ns]	20	25	40

Furthermore, we assume that our task needs to execute 10^9 cycles within 25 s. There are several ways of doing this, as can be seen from Figs. 7.18, 7.19, and 7.20. Using the maximum voltage (see Fig. 7.18), it is possible to shut down the processor during the slack time of 5 s (we assume the power consumption to be zero during this time).

Fig. 7.18 Possible voltage schedule

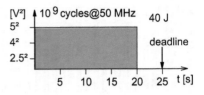

Another option is to initially run the processor at full speed and then reduce the voltage when the remaining cycles can be completed at the lowest voltage (see Fig. 7.19).

Fig. 7.19 Second voltage schedule

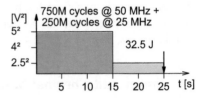

Finally, we can run the processor at a clock rate just large enough to complete the cycles within the available time (see Fig. 7.20).

Fig. 7.20 Third voltage schedule

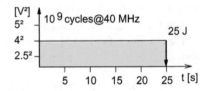

The corresponding energy consumptions can be calculated as

$$E_a = 10^9 * 40 * 10^{-9} = 40 \, \text{J} \tag{7.14}$$
$$E_b = 750 * 10^6 * 40 * 10^{-9} + 250 * 10^6 * 10 * 10^{-9} = 32.5 \, \text{J} \tag{7.15}$$
$$E_c = 10^9 * 25 * 10^{-9} = 25 \, \text{J} \tag{7.16}$$

A minimum energy consumption is achieved for the ideal supply voltage of 4 V. ∇

In the following, we use the term **variable voltage processor** only for processors that allow **any** supply voltage up to a certain maximum. It is expensive to support truly variable voltages, and therefore, actual processors support only a few fixed voltages.

The observations made for the above example can be generalized into the following statements. The proofs of these statements are given in the paper by Ishihara and Yasuura [242].

• If a variable voltage processor completes a task before the deadline, the energy consumption can be reduced[3].

---

[3]This formulation makes an implicit assumption in Lemma 1 of the paper by Ishihara and Yasuura explicit.

- If a processor uses a single supply voltage V_s and completes a task τ just at its deadline, then V_s is the unique supply voltage which minimizes the energy consumption of τ.

If a processor can only use a number of discrete voltage levels, then a voltage schedule using the two voltages which are the two immediate neighbors of the ideal voltage V_{ideal} can be chosen. These two voltages lead to the minimum energy consumption except if the need to use an integer number of cycles results in a small deviation from the minimum[4].

The statements can be used for allocating voltages to tasks. Next, we will consider such an allocation. We will use the following notation:

n : the number of tasks
EC_j : the number of executed cycles of task j
L : the number of voltages of the target processor
V_i : the ith voltage, where $1 \leq i \leq L$
f_i : the clock frequency for supply voltage V_i
d : the global deadline at which all tasks must have been completed
SC_j : the average switching capacitance during the execution of task j (SC_j comprises the actual capacitance C_L and the switching activity α (see Eq. (3.14) on p. 141))

The voltage scaling problem can then be formulated as an integer linear programming (ILP) problem (see p. 379). Toward this end, we introduce variables $X_{i,j}$ denoting the number of cycles executed at a particular voltage:

$X_{i,j}$: the number of clock cycles task j is executed at voltage V_i

Simplifying assumptions of the ILP model include the following:

- There is one processor that can be operated at a limited number of discrete voltages.
- The time for voltage and frequency switches is negligible.
- The worst case number of cycles for each task is known.

Using these assumptions, the ILP problem can be formulated as follows: Minimize

$$E = \sum_{j=1}^{n} \sum_{i=1}^{L} SC_j * X_{i,j} * V_i^2 \tag{7.17}$$

subject to

$$\forall j : \sum_{i=1}^{L} X_{i,j} = EC_j \tag{7.18}$$

---

[4]This need is not considered in the original paper.

and

$$\sum_{j=1}^{n}\sum_{i=1}^{L}\frac{X_{i,j}}{f_i} \leq d \qquad (7.19)$$

The goal is to find the number $X_{i,j}$ of cycles that each task τ_j is executed at a certain voltage V_i. According to the statements made above, no task will ever need more than two voltages. Using this model, Ishihara and Yasuura show that efficiency is typically improved if tasks have a larger number of voltages to choose from. If large amounts of slack time are available, many voltage levels help to find close to optimal voltage levels. However, four voltage levels do already give good results quite frequently.

There are many cases in which tasks actually run faster than predicted by their worst case execution times. This cannot be exploited by the above algorithm. This limitation can be removed by using checkpoints at which actual and worst case execution times are compared, and then to use this information to potentially scale down the voltage [28]. Also, voltage scaling in multi-rate task graphs was proposed [454]. DVFS can be combined with other optimizations such as body biasing [353]. Body biasing is a technique for reducing leakage currents.

7.4.2 Dynamic Power Management (DPM)

In order to reduce the energy consumption, we can also take advantage of power saving states, as introduced on p. 143. The essential question for exploiting DPM is: When should we go to a power-saving state? Straightforward approaches just use a simple timer to transition into a power-saving state. More sophisticated approaches model the idle times by stochastic processes and use these to predict the use of subsystems with more accuracy. Models based on exponential distributions have been shown to be inaccurate. Sufficiently accurate models include those based on renewal theory [465].

A comprehensive discussion of power management was published (see, e.g., [48, 339]). There are also advanced algorithms which integrate DVS and DPM into a single optimization approach for saving energy [466].

Allocating voltages and computing transition times for DPM may be two of the last steps of optimizing embedded software.

Power management is also linked to thermal management.

7.4.3 Thermal Management for MPSoCs

Design-time planning of the thermal behavior would need to leave large margins in terms of available performance. Hence, it is necessary to use run-time monitoring of temperatures. This means that thermal sensors must be available in systems which potentially could get too hot. This information is then used to control the generation of additional heat, and possibly has an impact on cooling mechanisms as well. Many users of mobile phones may already have observed this: It is, for example, very common to stop charging a mobile phone when it is already too hot. Controlling fans (when available) can be considered as another case of thermal management. Also, systems may be shutting down completely if temperatures are exceeding maximum thresholds. Some systems may be reducing the clock frequencies and voltages. There are also other options like a reduction of the performance by intentionally not using some of the available hardware. It is possible, for example, to issue less instructions per clock cycle or not to use some of the processor pipelines. For multiprocessor systems, tasks may be automatically migrated between various processors. In all of these cases, the objective "temperature" is evaluated at run-time and used to have an impact at run-time. Avoiding overheating is the goal of the work reported by Merkel et al. [373] and by Donald et al. [130]. Using temperature sensors to control the system means that control loops are being created. Potentially, such loops could start to oscillate. Atienza et al. have compared the behavior of various control strategies and came to the conclusion that an advanced control loop algorithm provides the best results, with a higher computing performance at a lower temperature, compared to standard approaches [582]. The details of this control loop design would be beyond the scope of a textbook useful for undergraduate students.

7.5 Problems

We suggest solving the following problems either at home or during a flipped classroom session:

7.1: Loop unrolling is one of the potentially useful optimizations. Please name two potential benefits and two potential problems!

7.2: Consider the following program:

```
1      #include <stdio.h>
2      #define DATALEN 15
3      #define FILTERTAPS 5
4      double x[DATALEN] = { 128.0, 130.0, 180.0, 140.0, 120.0,
5                            110.0, 107.0, 103.5, 102.0,  90.0,
6                             84.0,  70.0,  30.0,  77.3,  95.7 };
7      const double h[FILTERTAPS]={0.125,-0.25,0.5,-0.25,0.125};
8      double y[DATALEN]; // result;
9      int main(void)
10     {int i,n;
11      for(i=0;i<DATALEN;++i)
12     {y[i] = 0;
13          for(n=0; n < FILTERTAPS; ++n)
14            if ((i-n) > = 0) y[i] += h[n]*x[i-n];
15          }
16      for(i = 0; i < DATALEN; ++i) printf("%.2f ",y[i]);
17      return 0;
18  }
```

Perform at least the following optimizations:

- Removal of the if in the innermost loop (line 14)
- Loop unrolling (line 13)
- Constant propagation
- Floating-point to fixed-point conversion
- Avoidance of all accesses to arrays.

Please provide the optimized version of the program after each of the transformations and do also check for consistent results!

7.3: Suppose that your computer is equipped with a main memory and a scratchpad memory. Sizes and the required energy per access are shown in Table 7.2.

Table 7.2 Memory characteristics

Memory	Size in bytes	Energy per access (nJ)
Scratchpad	4096 (4k)	1.3
Main memory	262,144 (256 k)	31

Also, let us assume that we are accessing variables as shown in the Table 7.3.

Table 7.3 Variable characteristics

Variable	Size in bytes	Number of accesses
a	1024	16
b	2048	1024
c	512	2048
d	256	512
e	128	256
f	1024	512
g	512	64
h	256	512

Which of those variables should be allocated to the scratchpad memory, provided that we use a static, non-overlaying allocation of variables? Use the integer linear problem (ILP) model to select the variables. Your result should include the ILP model as well as the results. You may use the *lp_solve* program [15] to solve your ILP problem.

Chapter 8
Test

8.1 Scope

The purpose of testing is to make sure that a manufactured embedded system behaves as intended. Testing can be done during or after the fabrication (manufacturing test) and also after the system has been delivered to the customer (field testing). Testing of embedded systems contained in a cyber-physical or IoT system needs special attention for several reasons:

- Embedded systems integrated into a physical environment may be safety-critical. Therefore, their malfunctioning can be much more dangerous than, say, the malfunctioning of office equipment. As a result, expectations for the product quality are higher than for non-safety critical systems.
- Testing of timing-critical systems has to validate the correct timing behavior. This means that just testing the functional behavior is not sufficient.
- Testing embedded/cyber-physical systems in their real environment may be dangerous. For example, testing control software in a nuclear power plant can be a source of serious, far-reaching problems.

Preparations for testing should be done no later than at the end of the design phase. Preferably, necessary support for testing should even be considered earlier, intertwined with the design process and using testability as one of the objectives for evaluating designs. In order not to overload Chap. 5, we have moved all aspects of testing into this separate chapter. The presentation corresponds to considering testing only at the very end of the design flow (see Fig. 8.1), even though an earlier consideration during an actual design would be advisable. However, an early consideration is not always common practice, and, therefore, Fig. 8.1 might also correspond to an actual design flow.

© Springer International Publishing AG 2018
P. Marwedel, *Embedded System Design*, Embedded Systems,
DOI 10.1007/978-3-319-56045-8_8

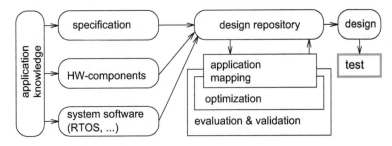

Fig. 8.1 Design flow with testing at its very end

In testing, we are typically denoting the system under design (SUD) as the **device under test** (DUT). We are applying a set of specially selected input patterns, so-called **test patterns** to the input(s) of the DUT, observe its behavior, and compare the behavior with the expected behavior. Test patterns are normally applied to the real, already manufactured system. The main purpose of testing is to identify systems that have not been correctly manufactured (manufacturing test) and to identify systems that fail later (field test).

Testing includes a number of different actions:

1. **test pattern generation,**
2. **test pattern application,**
3. **response observation,** and
4. **result comparison**.

8.2   Test Procedures

8.2.1   Test Pattern Generation for Gate Level Models

In test pattern generation, we try to identify a set of test patterns which distinguishes a correctly working from an incorrectly working system. Test pattern generation is usually based on **fault models**. Such fault models are models of possible faults. Test pattern generation tries to generate tests for all faults that are possible according to a certain fault model.

The stuck-at-fault model is a frequently used fault model. It is based on the assumption that any internal wire of an electronic circuit is either permanently connected to '0' or '1'. It has been observed that many faults actually behave as if some wire was permanently connected that way.

Example 8.1: As an example, consider the circuit shown in Fig. 8.2[1].

Fig. 8.2 Test pattern at the gate level

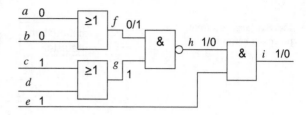

Suppose that we would like to check whether there is a stuck-at-1 fault for signal f. Toward this end, we try to set f to '0' by setting $a = b = $'0'. As a result, f should be '1' if there is this fault, and otherwise, it should be '0'. In order to observe this difference, we must propagate it to the output signal i. For this to happen, we must set e to '1' and set either c or d to '1'. h and i will be '1' if there is no fault and '0' otherwise. The test pattern comprises all values of inputs a to e. The D-algorithm can be used to generate this test pattern [304]. ▽

Many techniques for test pattern generation are based on the stuck-at-fault model. However, CMOS technologies require more comprehensive fault models. In CMOS technologies, faults can turn combinatorial devices into devices having internal states. This problem can occur if wires are broken (this case is known as **stuck-at-open fault**). As a result of this, gates of transistors can become disconnected. Such transistors will be conducting or non-conducting, depending on the charge stored on the gate before the wire was broken. In this way, the gate "remembers" the input signal due to stored charges. Furthermore, there may be transient faults and delay faults (faults changing the delay of a circuit). Delay faults may be the result of cross talk between adjacent wires. Fault models exist which take such hardware faults into account [298].

While good fault models exist for hardware testing, the same is not true for software testing.

8.2.2 Self-Test Programs

One of the key problems of testing modern integrated circuits is their limited number of pins, making it more and more difficult to access internal components. Also, it is getting very difficult to test these circuits at full speed, since testers must be at least as fast as the circuits themselves. The fact that many embedded systems are based on processors provides a way out of this dilemma: Processors are capable of running test programs or *diagnostics*. Such diagnostics have been used to test mainframe machines for decades.

---

[1]Please remember: consistent with standard ANSI/IEEE 91, the symbols ≥ 1 and & denote OR and AND gates, respectively.

Example 8.2: Figure 8.3 shows some components that might be contained in some processors.

Fig. 8.3 Segment from processor hardware

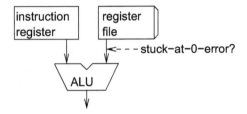

In order to test for stuck-at-faults at the input of the ALU, we can execute a small test program:

```
store pattern of all '1's in the register file;
perform xor between constant "0000...00" and register;
test if result contains a '0' bit;
if yes, report error;
otherwise start test for next fault;
```

∇

Similar small programs can be generated for other faults. Unfortunately, the process of generating diagnostics for mainframes has mostly been a manual one. Some researchers have proposed to generate diagnostics automatically [51, 56, 66, 295, 299, 300].

8.3 Evaluation of Test Pattern Sets and System Robustness

8.3.1 Fault Coverage

The quality of test pattern sets can be evaluated using **fault coverage** as a metric.

Definition 8.1: Fault coverage is the percentage of potential faults that can be found for a given test pattern set:

$$Coverage = \frac{\text{Number of detectable faults for a given test pattern set}}{\text{Number of faults possible due to the fault model}}$$

In practice, achieving a good product quality requires fault coverages in the area of at least 98 to 99%. The requirements may be higher for particular systems. Also, special fault models may be necessary for certain hardware components (e.g., for batteries).

In addition to achieving a high coverage, we must also achieve a high **correctness coverage**. This means that a fault-free system must be recognized as such. Otherwise, it would be possible to achieve a 100% coverage by classifying all systems as faulty.

In order to increase the number of options that exist for system validation, it has been proposed to use test methods already during the design phase. For example, test pattern sets can be applied to software models of systems in order to check whether two software models behave in the same way. More time-consuming formal methods need to be applied only to those cases in which the system passed this test-based equivalence check.

8.3.2 Fault Simulation

It is currently not feasible (and it will probably not be feasible) to completely predict the behavior of systems in the presence of faults or to analytically compute the coverage. Therefore, the behavior of systems in the presence of faults is frequently simulated. This type of simulation is called **fault simulation**. In fault simulation, system models are modified to reflect the behavior of the system in the presence of a certain fault.

The goals of fault simulation include:

- to know the effect of a fault of the components at the system level (i.e., to check whether faults are redundant) and
- to know whether or not mechanisms for improving fault tolerance actually help.

Definition 8.2: Faults are called **redundant** if they do not affect the observable behavior of the system.

Fault simulation requires the simulation of the system for all faults feasible for the fault model and also for a possibly large number of different input patterns. Accordingly, fault simulation is an extremely time-consuming process. Different techniques have been proposed to speed up fault simulation.

One such technique applies to fault simulation at the gate level. In this case, internal signals are single bit signals. This fact enables the mapping of a signal to a single bit of some machine word of a simulating host machine. AND- and OR-machine instructions can then be used to simulate Boolean networks. However, only a single bit would be used per machine word. Efficiency is improved with parallel fault simulation.

Definition 8.3: Fault simulation is called **parallel fault simulation** if $n > 1$ different test patterns are simulated at the same time where n is the length of a bit vector supported as a machine data type of the simulating processor.

The values of each of the n test patterns are mapped to a different bit position in the machine. Executing the same set of AND- and OR-instructions will then simulate the behavior of the Boolean network for n test patterns instead of for just one. AVX instructions mentioned on p. 152 are very useful for this.

8.3.3   Fault Injection

Fault simulation may be too time-consuming for real systems. If actual systems are available, fault injection can be used instead. In fault injection, real existing systems are modified and the overall effect on the system behavior is checked. Fault injection does not rely on fault models (even though they can be used). Hence, fault injection has the potential of generating faults that would not have been predicted by a fault model.

We can distinguish between two types of fault injection:

- local faults within the system and
- faults in the environment (behaviors which do not correspond to the specification). For example, we can check how the system behaves if it is operated outside the specified temperature or radiation ranges.

Several methods can be used for fault injection:

- fault injection at the hardware level: examples include pin manipulation, electromagnetic and nuclear radiation and
- fault injection at the software level: examples include toggling some memory bits.

The quality of fault injection depends on the "probe effect": Probing might have an impact on the behavior of the system. This impact should be as small as possible and essentially be negligible.

According to experiments reported by Kopetz [292], software-based fault injection was essentially as effective as hardware-based fault injection. Nuclear radiation was a noticeable exception; in that, it generated errors which were not generated with other methods.

8.4   Design for Testability

8.4.1   Motivation

Ideas for test pattern generation for Boolean circuits have been presented in Sect. 8.2.1. For circuits implementing state machines (automata), test pattern generation is more difficult. Verifying whether or not two finite state machines are equivalent may require complex input sequences [290].

Example 8.3: The state chart of Fig. 2.25 is shown again in Fig. 8.4 for convenience: Suppose that we would like to test the transition from state C to state D. This requires us to get into state C first, by applying an appropriate sequence of input patterns. Assuming that we start from the default state, we have to generate a sequence comprising signals g and h. Next, we must generate input event i and check if output y is generated. Also, we need to check whether we reached state D. We could apply

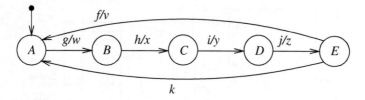

Fig. 8.4 Finite state machine to be tested

input signal j and check if output z is emitted. Still, we would not be sure that we actually had reached state D. There could be a fault resulting in the generation of z for a transition from a different state. This procedure is rather complicated, takes a lot of time, and is susceptible to interference with other errors. Nevertheless, the procedure could even be more complicated since the overall test in our example is simplified by the fact that the FSM contains a linear chain of transitions (see the assignments of this chapter). ∇

This example demonstrates that if testing comes in only as an afterthought, it may be very difficult to test a system. In order to simplify tests, special hardware can be added such that testing becomes easier. The process of designing for better testability is called **design for testability**, or **DfT**. Special purpose hardware for testing finite state machines is a prominent example of this.

8.4.2   Scan Design

Reaching certain states and observing states resulting from the application of input patterns is very much simplified with **scan design**. In scan design, all flip-flops storing states are connected to form serial shift registers (see Fig. 8.5).

The circuit contains three D-type flip-flops DFF and one multiplexer at each of the flip-flop inputs. Using the control input of the multiplexers (shown at the bottom of the multiplexer inputs), we can either connect the flip-flops to the network generating the next state from the current state and the current input or we can connect flip-flops to form a serial chain. Setting the multiplexers to scan mode, we can load state bit after state bit into the scan chain (one bit at every clock tick). This way, we can load any state into the three flip-flops serially. In a second phase, we can apply an input pattern to the FSM while the multiplexers are set to normal mode. After the next clock tick, the FSM will be in a new state. This new state can be serially shifted out in the third and final phase, using the serial mode again (one bit per clock tick). The net effect is that we do not need to worry about how to get into certain states and how to observe whether or not the Boolean function δ for computing the next state has been correctly implemented while we are generating tests for the FSM. Effectively, the fact that we are dealing with state-based systems has an impact only on the two (simple) shift phases, and test pattern generation for (stateless) Boolean networks

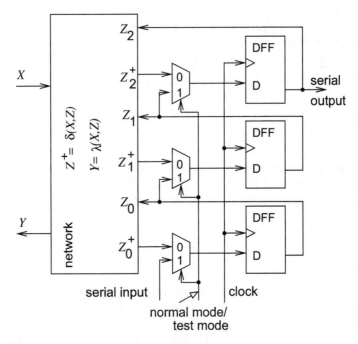

Fig. 8.5 Scan path design

can be used for checking for correct outputs. This means that it is sufficient to use test pattern generation methods for Boolean functions (stateless networks) instead of caring about complex input sequences, etc.

Scan design is a technique which works well for single chips. For board-level integration, it is necessary to have some technique for connecting scan chains of several chips. **JTAG** is a standard designed for this. The standard defines registers at the boundaries of all chips and a number of test pins and control commands such that all chips can be connected in scan chains. JTAG is also known as [425] boundary scan.

8.4.3   Signature Analysis

In order to also avoid shifting out the response of the device under test (DUT), responses can be compacted. A setup like the pipeline shown in Fig. 8.6 can be used for this purpose.

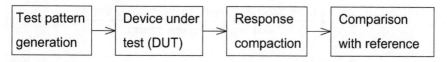

Fig. 8.6 Testing a device under test (DUT)

Generated test patterns are used as inputs (or so-called stimuli) to the DUT. The response of the DUT is then compacted to form a **signature**, which characterizes the response. This response is later compared to the expected response. The expected response can be computed by simulation.

The compaction is typically performed with linear-feedback shift registers (LFSRs), shift registers with an XOR-feedback.

Example 8.4: Figure 8.7 shows a 4-bit LFSR (**left**) and the associated state diagram (**right**) [304].

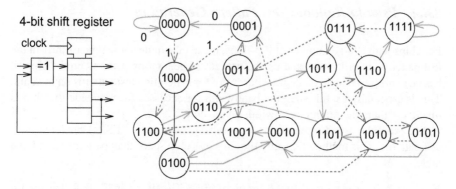

Fig. 8.7 LFSR for response compaction: **left**: schematic; **right**: state diagram

Blue dashed lines denote an input of '1', red solid lines denote an input of '0'. The selected feedback yields all possible signatures. During testing, the response of the system tested is sent to the input of the LFSR. The LFSR will then generate a signature reflecting the response. ∇

Due to storing the signature instead of the full response, several response patterns can be mapped to the same signature. What is the probability of obtaining a correct signature from an incorrect response?

In general, an n-bit signature generator can generate 2^n signatures. For an m-bit response of the DUT, the best that we can do is to evenly map $2^{(m-n)}$ responses to the same signature. Suppose that we expect a certain signature to be generated for the correct response of the system. Then, $2^{(m-n)} - 1$ incorrect responses would also map to the same signature. There is a total of $2^m - 1$ incorrect responses if responses are m-bit long. Hence, the probability of an incorrect response to map to the correct signature (provided patterns map evenly to signatures) is

$$P = Pr\left(\frac{\text{other patterns mapping to the same signature}}{\text{total number of other patterns}}\right) \tag{8.1}$$

$$= \frac{2^{(m-n)} - 1}{2^m - 1} \tag{8.2}$$

$$\approx \frac{2^{(m-n)}}{2^m} \text{ for } m \gg n \tag{8.3}$$

$$\approx \frac{1}{2^n} \text{ for } m \gg n \tag{8.4}$$

This means that the probability of generating correct signatures from an incorrect test response is very small if the shift register is long. For example, actual shift registers may be 32 bits long. Nevertheless, it is still feasible to have the correct signature for wrong inputs. The corresponding effect is called **aliasing**. A careful analysis of aliasing is recommended at least for critical applications.

8.4.4 Pseudo-Random Test Pattern Generation

For chips with a large number of flip-flops, it can take quite some time to shift-in the test patterns. In order to speed up the process of generating patterns on the chip, it has been proposed to also integrate hardware for generating test patterns on the chip. This is especially useful when the bandwidth for accesses from outside the chip is much less than the internal bandwidth on the chip.

For example, pseudo-random patterns (also generated by LFSRs) can be used as test patterns. This method typically requires less chip space than patterns stored in a table.

Example 8.5: We can modify the circuit of Fig. 8.7 as shown in Fig. 8.8. The circuit generates all possible test patterns, except the pattern consisting of all zeros.

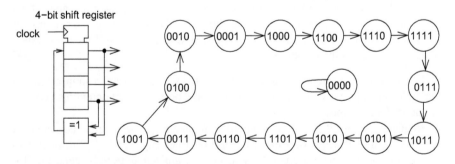

Fig. 8.8 Linear-feedback shift register for test pattern generation

∇

The pattern consisting of all zeros has to be avoided, since the generator would get stuck once it arrives at this pattern. The generated patterns are typically exercising systems to be tested much better than simple counters.

8.4.5 The Built-In Logic Block Observer (BILBO)

The **built-in logic block observer** (BILBO) [286] has been proposed as a circuit combining test pattern generation, test response compaction, and serial scan capabilities. A BILBO with three D-type flip-flops is shown in Fig. 8.9.

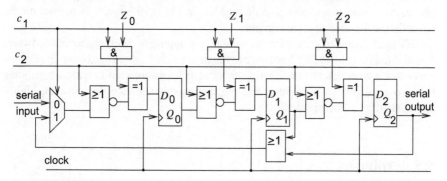

Fig. 8.9 BILBO

Modes of BILBO registers are shown in Table 8.1. The 3-bit register shown in Fig. 8.9 can be in scan path, reset, linear-feedback shift register (LFSR), and normal mode. In LFSR mode, it can be used for either generating pseudo-random patterns or for compacting responses from inputs (Z_0 to Z_2). In this case, compaction is based on parallel inputs instead of the sequential inputs we have considered so far. Purpose and behavior of compaction from parallel inputs are similar to that for serial inputs.

Table 8.1 Modes of BILBO registers

c_1	c_2	D_i	Comment
'0'	'0'	$'0' \oplus \overline{Q_{i-1}} = \overline{Q_{i-1}}$	Scan path mode
'0'	'1'	$'0' \oplus \overline{'1'} = '0'$	Reset
'1'	'0'	$Z_i \oplus \overline{Q_{i-1}}$	LFSR mode
'1'	'1'	$Z_i \oplus \overline{'1'} = Z_i$	Normal mode

Typically, BILBOs are used in pairs (see Fig. 8.10).

Fig. 8.10 Cross-coupled BILBOs

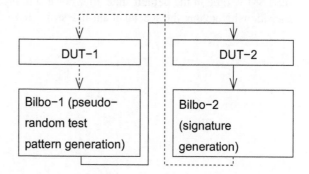

One BILBO generates pseudo-random test patterns, feeding some Boolean network with these patterns. The response of the Boolean network is then compressed by a second BILBO connected to the output of the network. At the end of the test sequence, the compacted response is serially shifted out and compared with the expected response. The expected response can be computed by simulation.

During a second phase, the roles of the two BILBOs can be swapped. During this phase, the connection shown as a dashed line in Fig. 8.10 is used. In normal mode, BILBOs can be used as state registers.

DfT hardware is of great help during the prototyping and debugging of hardware. It is also useful to have DfT hardware in the final product, since hardware fabrication never has a zero defect rate. Testing fabricated hardware significantly contributes to the overall cost of a product, and mechanisms that reduce this cost are highly appreciated by all companies.

8.5   Problems

8.1: Consider the circuit shown in Fig. 8.2. Generate a test pattern for a stuck-at-0 fault at signal h!

8.2: Which state diagram corresponds to the LFSR shown in Fig. 8.11?

Fig. 8.11 LFSR

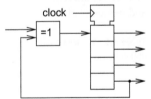

8.3: Specify test patterns and expected responses for the FSM shown in Fig. 8.4. These patterns must be specified as a sequence of pairs (test pattern, expected response). Events shown in Fig. 8.4 can be used as test patterns. We assume that the FSM will be in the default state after power-on. Provide a complete test for all transitions! Note that the special chain-like structure of the FSM simplifies testing.

Appendix A
Integer Linear Programming

Integer linear programming (ILP) is a mathematical optimization technique applicable to a large number of optimization problems.

ILP models provide a general approach for modeling optimization problems. ILP models consist of two parts: a cost function and a set of constraints. Both parts involve references to a set $X = \{x_i\}$ of integer-valued variables. Cost functions must be linear functions of those variables. So, they must be of the general form

$$C = \sum_i a_i x_i, \text{ with } a_i \in \mathbb{R}, x_i \in \mathbb{N}_0 \qquad \text{(A.1)}$$

The set J of constraints must also consist of linear functions of integer-valued variables. They must be of the form

$$\forall j \in J : \sum_i b_{i,j} x_i \geq c_j \text{ with } b_{i,j}, c_j \in \mathbb{R} \qquad \text{(A.2)}$$

Definition A.1: The **integer linear programming (ILP-) problem** is the problem of minimizing the cost function of equation (A.1) subjects to the constraints given in Eq. (A.2). If all variables are constrained to being either 0 or 1, the corresponding model is called a **0/1-integer linear programming model**. In this case, variables are also denoted as **(binary) decision variables**.

Note that $\geq$ can be replaced by $\leq$ in Eq. (A.2) if constants $b_{i,j}$ are modified accordingly. Also, the case of negative variables x_i (i.e., allowing x_i to have any integer value) can be transformed into the case of nonnegative variables shown above by multiplying constants by -1. Applications requiring **maximizing** some **gain** function C' can be changed into the above form by setting $C = -C'$. Equations may be represented by pairs of constraints, but they are typically used to eliminate some variables.

© Springer International Publishing AG 2018
P. Marwedel, *Embedded System Design*, Embedded Systems,
DOI 10.1007/978-3-319-56045-8

For example, assuming that x_1, x_2, and x_3 must be integers, the following set of equations represent a 0/1-IP model:

$$C = 5x_1 + 6x_2 + 4x_3 \tag{A.3}$$
$$x_1 + x_2 + x_3 \geq 2 \tag{A.4}$$
$$x_1 \leq 1 \tag{A.5}$$
$$x_2 \leq 1 \tag{A.6}$$
$$x_3 \leq 1 \tag{A.7}$$

Due to the constraints, all variables are either 0 or 1. There are four possible solutions. These are listed in Table A.1. The solution with a cost of 9 is optimal.

Table A.1 Possible solutions of the presented ILP problem

x_1	x_2	x_3	C
0	1	1	10
1	0	1	9
1	1	0	11
1	1	1	15

ILP is a variant of linear programming (LP). For linear programming, variables can take any real values. ILP and LP models can be solved optimally using mathematical programming techniques. Unfortunately, ILP is NP-complete (but LP is not) and ILP execution times may become very large.

Nevertheless, ILP models are useful for modeling optimization problems as long as the model sizes are not extremely large. Modeling optimization problems as integer linear programming problems make sense despite the complexity of the problem: Many problems can be solved in acceptable execution times and if they cannot, ILP models provide a good starting point for heuristics. Execution times depend on the number of variables and on the number and structure of the constraints. Good ILP solvers (like lp_solve [15] or CPLEX) can solve well-structured problems containing a few thousand variables in acceptable computation times (e.g., minutes). For more information on ILP and LP, refer to books on the topic (e.g., to Wolsey [566]).

Appendix B
Kirchhoff's Laws and Operational Amplifiers

Our presentation of D/A converters on page 188 assumes some basic knowledge about operational amplifiers. This knowledge is frequently lacking among computer science students, and therefore, the necessary fundamentals are presented in this Appendix. These fundamentals require an understanding of Kirchhoff's laws, of which students will also be reminded in this Appendix.

Kirchhoff's Laws

Kirchhoff's laws provide a means for analyzing electrical circuits. The first rule is Kirchhoff's Current Law, also called Kirchhoff's Junction Rule, or Kirchhoff's First Law. The rule applies to junctions such as the one shown in Fig. B.1.

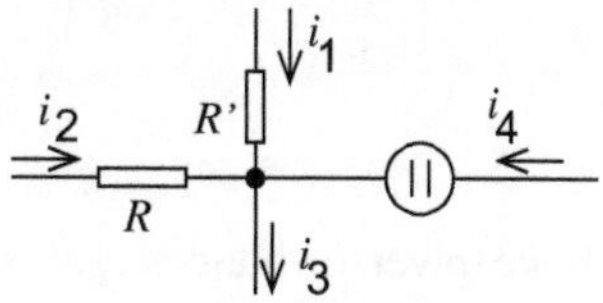

Fig. B.1 Junction in an electrical circuit

Theorem B.1 (Kirchhoff's Current Law): *At any point in an electrical circuit, the sum of currents flowing toward that point is equal to the sum of currents flowing away from that point [261]. Formally, for any node in a circuit we have:*

$$\sum_k i_k = 0 \qquad (B.1)$$

If Kirchhoff's law is used in the form of equation (B.1), currents denoted by arrows pointing away from the node must be counted as negative, and this counting is independent of the direction into which electrons are actually flowing.

© Springer International Publishing AG 2018
P. Marwedel, *Embedded System Design*, Embedded Systems,
DOI 10.1007/978-3-319-56045-8

Example B.1: For the currents of Fig. B.1, we have

$$i_1 + i_2 - i_3 + i_4 = 0 \qquad (B.2)$$
$$i_1 + i_2 + i_4 = i_3 \qquad (B.3)$$

This invariance exists due to the conservation of electrical charge. Without this rule, the total electrical charge would not remain constant, and the voltage would increase.

Kirchhoff's second rule applies to loops in a circuit. It is known as Kirchhoff's Voltage Law, Kirchhoff's Loop Rule, or Kirchhoff's Second Law. Figure B.2 shows an example.

Fig. B.2 Loop in an electrical circuit

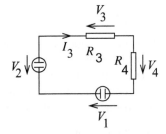

Theorem B.2 (Kirchhoff's Voltage Law): *The sum of the potential differences (voltages) across all elements around any closed circuit must be zero [261]. Formally, for any loop in a circuit we have:*

$$\sum_k V_k = 0 \qquad (B.4)$$

If we traverse voltages against the arrow direction, we have to count them as negative.

Example B.2: For the schematic of Fig. B.2, we have

$$V_1 - V_2 - V_3 + V_4 = 0 \qquad (B.5)$$

The underlying reason for this invariance is the conservation of energy. Without this rule, we could accelerate charge in the loop and the charge would accumulate energy without any energy consumption elsewhere.

In general, it is not relevant into which direction electrons are actually flowing and which of two terminals are actually positive with respect to some other terminal. Arrows can be selected in an arbitrary way. We just have to make sure that we respect the direction of the arrows when we apply Kirchhoff's laws. If arrows for voltages and currents across components are pointing in opposite directions, the equation for that component has to take that into account.

Example B.3: Ohm's law for resistor R_3 in Fig. B.2 reads as follows, due to the opposite directions of voltage and current arrows:

$$I_3 = -\frac{V_3}{R_3} \tag{B.6}$$

Of course, we will typically try to define the direction of voltages and currents such that we avoid having too many minus signs.

Operational Amplifiers

In electronics, there is frequently the need to amplify some signal $x(t)$ in order to obtain some amplified signal $y(t) = a \cdot x(t)$, with $a > 1$. a is called the **gain**. Designing different circuits for each and every gain would be a laborious task. Therefore, designers are frequently using a general amplifier which can be easily configured to have the required gain. Such a general amplifier is called **operational amplifier**, or op-amp for short. Op-amps are designed for a very large maximum gain. The required actual gain can be adjusted with a proper selection of a few hardware components in the circuit surrounding the op-amp.

More precisely, an operational amplifier is a component having two signal inputs and one signal output. In addition, there are at least two power supply inputs (see Fig. B.3).

Fig. B.3 Operational amplifier

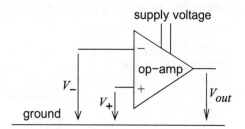

Op-amps amplify the difference between the voltages at the two signal inputs with respect to ground by a gain g:

$$V_{out} = g * (V_+ - V_-) \tag{B.7}$$

g is called the **open loop gain** and is typically very large ($10^4 < g < 10^6$). For an ideal op-amp, g would approach infinity. Furthermore, op-amps usually come with a very high input impedance ($> 1\,M\Omega$). Hence, we can frequently ignore signal input currents. For an ideal op-amp, the input impedance would be infinity and input currents would be zero.

Op-amps have been commercially available for decades, both as separate integrated circuits and within other circuits. They differ by their speed, their voltage ranges, their current drive capability, and other characteristics. The actual gain of the circuit is selected with external resistors. Figure B.4 shows how this can be done.

Fig. B.4 Op-amp with feedback

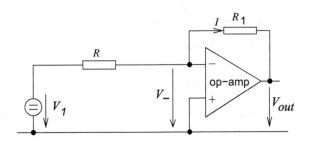

Any small voltage between the two signal inputs is amplified by a large factor. Via resistor R_1, the resulting output voltage is fed back. feedback is to the inverting input and therefore, any positive voltage V_- results in a negative voltage V_{out} and vice versa. This means that the feedback will work against the input voltage, and it does so very strongly, due to the large amplification. Therefore, the feedback will reduce the voltage at the input pin. The question is: By how much? We can use Kirchhoff's rules to find the resulting voltage V_- (see Fig. B.5).

Fig. B.5 Op-amp with feedback (loop highlighted)

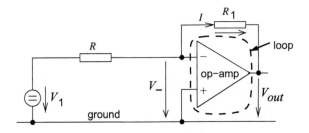

Due to the characteristics of op-amps, we have

$$V_{out} = -g * V_-$$ (B.8)

Due to Kirchhoff's law for the loop shown by a dashed line in Fig. B.5, we have

$$I * R_1 + V_{out} - V_- = 0$$ (B.9)

Note that we include a minus sign for V_-, since we are traversing a segment of the loop against the direction of the arrow. From Eqs. (B.8) and (B.9), we get

$$I * R_1 + (-g) * V_- - V_- = 0 \tag{B.10}$$

$$(1 + g) * V_- = I * R_1 \tag{B.11}$$

$$V_- = \frac{I * R_1}{1 + g} \tag{B.12}$$

$$V_{-,ideal} = \lim_{g \to \infty} \frac{I * R_1}{1 + g} \tag{B.13}$$

$$= 0 \tag{B.14}$$

This means that, for an ideal op-amp, V_- is 0. Due to this, the inverting signal input is called **virtual ground**. Nevertheless, this input cannot be connected to ground, since this would change the currents.

Computing the actual gain of the circuit in Fig. B.4 is left as an exercise for Chap. 3.

Appendix C
Paging and Memory Management Units

In this Appendix, we are discussing a basic technique for managing memories. In simple systems, physical memories are actually addressed by the addresses which are seen by (assembly language) programmers. This approach is very easy to implement from a hardware technology point of view. However, this approach has disadvantages for using the memory. For example, the allocation of objects to memory is very static. The size of memory objects needs to be estimated before actually allocating memory.

More flexibility is obtained when we distinguish between the memory addresses as seen by the (assembly language) programmer and the ones used to address physical memory. Addresses as seen by the programmer are called **virtual addresses**, and addresses seen at the memory are called **real** or **physical addresses**.

For a memory organization called **paging**, we partition the space of virtual addresses into chunks of equal size, called pages. The size of these pages is a power of two, such as 2 k bytes or 4 k bytes. As a result, virtual addresses consist of those bits addressing a particular page and those addressing a word or byte within a page. The first set of bits is called page number, the second the offset.

Physical memory is partitioned into **page frames** of the same size.

Then, a mapping table – called **page table** – contains the information needed to map page numbers to the corresponding start address in physical memory. The offset is identical for virtual and real addresses (see Fig. C.1 (**left**)).

This allows for a more dynamic allocation to memory. Contiguous ranges of virtual addresses do not need to be allocated to contiguous ranges in real memory, offering much more allocation freedom (see Fig. C.1 (**right**)). Certain memory objects (such as stacks) can grow and shrink in multiples of page sizes.

There may be more than one virtual address space, such as one address space per process managed by the operating system. In this case, the relevant page table has to be set during context switches. The actual mapping from virtual to real addresses is typically performed in a **memory management unit** (MMU). The MMU is placed between processors and the memory and converts virtual addresses into real addresses. During its operation, the MMU needs to know the contents of the page table. Page tables may be large and, hence, fast buffers can be used as

P. Marwedel, *Embedded System Design*, Embedded Systems,
DOI 10.1007/978-3-319-56045-8

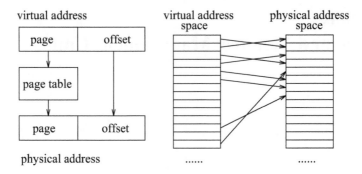

Fig. C.1 Paging: *left* address translation; *right*: impact on mapping of address spaces

caches specialized for accesses to the page table. These fast buffers are called **translation look-aside buffers** (TLB) or **address translation memories** (ATM). TLBs are assumed to contain copies of frequently used correspondences between virtual and real addresses.

For PCs, the presented approach is frequently combined with **demand paging**, i.e., fetching page frames currently not in main memory from a slower background memory on demand. Demand paging is less popular in embedded systems, frequently due to the nonavailability of a background memory. The term "paging" is often used for what we have called demand paging. However, the distinction between paging as a method for mapping virtual addresses to real addresses and paging fetching information from a background memory automatically is important for embedded systems.

In addition to the reasons mentioned above, paging is also useful for memory protection. Page table entries commonly contain bits which indicate access permissions to the memory represented by this entry. Common permission bits include read, write, and execute permissions. These protection bits enable the system designer to isolate memory spaces of different tasks or processes against each other and also help to protect the operating system from erroneous memory accesses by tasks or even rogue processes which try to subvert a system's security. The latter aspect is gaining relevance especially in the context of networked embedded systems, as used in Internet of Things applications.

Please refer to books on computer architecture for more information on memory management [205].

References

1. Aamodt, T., Chow, P.: Embedded ISA support for enhanced floating-point to fixed-point ANSI C compilation. In: Proceedings of the International Conference on Compilers, Architectures, and Synthesis for Embedded Systems (CASES), pp. 128–137 (2000)
2. Abella, J., Hardy, D., Puaut, I., Quiñones, E., Cazorla, F.J.: On the comparison of deterministic and probabilistic wcet estimation techniques. In: Euromicro Conference on Real-Time Systems (ECRTS), pp. 266–275 (2014). doi:10.1109/ECRTS.2014.16
3. Abeni, L., Buttazzo, G.: Integrating multimedia applications in hard real-time systems. In: Proceedings of the Real-Time Systems Symposium (RTSS), pp. 4–13 (1998)
4. Absint: aiT worst-case execution time analyzers. http://www.absint.de/ait (2016)
5. AbsInt Angewandte Informatik GmbH: Stack overflow is a thing of the past. https://www.absint.com/stackanalyzer/index.htm (2016)
6. Accellera Systems Initiative™: Core SystemC language and examples. http://accellera.org/downloads/standards/systemc/files (2014)
7. Accellera Systems Initiative™: SystemC synthesizable subset – version 1.4.7. http://accellera.org/images/downloads/standards/systemc/SystemC_Subset_1_4_7-Apache.pdf (2016)
8. ACM SIGBED: Home page. http://www.sigbed.org (2016)
9. ACM/IEEE: Computer science curricula 2013: Curriculum guidelines for undergraduate degree programs in computer science. The Joint Task Force on Computing Curricula Association for Computing Machinery (ACM) IEEE Computer Society (2013). http://www.acm.org/education/CS2013-final-report.pdf
10. Ahmad I, Ranka S (2016) Handbook of Energy-Aware and Green Computing - Two. Volume. Set. CRC Press, Boca Raton
11. Akatech: Cyber-physical systems. Driving force for innovation in mobility, health, energy and production. http://www.acatech.de/de/publikationen/stellungnahmen/kooperationen/detail/artikel/cyber-physical-systems-innovationsmotor-fuer-mobilitaet-gesundheit-energie-und-produktion.html (2011)
12. Ambler, S.: The diagrams of UML 2.0. http://www.agilemodeling.com/essays/umlDiagrams.htm (2003)
13. Analog Devices Inc. Eng.: Data Conversion Handbook (Analog Devices). Newnes (2004)
14. Andersson, B., Baruah, S., Jonsson, J.: Static-priority scheduling on multiprocessors. In: Proceedings of the Real-Time Systems Symposium (RTSS), pp. 193–202. IEEE Computer Society, Washington, DC, USA (2001). http://dl.acm.org/citation.cfm?id=882482.883823
15. Anonymous: Introduction to lp_solve 5.5.2.0. http://lpsolve.sourceforge.net (2016)
16. Anonymous: RTJS home page. http://www.rtsj.org (2016)

© Springer International Publishing AG 2018

P. Marwedel, *Embedded System Design*, Embedded Systems,

DOI 10.1007/978-3-319-56045-8

17. Anonymus: OSEK. https://en.wikipedia.org/wiki/OSEK (2016)
18. ARM Ltd.: AMBA specifications. http://www.arm.com/products/system-ip/amba-specifications.php (2016)
19. ARM Ltd.: big.LITTLE Technology. http://www.arm.com/products/processors/technologies/biglittleprocessing.php (2016)
20. ARM Ltd.: Mali-T860 & Mali-T880. http://www.arm.com/products/multimedia/mali-gpu/high-performance/mali-t860-t880.php (2016)
21. ARM Ltd.: #pragma arm section [section_type_list]. http://www.keil.com/support/man/docs/armcc/armcc_chr1359124985290.htm (2016)
22. Arnaud, F., Colquhoun, S., Mareau, A., Kohler, S., Jeannot, S., Hasbani, F., Paulin, R., Cremer, S., Charbuillet, C., Druais, G., Scheer, P.: Technology-Circuit Convergence for Full-SOC Platform in 28 nm and Beyond. International Electron Devices Meeting (2011)
23. Artist Consortium: Home page. http://www.artist-embedded.org (2016)
24. Atienza, D., Baloukas, C., Papadopoulos, L., Poucet, C., Mamagkakis, S., Hidalgo, J.I., Catthoor, F., Soudris, D., Lanchares, J.: Optimization of dynamic data structures in multimedia embedded systems using evolutionary computation. In: Proceedings of the International Workshop on Software and Compilers for Embedded Systems (SCOPES), pp. 31–40 (2007). doi:10.1145/1269843.1269849
25. Atmel: 32-Bit AVR Microcontroller. http://www.atmel.com/Images/doc32058.pdf (2012)
26. AUTOSAR: Automotive Open System Architecture. http://www.autosar.org (2016)
27. Avižienis A, Laprie JC, Randell B, Landwehr C (2004) Basic concepts and taxonomy of dependable and secure computing. IEEE Trans. Dependable Secur. Comput. 1(1):11–33
28. Azevedo, A., Issenin, I., Cornea, R., Gupta, R., Dutt, N., Veidenbaum, A., Nicolau, A.: Profile-based dynamic voltage scheduling using program checkpoints. In: Proceedings of Design, Automation and Test in Europe (DATE), pp. 168–175 (2002)
29. Bäck T, Fogel D, Michalewicz Z (1997) Handbook of Evolutionary Computation. Oxford University Press, Oxford
30. Bäck, T., Schwefel, H.P.: An overview of evolutionary algorithms for parameter optimization. In: Evolutionary computation, pp. 1–23 (1993)
31. Bai, K., Shrivastava, A.: Heap data management for limited local memory (LLM) multi-core processors. In: Proceedings of the International Conference on Hardware/Software Codesign and System Synthesis (CODES+ISSS), pp. 317–325 (2010)
32. Baker, T.: Rate monotone scheduling. http://www.cs.fsu.edu/~baker/realtime/restricted/notes/rmscheduling.html (2008)
33. Baker, T.P.: Stack-based scheduling of real-time processes. J. Real-Time Syst. 3 (1991)
34. Ball, S.R.: Embedded Microprocessor Systems - Real World Designs. Newnes (1996)
35. Ball, S.R.: Debugging Embedded Microprocessor Systems. Newnes (1998)
36. Banakar, R., Steinke, S., Lee, B.S., Balakrishnan, M., Marwedel, P.: Scratchpad memory: a design alternative for cache on-chip memory in embedded systems. In: Proceedings of the International Symposium on Hardware-Software Codesign (CODES), pp. 73–78 (2002). doi:10.1145/774789.774805
37. Barney, B.: POSIX threads programming. https://computing.llnl.gov/tutorials/pthreads (2015)
38. Barr, M.: Programming Embedded Systems. O'Reilly (1999)
39. Barrett S, Pack D (2005) Embedded Systems - Design and Applications with the 68HC12 and HCS12. Prentice Hall, Englewood Cliffs
40. Baruah S, Bertogna M, Buttazzo G (2015) Multiprocessor Scheduling for Real-Time Systems. Springer, Berlin
41. Baruah SK, Cohen NK, Plaxton CG, Varvel DA (1996) Proportionate progress: A notion of fairness in resource allocation. Algorithmica 15(6):600–625. doi:10.1007/BF01940883
42. Baruah, S.K., Mok, A.K., Rosier, L.E.: Preemptively scheduling hard-real-time sporadic tasks on one processor. In: Proceedings of the Real-Time Systems Symposium (RTSS), pp. 182–190 (1990). doi:10.1109/REAL.1990.128746

43. Basten, T.: Opening remarks, 2nd Artist workshop on models of computation and communication. Eindhoven. http://www.es.ele.tue.nl/~tbasten/mocc2008/presentations/mocc.pdf (2008)

44. Benavides T, Treon J, Hulbert J, Chang W (2008) The enabling of an execute-in-place architecture to reduce the embedded system memory footprint and boot time. JCP 3(1):79–89. doi:10.4304/jcp.3.1.79-89

45. Bender, A.: Milp based task mapping for heterogeneous multiprocessor systems. In: Design Automation Conference, 1996, with EURO-VHDL '96 and Exhibition, Proceedings EURO-DAC '96, European, pp. 190–197 (1996). doi:10.1109/EURDAC.1996.558204

46. Bengtsson J, Yi W (2004) Timed automata: semantics, algorithms and tools. In: Desel J, Reisig W, Rozenberg G (eds) ACPN 2003, vol 3098. LNCS. Springer, Berlin, pp 87–124

47. Benini L, Bogliolo A, De Micheli G (2000) A survey of design techniques for system-level dynamic power management. IEEE Trans. VLSI Syst. 8(3):299–316

48. Benini L, De Micheli G (1998) Dynamic Power Management - Design Techniques and CAD Tools. Kluwer Academic Publishers, Norwell

49. Berger H (2001) Automating with STEP 7 in LAD and FBD: SIMATIC S7–300/400 Programmable Controllers. Wiley, New York

50. van Berkel, C.H.K.: Multi-core for mobile phones. In: Proceedings of the Conference on Design, Automation and Test in Europe, DATE '09, pp. 1260–1265. European Design and Automation Association, 3001 Leuven, Belgium, Belgium (2009). http://dl.acm.org/citation.cfm?id=1874620.1874924

51. Bernardi, P., Rebaudengo, M. Reorda, S.: Using infrastructure IPs to support SW-based self-test of processor cores. In: Workshop on Fibres and Optical Passive Components, pp. 22–27 (2005)

52. Bernstein JB, Gurfinkel M, Li X, Walters J, Shapira Y, Talmor M (2006) Electronic circuit reliability modeling. Microelectron. Reliab. 46(12), 1957–1979. doi:10.1016/j.microrel.2005.12.004. http://www.sciencedirect.com/science/article/pii/S0026271406000023

53. Bertogna, M., Cirinei, M., Lipari, G.: Improved schedulability analysis of EDF on multiprocessor platforms. In: Euromicro Conference on Real-Time Systems (ECRTS), pp. 209–218 (2005). doi:10.1109/ECRTS.2005.18

54. Bertolotti IC (2006) Real-time embedded operating systems: standards and perspectives. In: Zurawski R (ed) Embedded Systems Handbook. CRC Press, Boca Raton

55. Beszedes, A.: Survey of code size reduction methods. ACM Computing Surveys, pp. 223–267 (2003)

56. Bieker, U., Marwedel, P.: Retargetable self-test program generation using constraint logic programming. In: Proceedings of the Design Automation Conference (DAC), pp. 605–611 (1995)

57. Bini, E., Buttazzo, G., Buttazzo, G.: A hyperbolic bound for the rate monotonic algorithm. Euromicro Conference on Real-Time Systems (ECRTS), pp. 59–73 (2001)

58. Black JR (1969) Electromigration failure modes in aluminum metallization for semiconductor devices. Proc. IEEE 57(9):1587–1594. doi:10.1109/PROC.1969.7340

59. Boldt, M., Traulsen, C., von Hanxleden, R.: Compilation and worst-case reaction time analysis for multithreaded esterel processing. EURASIP J. Embed. Syst. 2008, 4:1–4:21 (2008). doi:10.1155/2008/594129

60. Bonfietti, A., Benini, L., Lombardi, M., Milano, M.: An efficient and complete approach for throughput-maximal SDF allocation and scheduling on multi-core platforms. In: Proceedings of Design, Automation and Test in Europe (DATE), pp. 897–902. European Design and Automation Association, 3001 Leuven, Belgium, Belgium (2010). http://dl.acm.org/citation.cfm?id=1870926.1871143

61. Bonny, T., Henkel, J.: Huffman-based code compression techniques for embedded processors. ACM Trans. Des. Autom. Electron. Syst. 15(4), 31:1–31:37 (2010). doi:10.1145/1835420.1835424

62. Boulanger JL (2015) CENELEC 50128 and IEC 62279 Standards. Wiley, New York

63. Boussinot F, de Simone R (1991) The Esterel language. Proc. IEEE 79(9):1293–1304

64. Bouwmeester, D., Ekert, A., (eds.), A.Z.: The Physics of Quantum Information: Quantum Cryptography, Quantum Teleportation, Quantum Computation. Springer, Berlin (2000)
65. Bouyssounouse, B., Sifakis, J. (eds.): Embedded Systems Design, The ARTIST Roadmap for Research and Development. Lecture Notes in Computer Science, vol. 3436. Springer, Berlin (2005)
66. Brahme, D., Abraham, J.A.: Functional testing of microprocessors. IEEE Trans. Comput. 475–485 (1984)
67. Brand, D., Bergamaschi, R.A., Stok, L.: Don't cares in synthesis: theoretical pitfalls and practical solutions. IEEE Trans. Comput. Aided Des. Integr. Circuits Syst. **17**(4), 285–304 (1998). doi:10.1109/43.703819
68. Braun, A., Bringmann, O., Lettnin, D., Rosenstiel, W.: Simulation-based verification of the MOST net interface specification revision 3.0. In: Proceedings of Design, Automation and Test in Europe (DATE) (2010)
69. Bremaud P (1999) Markov Chains. Springer, Berlin
70. Brettel, M., Friederichsen, N., Keller, M., Rosenberg, M.: How virtualization, decentralization and network building change the manufacturing landscape: an industry 4.0 perspective. Int. J. Mech. Indust. Sci. Eng. **8**(1), 37–44 (2014)
71. Bril, R.J.: Real-time scheduling for media processing using conditionally guaranteed budgets. Ph.D. thesis, TU Eindhoven (2004)
72. Brooks, D., Tiwari, V., Martonosi, M.: Wattch: a framework for architectural-level power analysis and optimizations. In: 27th International Symposium on Computer Architecture (ISCA), pp. 83–94 (2000)
73. Bruno E, Bollella G (2009) Real-Time Java Programming: With Java RTS. Prentice Hall, Englewood Cliffs
74. Bryant, R.: A switch-level model and simulator for MOS digital circuits. IEEE Trans. Comput. **33** (1984)
75. Buck, J.T.: Scheduling dynamic dataflow graphs with bounded memory using the token flow model. Ph.D. thesis, University of California at Berkeley (1993)
76. Budkowski S, Dembinski P (1987) An introduction to estelle: a specification language for distributed systems. Comput. Netw. ISDN Syst. **14**(1):3–23. doi:10.1016/0169-7552(87)90084-5. http://www.sciencedirect.com/science/article/B6TYT-48V22NJ-5C/2/611631c58f275c04f464fba932b8b699
77. Burd, T., Brodersen, R.: Design issues for dynamic voltage scaling. In: International Symposium on Low Power Electronics and Design (ISLPED), pp. 9–14 (2000)
78. Burd T, Brodersen RW (2003) Energy Efficient Microprocessor Design. Kluwer Academic Publishers, Norwell
79. Burks, A., Goldstine, H., von Neumann, J.: Preliminary discussion of the logical design of an electronic computing element. Report to U.S. Army Ordnance Department. Reprinted at https://www.cs.princeton.edu/courses/archive/fall10/cos375/Burks.pdf (1946)
80. Burns, A., Wellings, A.: Real-Time Systems and Programming Languages. Addison-Wesley, Reading (1990)
81. Burns, A., Wellings, A.: Real-Time Systems and Programming Languages, 4th edn. Addison Wesley, Reading (2009)
82. Buttazzo G (2011) Hard Real-Time Computing Systems, 4th printing. Springer, Berlin
83. Cadence Design Systems Inc.: Tensilica Customizable Processor IP. http://ip.cadence.com/ipportfolio/tensilica-ip (2016)
84. Cai, L., Gajski, D.: Transaction level modeling: An overview. In: Proceedings of the International Conference on Hardware/Software Codesign and System Synthesis (CODES+ISSS), pp. 19–24. ACM, New York, NY, USA (2003). doi:10.1145/944645.944651
85. Carroll, A., Heiser, G.: An analysis of power consumption in a smartphone. In: Proceedings of the 2010 USENIX Conference on USENIX Annual Technical Conference, USENIXATC'10, pp. 21–21. USENIX Association, Berkeley, CA, USA (2010). http://dl.acm.org/citation.cfm?id=1855840.1855861

86. Caspi, P., Sangiovanni-Vincentelli, A., Almeida, L., et al: Guidelines for a graduate curriculum on embedded software and systems. ACM Trans. Embed. Comput. Syst. 587–611 (2005)

87. Castrillon, J., Tretter, A., Leupers, R., Ascheid, G.: Communication-aware mapping of KPN applications onto heterogeneous MPSoCs. In: Proceedings of the Design Automation Conference (DAC), pp. 1266–1271. ACM, New York, NY, USA (2012). doi:10.1145/2228360. 2228597

88. Cederqvist, P.: The CVS Manual - Version Management with CVS. Network Theory Ltd. (2006)

89. Cena, G., Valenzano, A.: Performance analysis of byteflight networks. In: Proceedings of the 2004 IEEE International Workshop on Factory Communication Systems, 2004, pp. 157–166 (2004). doi:10.1109/WFCS.2004.1377701

90. Ceng, J., Castrillón, J., Sheng, W., Scharwächter, H., Leupers, R., Ascheid, G., Meyr, H., Isshiki, T., Kunieda, H.: MAPS: an integrated framework for MPSoC application parallelization. In: Proceedings of the Design Automation Conference (DAC), pp. 754–759 (2008)

91. Chandrakasan AP, Sheng S, Brodersen RW (1992) Low-power CMOS digital design. IEEE J. Solid-State Circuits 27(4):119–123

92. Chandrakasan AP, Sheng S, Brodersen RW (1995) Low power CMOS digital design. Kluwer Academic Publishers, Norwell

93. Chanet, D., Sutter, B.D., Bus, B.D., Put, L.V., Bosschere, K.D.: Automated reduction of the memory footprint of the Linux kernel. ACM Trans. Embed. Comput. Syst. **6**(4), 23 (2007). doi:10.1145/1274858.1274861

94. Chang, D.W., Lin, I.C., Chien, Y.S., Lin, C.L., Su, A.W.Y., Young, C.P.: CASA: contention-aware scratchpad memory allocation for online hybrid on-chip memory management. IEEE Trans. Comput. Aided Des. Integr. Circuits Syst. **33**(12), 1806–1817 (2014). doi:10.1109/TCAD.2014.2363385

95. Chen, J., Huang, W., Liu, C.: k2q: A quadratic-form response time and schedulability analysis framework for utilization-based analysis. In: CoRR (2015). http://arxiv.org/abs/1505.03883

96. Chen, K., Sztipanovits, J., Neema, S.: Compositional specification of behavioral semantics. In: Design, Automation and Test in Europe (DATE), pp. 906–911 (2007)

97. Chen, M., Rincón-Mora, G.: Accurate electrical battery model capable of predicting runtime and i-v performance. IEEE Trans. Energy Convers. 504–511 (2006)

98. Chen, X., Dick, R., Shang, L.: Properties of and improvements to time-domain dynamic thermal analysis algorithms. In: Proceedings of Design, Automation and Test in Europe (DATE) (2010)

99. Chetto, H., Silly, M., Bouchentouf, T.: Dynamic scheduling of real-time tasks under precedence constraints. J. Real-Time Syst. **2** (1990)

100. Cheung, E., Hsieh, H., Balarin, F.: Automatic buffer sizing for rate-constrained KPN applications on multiprocessor system-on-chip. In: IEEE International High Level Design Validation and Test Workshop, 2007. HLVDT 2007, pp. 37–44 (2007). doi:10.1109/HLDVT.2007. 4392782

101. Chiu, Y.: Folding and interpolating ADC. University of Texas at Dallas, EECT 7327. http://www.utdallas.edu/~yxc101000/courses/7327/slides/intp (2014)

102. Cho, H., Ravindran, B., Jensen, E.: An optimal real-time scheduling algorithm for multiprocessors. In: Proceedings of the Real-Time Systems Symposium (RTSS) (2006)

103. Chung, E.Y., Benini, L., De Micheli, G.: Source code transformation based on software cost analysis. In: Proceedings of the International Symposium on System Synthesis (ISSS), pp. 153–158 (2001)

104. Clarke EM, Grumberg O, Hiraishi H, Jha S, Long DE, McMillan KL, Ness LA (2005) Verification of the Futurebus+ cache coherence protocol. Formal Methods Syst. Des. 6(2):217–232

105. Clouard, A., Jain, K., Ghenassia, F., Maillet-Contoz, L., Strassen, J.: Using transactional models in SoC design flow. In: [390], pp. 29–64 (2003)

106. Coelho DR (1989) The VHDL Handbook. Kluwer Academic Publishers, Norwell

107. Coello, C.A.C., Lamont, G.B., Veldhuizen, D.A.v.: Evolutionary Algorithms for Solving Multi-Objective Problems. Springer, Berlin (2007)

108. Collins-Sussman, B., Fitzpatrick, B., Pilato, C.: Version control with subversion – for subversion 1.5. http://svnbook.red-bean.com/en/1.5/svn-book.pdf (2008)
109. Cooling, J.: Software Engineering for Real-Time Systems. Addison Wesley, Reading (2003)
110. Cortadella, J., Kondratyev, A., Lavagno, L., Massot, M., Moral, S., Passerone, C., Watanabe, Y., Sangiovanni-Vincentelli, A.: Task generation and compile-time scheduling for mixed data-control embedded software. In: Proceedings of the Design Automation Conference (DAC), pp. 489–494 (2000)
111. Coskun, A.K., Rosing, T.S., Whisnant, K.A., Gross, K.C.: Temperature-aware mpsoc scheduling for reducing hot spots and gradients. In: Proceedings of the 2008 Asia and South Pacific Design Automation Conference, ASP-DAC '08, pp. 49–54. IEEE Computer Society Press, Los Alamitos, CA, USA (2008). http://dl.acm.org/citation.cfm?id=1356802.1356815
112. Coussy P, Morawiec A (2008) High-Level Synthesis - From Algorithm to Digital Circuit. Springer, Berlin
113. Craig ID (2006) Virtual Machines. Springer, Berlin
114. Damm, W., Harel, D.: LSCs: Breathing Life into Message Sequence Charts. Formal Methods in System Design (2001)
115. Dasgupta S (1979) The organization of microprogram stores. ACM Comput. Surv. 11:39–65
116. Davis, R.I., Burns, A.: A survey of hard real-time scheduling for multiprocessor systems. ACM Comput. Surv. **43**(4), 35:1–35:44 (2011). doi:10.1145/1978802.1978814
117. De Greef, E., Catthoor, F., Man, H.: Memory size reduction through storage order optimization for embedded parallel multimedia applications. In: Proceedings of the Workshop on Parallel Processing and Multimedia, pp. 84–98 (1997)
118. De Greef, E., Catthoor, F., Man, H.D.: Array placement for storage size reduction in embedded multimedia systems. In: IEEE International Conference on Application-Specific Systems, Architectures and Processors (ASAP), pp. 66–75 (1997)
119. De Micheli G, Ernst R, Wolf W (2002) Readings in Hardware/Software Co-Design. Academic Press, Dublin
120. Derin, O.: Self-adaptivity of applications on network on chip multiprocessors – the case of fault-tolerant Kahn Process Networks. Ph.D. thesis, Università della Svizzera Italiana de Lugano (2013)
121. Derler P, Lee EA, Vincentelli AS (2012) Modeling cyber-physical systems. Proc. IEEE 100(1):13–28. doi:10.1109/JPROC.2011.2160929
122. Deutsches Institut für Normung: DIN 66253, Programmiersprache PEARL, Teil 2 PEARL 90. Beuth-Verlag (1997). http://www.din.de
123. Devi, U.M.C., Anderson, J.H.: Tardiness bounds under global EDF scheduling on a multiprocessor. In: Proceedings of the Real-Time Systems Symposium (RTSS), pp. 312–341 (2005). doi:10.1109/RTSS.2005.39
124. Devillers, R.R., Goossens, J.: Liu and Layland's schedulability test revisited. Inf. Process. Lett. 157–161 (2000)
125. Dhall K, Liu C (1978) On a real-time scheduling problem. Oper. Res. 26(1):127–140. doi:10.1287/opre.26.1.127
126. Dibble PC (2008) Real-Time Java Platform Programming, 2nd edn. BookSurge Publishing, Charleston
127. Diederichs, C., Margull, U., Slomka, F., Wirrer, G.: An application-based EDF scheduler for OSEK/VDX. In: Proceedings of Design, Automation and Test in Europe (DATE), pp. 1045–1050 (2008)
128. Dill, D., Alur, R.: A theory of timed automata. In: Theoretical Computer Science, pp. 183–235 (1994)
129. Dominguez A, Udayakumaran S, Barua R (2005) Heap data allocation to scratch-pad memory in embedded systems. J. Embed. Comput. 1(4):521–540
130. Donald, J., Martonosi, M.: Techniques for multicore thermal management: classification and new exploration. SIGARCH Comput. Archit. News **34**(2), 78–88 (2006). doi:10.1145/1150019.1136493

131. Dósa, G.: The Tight Bound of First Fit Decreasing Bin-Packing Algorithm Is FFD(I) ? 11/9OPT(I) + 6/9, pp. 1–11. Springer, Berlin (2007). doi:10.1007/978-3-540-74450-4_1
132. Douglass, B.P.: Real-Time UML, 3rd edn. Addison Wesley, Reading (2004)
133. Dozio, L., Mantegazza, P.: Real time distributed control systems using RTAI. In: Proceedings of the Sixth IEEE International Symposium on Object-Oriented Real-Time Distributed Computing (2003)
134. Drepper, U.: What every programmer should know about memory. http://www.akkadia.org/drepper/cpumemory.pdf (2007)
135. Drusinsky, D., Harel, D.: Using statecharts for hardware description and synthesis. IEEE Trans. Comput. Design 798–807 (1989)
136. Dunkels A (2001) Design and Implementation of the lwIP TCP/IP Stack. Swedish Inst. Comput. Sci. 2:77
137. Dunn, W.: Practical Design of Safety-Critical Computer Systems. Reliability Press (2002)
138. Dusza, B., Marwedel, P., Spinczyk, O., Wietfeld, C.: A context-aware battery lifetime model for carrier aggregation enabled LTE-A systems. In: IEEE Consumer Communications and Networking Conference (CCNC) (2014)
139. Ecker W, Müller W, Dömer R (2009) Hardware-dependent software - Principles and practice. Springer, Berlin
140. Edwards, S.: Dataflow languages. http://www.cs.columbia.edu/~sedwards/classes/2001/w4995-02/presentations/dataflow.ppt (2001)
141. Edwards S (2006) Languages for embedded systems. In: Zurawski R (ed) Embedded Systems Handbook. CRC Press, Boca Raton
142. efunda: Materials home. http://www.efunda.com/materials/elements/element_info.cfm?Element_ID=Si (2017)
143. Egger, B., Lee, J., Shin, H.: Scratchpad memory management for portable systems with a memory management unit. In: Proceedings of the International Conference on Compilers, Architectures, and Synthesis for Embedded Systems (CASES), pp. 321–330 (2006)
144. Eggermont, L.: Embedded systems roadmap. STW. http://citeseerx.ist.psu.edu/viewdoc/download?doi=10.1.1.119.6407&rep=rep1&type=pdf (2002)
145. Elsevier B.V.: Sensors and actuators a: physical. Int. J. (2016)
146. Elsevier B.V.: Sensors and actuators b: chemical. Int. J. (2016)
147. Esterel Technologies: SCADE Suite™ - control software design. http://www.esterel-technologies.com/products/scade-suite (2016)
148. Esterel Technologies SA: Homepage. http://www.esterel-technologies.com (2014)
149. European Commission: Topic: Smart cyber-physical systems. http://ec.europa.eu/research/participants/portal/desktop/en/opportunities/h2020/topics/ict-01-2014.html (2013)
150. Evidence: Erika enterprise. http://erika.tuxfamily.org/drupal/ (2016)
151. Falk, H.: WCET-aware register allocation based on graph coloring. In: Proceedings of the Design Automation Conference (DAC), pp. 726–731 (2009)
152. Falk, H., Marwedel, P.: Control flow driven splitting of loop nests at the source code level. In: Proceedings of Design, Automation and Test in Europe (DATE), pp. 410–415 (2003)
153. Falk, H., Verma, M.: Combined data partitioning and loop nest splitting for energy consumption minimization. In: Proceedings of the International Workshop on Software and Compilers for Embedded Systems (SCOPES), pp. 137–151. Amsterdam/The Netherlands (2004)
154. Fard, H.M., Prodan, R., Barrionuevo, J.J.D., Fahringer, T.: A multi-objective approach for workflow scheduling in heterogeneous environments. In: Proceedings of the 2012 12th IEEE/ACM International Symposium on Cluster, Cloud and Grid Computing (Ccgrid 2012), CCGRID '12, pp. 300–309. IEEE Computer Society, Washington, DC, USA (2012). doi:10.1109/CCGrid.2012.114
155. Ferrell, T.K., Ferrell, U.D.: RTCA DO-178-B/EUROCAE ED-12B. In: Spitzer, C. (ed.) The Avionics Handbook (Electrical Engineering Handbook), Chap. 27. CRC Press, Boca Raton (2001)

156. Fettweis, G., Weiss, M., Drescher, W., Walther, U., Engel, F., Kobayashi, S., Richter, T.: Breaking new grounds over 3000 MMAC/s: a broadband mobile multimedia modem DSP. In: International Conference on Signal Processing Application and Technology (ICSPA) (1998). http://citeseerx.ist.psu.edu/viewdoc/summary?doi=10.1.1.50.9340

157. Feucht, J.: The mystery of the bouncing ball. http://feucht.us/writings/bouncing_ball.php (2016)

158. Fiorin, L., Palermo, G., Lukovic, S., Silvano, C.: A data protection unit for NoC-based architectures. In: Proceedings of the International Conference on Hardware/Software Codesign and System Synthesis (CODES+ISSS), pp. 167–172 (2007)

159. Flautner, K.: Heterogeneity to the rescue. https://www.bscmsrc.eu/sites/default/files/media/arm-heterogenous-mp-november-2011.pdf (2011)

160. FlexRay Consortium: Flexray® requirement specification. version 2.01. http://www.flexray.de (2002)

161. Fowler, M., Scott, K.: UML Distilled - Applying the Standard Object Modeling Language. Addison-Wesley, Reading (1998)

162. Franke, B.: Fast cycle-approximate instruction set simulation. In: Proceedings of the International Workshop on Software and Compilers for Embedded Systems (SCOPES), pp. 69–78 (2008)

163. Franke B, O'Boyle MF (2005) A complete compiler approach to auto-parallelizing C programs for multi-DSP systems. IEEE Trans. Parallel Distrib. Syst. 16:234–245

164. Freescale semiconductor/NXP: ColdFire® family programmer's reference manual. http://www.nxp.com/assets/documents/data/en/reference-manuals/CFPRM.pdf (2005)

165. Gajski, D., Kuhn, R.: New VLSI tools. IEEE Comput. 11–14 (1983)

166. Gajski D, Vahid F, Narayan S, Gong J (1994) Specification and Design of Embedded Systems. Prentice Hall, Englewood Cliffs

167. Gajski D, Zhu J, Dömer R, Gerstlauer A, Zhao S (2000) SpecC: Specification Language Methodology. Kluwer Academic Publishers, Norwell

168. Gajski DD, Abdi S, Gerstlauer A, Schirner G (2009) Embedded System Design. Springer, Heidelberg

169. Ganssle, J. (ed.): Embedded Systems (World Class Designs). Newnes (2008)

170. Ganssle, J.G.: The Art of Designing Embedded Systems. Newnes (2000)

171. Ganssle, J.G., Noergaard, T., Eady, F., Edwards, L., Katz, D.J., Gentile, R., Arnold, K., Hyder, K., Perrin, B.: Embedded Hardware - Know it all. Newnes (2008)

172. Garey MR, Johnson DS (1979) Computers and Intractability. Bell Labaratories, Murray Hill, New Jersey

173. Garg R, Khatri S (2009) Analysis and Design of Resilient VLSI Circuits. Springer, Berlin

174. Gebotys C (2010) Security in Embedded Devices. Springer, Berlin

175. Geffroy JC, Motet G (2002) Design of Dependable computing Systems. Kluwer Academic Publishers, Norwell

176. Gerum, P.: Xenomai-implementing a RTOS emulation framework on GNU Linux. Xenomai White Paper (2004)

177. Gierlichs, B., Poschmann, A.: International workshop on cryptographic hardware and embedded systems (CHES). http://www.chesworkshop.org/ (2016)

178. Girard A, Pappas GJ (2007) Approximation metrics for discrete and continuous systems. IEEE Trans. Autom. Control 52(5):782–798. doi:10.1109/TAC.2007.895849

179. Giusto D, Iera A, Morabito G, Atzori L (eds) (2010) The Internet of Things, 20th Tyrrhenian Workshop on Digital Communications. Springer, Berlin

180. Gomez, L., Fernandes, J.: Behavioral Modeling for Embedded Systems and Technologies. IGI Global (2010)

181. Goossens, J., Funk, S., Baruah, S.: EDF scheduling on multiprocessor platforms: some (perhaps) counterintuitive observations. In: RealTime Computing Systems and Applications Symposium (2002)

182. Graham, R., Lawler, E., Lenstra, J.K., Kan, A.H.G.: Optimization and approximation in deterministic sequencing and scheduling: a survey. Ann. Discret. Math. 287–326 (1979)

183. Grimheden, M.E. (ed.): WESE'15: Proceedings of the WESE'15: Workshop on Embedded and Cyber-Physical Systems Education. ACM, New York, NY, USA (2015)
184. Grötker T, Liao S, Martin G (2002) System design with SystemC. Springer, Berlin
185. Gubbia J, Buyyab R, Marusica S, Palaniswamia M (2013) Internet of things (IoT): a vision, architectural elements, and future directions. Future Gener. Comput. Syst. 29:1645–1660
186. Guerrieri A, Loscri V, Rovella A, Fortino G (2016) Management of Cyber Physical Objects in the Future Internet of Things: Methods. Architectures and Applications. Springer, Berlin
187. Gupta, R.: Tasks and task management. Course ICS 212, Winter 2002, UC Irvine. http://citeseerx.ist.psu.edu/viewdoc/download?doi=10.1.1.7.8704&rep=rep1&type=pdf (2002)
188. Ha, S.: Model-based programming environment of embedded software for MPSoC. In: Proceedings of the Asia and South Pacific Design Automation Conference (ASPDAC), pp. 330–335 (2007). doi:10.1109/ASPDAC.2007.358007
189. de Haan L, Ferreira A (2006) Extreme Value Theory - An Introduction. Springer, Berlin
190. Hahn, S., Reineke, J., Wilhelm, R.: Toward compact abstractions for processor pipelines. In: Correct System Design, pp. 205–220 (2015)
191. Halbwachs, N.: Synchronous programming of reactive systems, a tutorial and commented bibliography. In: Tenth International Conference on Computer-Aided Verification, CAV'98. LNCS, vol. 1427. Springer, Berlin (1998). http://www.springerlink.com/content/5127074271136j71/fulltext.pdf
192. Halbwachs N (2008) Personal Communication. South American Artist School on Embedded Systems, Florianopolis
193. Halbwachs N, Caspi P, Raymond P, Pilaud D (1991) The synchronous dataflow language LUSTRE. Proc. IEEE Trans. Softw. Eng. 79:1305–1320
194. Hamada, T., Benkrid, K., Nitadori, K., Taiji, M.: A comparative study on ASIC, FPGAs, GPUs and general purpose processors in the $O(N^2)$ gravitational N-body simulation. In: Proceedings of the 2009 NASA/ESA Conference on Adaptive Hardware and Systems, AHS '09, pp. 447–452. IEEE Computer Society, Washington, DC, USA (2009). doi:10.1109/AHS.2009.55
195. Harbour, M.G.: RT-POSIX: An overview. http://www.ctr.unican.es/publications/mgh-1993a.pdf (1993)
196. Hardkernel co, Ltd.: Odroid XU3. http://www.hardkernel.com/main/products/prdt_info.php?g_code=G140448267127 (2013)
197. Harel, D.: StateCharts: A visual formalism for complex systems. In: Science of Computer Programming, pp. 231–274 (1987)
198. Hattori, T.: MPSoC approaches for low-power embedded SoC's. http://www.mpsoc-forum.org/2007/slides/Hattori.pdf (2007)
199. Haugen O, Moller-Pedersen B (2006) Introduction to UML and the modeling of embedded systems. In: Zurawski R (ed) Embedded Systems Handbook. CRC Press, Boca Raton
200. Hayes J (1982) A unified switching theory with applications to VLSI design. Proc. IEEE 70:1140–1151
201. Heath, S.: Embedded System Design. Newnes (2000)
202. Heemels, W., Johansson, K., Tabuada, P.: An introduction to event-triggered and self-triggered control. In: 2012 IEEE 51st Annual Conference on Decision and Control (CDC), pp. 3270–3285 (2012). doi:10.1109/CDC.2012.6425820
203. Henia, R., Hamann, A., Jersak, M., Racu, R., Richter, K., Ernst, R.: System level performance analysis - the SymTA/S approach. In: IEEE Computers and Digital Techniques, pp. 148–166 (2005)
204. Hennessy JL, Patterson DA (2008) Computer Organization - The Hardware/Software Interface. Morgan Kaufmann Publishers Inc., San Francisco
205. Hennessy JL, Patterson DA (2011) Computer Architecture - A Quantitative Approach, 5th edn. Morgan Kaufmann Publishers Inc., San Francisco
206. Herken R (1995) The Universal Turing Machine: A Half-Century Survey. Springer, Berlin
207. Herrera, F., Fernández, V., Sánchez, P., Villar, E.: Embedded software generation from SystemC for platform based design. In: [390], pp. 247–272 (2003)

208. Herrera, F., Posadas, H., Sánchez, P., Villar, E.: Systemic embedded software generation from SystemC. In: Proceedings of Design, Automation and Test in Europe (DATE), pp. 10142–10149 (2003)

209. Hoare C (1985) Communicating Sequential Processes. International Series in Computer Science. Prentice Hall, Englewood Cliffs

210. Hofer, W., Lohmann, D., Scheler, F., Schröder-Preikschat, W.: Sloth: threads as interrupts. In: Proceedings of the Real-Time Systems Symposium (RTSS), pp. 2004–2013 (2009)

211. Hofmann, M., Jost, S.: Static prediction of heap space usage for first-order functional programs. In: Proceedings of the Symposium on Principles of Programming Languages (POPL), pp. 185–197. ACM, New York, NY, USA (2003). doi:10.1145/604131.604148

212. Hopcroft, J., Motwani, R., Ullman, J.D.: Introduction to Automata Theory, Languages, and Computation. Addison Wesley, Reading (2006)

213. Horn W (1974) Some simple scheduling algorithms. Nav. Res. Logist. Q. 21:177–185

214. Huang, L., Xu, Q.: AgeSim: A simulation framework for evaluating the lifetime reliability of processor-based SoCs. In: Proceedings of Design, Automation and Test in Europe (DATE) (2010)

215. Huerlimann, D.: Opentrack home page. http://www.opentrack.ch (2016)

216. Hüls, T.: Optimizing the energy consumption of an MPEG application (in German). Master thesis, CS Dept., Univ. Dortmund. http://ls12-www.cs.uni-dortmund.de/publications/theses (2002)

217. Hunt VD, Puglia A, Puglia M (2007) RFID: A Guide to Radio Frequency Identification. Wiley, New York

218. McGregor, I.: The relationship between simulation and emulation. In: Winter Simulation Conference, pp. 1683–1688 (2002)

219. IBM: What's new in Rational Rhapsody 7.5.1. http://www.ibm.com/developerworks/rational/library/09/whatsnewinrationalrhapsody-7-5-1 (2009)

220. IBM: IBM Rational StateMate. http://www.ibm.com/developerworks/rational/products/statemate/ (2016)

221. IBM: Rational DOORS. http://www-01.ibm.com/software/awdtools/doors/ (2016)

222. ICD Staff: ICD-C compiler framework. http://www.icd.de/en/embedded-systems/compiler-tool-development/icd-c (2016)

223. IEC: IEC 60848 – GRAFCET specification language for sequential function charts. https://webstore.iec.ch/publication/3684 (2002)

224. IEC: IEC 61508-1:2010 functional safety of electrical/ electronic/ programmable electronic safety-related systems – part 1: General requirements. https://webstore.iec.ch/publication/5515 (2010)

225. IEC: IEC 61508-2:2010 functional safety of electrical/electronic/programmable electronic safety-related systems - part 2: Requirements for electrical/electronic/programmable electronic safety-related systems. https://webstore.iec.ch/publication/5516 (2010)

226. IEC: IEC 61508-3:2010 functional safety of electrical/electronic/programmable electronic safety-related systems - part 3: Software requirements. https://webstore.iec.ch/publication/5517 (2010)

227. IEC: IEC 61513:2011-nuclear power plants - instrumentation and control important to safety - general requirements for systems. https://webstore.iec.ch/publication/5532 (2011)

228. IEC: IEC 61511:2016 SER functional safety - safety instrumented systems for the process industry sector - all parts. https://webstore.iec.ch/publication/5527 (2016)

229. IEEE: 1076-1987 - IEEE standard VHDL language reference manual. http://standards.ieee.org/findstds/standard/1076-1987.html (1987)

230. IEEE: IEEE graphic symbols for logic functions Std 91a-1991. http://ieeexplore.ieee.org/stamp/stamp.jsp?tp=&arnumber=27895 (1991)

231. IEEE: 1076-1993 - IEEE standard VHDL language reference manual. http://standards.ieee.org/findstds/standard/1076-1993.html (1993)

232. IEEE: 1076-2000 - IEEE standard VHDL language reference manual. http://standards.ieee.org/findstds/standard/1076-2000.html (2002)

233. IEEE: 1076-2002 - IEEE standard VHDL language reference manual. http://standards.ieee.org/findstds/standard/1076-2002.html (2002)
234. IEEE: 1364-2005 - IEEE standard for Verilog hardware description language. https://standards.ieee.org/findstds/standard/1364-2005.html (2005)
235. IEEE: 1076-2008 - IEEE standard VHDL language reference manual. http://standards.ieee.org/findstds/standard/1076-2008.html, pp. c1–626 (2009)
236. IEEE: P1076.1 - standard VHDL analog and mixed-signal extensions. http://standards.ieee.org/develop/project/1076.1.html (2011)
237. IEEE: 1800-2012 - IEEE standard for SystemVerilog – unified hardware design, specification, and verification language. https://standards.ieee.org/findstds/standard/1800-2012.html (2012)
238. IMEC: ADRES multimedia processor & 3MF multimedia platform. http://www2.imec.be/content/user/File/ADRES_3MF.pdf (2010)
239. Intel: Motion estimation with Intel® streaming SIMD extensions 4 (Intel® SSE4). http://software.intel.com/en-us/articles/motion-estimation-with-intel-streaming-simd-extensions-4-intel-sse4 (2008)
240. Intel: Intel® AVX. http://software.intel.com/en-us/avx (2010)
241. Intel: Intel Itanium processor family. http://www.intel.com/itcenter/products/itanium (2016)
242. Ishihara, T., Yasuura, H.: Voltage scheduling problem for dynamically variable voltage processors. In: International Symposium on Low Power Electronics and Design (ISLPED), pp. 197–202 (1998)
243. ISO: Road vehicles – functional safety – part 3: concept phase. https://www.iso.org/obp/ui/#iso:std:iso:26262:-3:ed-1:v1:en (2011)
244. ISO: Road vehicles - FlexRay communications system – part 1: general information and use case definition. http://www.iso.org/iso/home/store/catalogue_tc/catalogue_detail.htm?csnumber=59804 (2013)
245. ISO: ISO 9001:2015(en)-Quality management systems – requirements. https://www.iso.org/obp/ui/#iso:std:iso:9001:ed-5:v1:en (2015)
246. ISO/IEC: ISO/IEC 15437:2001 - Information technology – enhancements to LOTOS (E-LOTOS). http://www.iso.org/iso/home/store/catalogue_tc/catalogue_detail.htm?csnumber=27680 (2001)
247. Israr, A., Huss, S.: Specification and design considerations for reliable embedded systems. In: Proceedings of Design, Automation and Test in Europe (DATE), pp. 1111–1116 (2008)
248. ITRS Organization: International technology roadmap for semiconductors (ITRS). http://public.itrs.net (2009)
249. ITRS Organization: International technology roadmap for semiconductors – 2013 edition – executive summary. http://www.itrs2.net/itrs-reports.html (2013)
250. Iyer, A., Marculescu, D.: Power and performance evaluation of globally asynchronous locally synchronous processors. In: International Symposium on Computer Architecture (ISCA), pp. 158–168 (2002)
251. Jackson, J.: Scheduling a production line to minimize maximum tardiness. Management Science Research Project 43, University of California, Los Angeles (1955)
252. Jacobs, M., Hahn, S., Hack, S.: Wcet analysis for multi-core processors with shared buses and event-driven bus arbitration. In: Proceedings of the 23rd International Conference on Real Time and Networks Systems, RTNS '15, pp. 193–202. ACM, New York, NY, USA (2015). doi:10.1145/2834848.2834872
253. Jacobs, M., Hahn, S., Hack, S.: A framework for the derivation of WCET analyses for multi-core processors. In: 28th Euromicro Conference on Real-Time Systems, ECRTS 2016, Toulouse, France, July 5–8, 2016, pp. 141–151. IEEE Computer Society (2016). doi:10.1109/ECRTS.2016.19
254. Jain, M., Balakrishnan, M., Kumar, A.: ASIP design methodologies: survey and issues. In: 14th International Conference on VLSI Design, pp. 76–81 (2001)
255. Janka R (2002) Specification and Design Methodology for Real-Time Embedded Systems. Kluwer Academic Publishers, Norwell

256. Jantsch A (2004) Modeling Embedded Systems and SoC's: Concurrency and Time in Models of Computation. Morgan Kaufmann, San Francisco
257. Jantsch A (2006) Models of embedded computation. In: Zurawski R (ed) Embedded Systems Handbook. CRC Press, Boca Raton
258. Java Community Process: JSR-1 – real-time specification for Java. http://www.jcp.org/en/jsr/detail?id=1 (2006)
259. Jayaseelan, R., Mitra, T., Li, X.: Estimating the worst-case energy consumption of embedded software. In: 12th IEEE Real-Time and Embedded Technology and Applications Symposium (RTAS'06), pp. 81–90. IEEE, New York (2006)
260. Jensen K (2013) Coloured Petri Nets: Basic Concepts, Analysis Methods and Practical Use, vol 1. Springer, Berlin
261. Jewett, J.W., Serway, R.A.: Physics for Scientists and Engineers with Modern Physics. Thomson Higher Education (2007)
262. Jha, P., Dutt, N.: Rapid estimation for parameterized components in high-level synthesis. IEEE Trans. VLSI Syst. 296–303 (1993)
263. Jones, M.: What really happened on Mars Rover Pathfinder. In: Neumann, P.G. (ed.) comp.risks, The Risks Digest, vol. 19(49) (1997). http://research.microsoft.com/en-us/um/people/mbj/mars_pathfinder/Mars_Pathfinder.html
264. Jones, N.D.: An introduction to partial evaluation. ACM Comput. Surv. 28(3), 480–503 (1996). doi:10.1145/243439.243447
265. JXTA Community: Home page. https://jxta.kenai.com/ (2016)
266. Kahn, G.: The semantics of a simple language for parallel programming. In: Proceedings of the International Federation for Information Processing (IFIP), pp. 471–475 (1974)
267. Kamal R (2003) Embedded Systems: Architecture. Programming and Design. Tata McGraw-Hill, New York
268. Kang, S., Dean, A.G.: Leveraging both data cache and scratchpad memory through synergetic data allocation. In: Proceedings of the Real Time and Embedded Technology and Applications Symposium (RTAS), pp. 119–128. IEEE Computer Society, Washington, DC, USA (2012). doi:10.1109/RTAS.2012.22
269. Kannan, A., Shrivastava, A., Pabalkar, A., Lee, J.E.: A software solution for dynamic stack management on scratch pad memory. In: Proceedings of the Asia and South Pacific Design Automation Conference (ASPDAC), pp. 612–617 (2009)
270. Karp RM, Miller RE (1966) Properties of a model for parallel computations: determinancy, termination, queueing. SIAM J. Appl. Math. 14:1390–1411
271. Keding, H., Willems, M., Coors, M., Meyr, H.: FRIDGE: a fixed-point design and simulation environment. In: Design, Automation and Test in Europe (DATE), pp. 429–435 (1998)
272. Keinert J, Streubühr M, Schlichter T, Falk J, Gladigau J, Haubelt C, Teich J, Meredith M (2009) SystemCodesigner - an automatic ESL synthesis approach by design space exploration and behavioral synthesis for streaming applications. ACM Trans. Des. Autom. Electron. Syst. 14:1–23
273. Kelter T, Marwedel P (2017) Parallelism analysis: precise WCET values for complex multi-core systems. Sci. Comput. Program. 133:175–193. doi:10.1016/j.scico.2016.01.007
274. Kempe, M.: Ada 95 reference manual, ISO/IEC standard 8652:1995. (HTML-version). http://www.adahome.com/rm95/ (1995)
275. Kempe Software Capital Enterprises (KSCE): Ada home: the web site for Ada. http://www.adahome.com (2010)
276. Kernighan BW, Ritchie DM (1988) The C Programming Language. Prentice Hall, Englewood Cliffs
277. Kerrison, S., Eder, K.: Energy modeling of software for a hardware multithreaded embedded microprocessor. ACM Trans. Embed. Comput. Syst. 14(3), 56:1–56:25 (2015). doi:10.1145/2700104
278. Kühn, R.: Analysis and evaluation of quality metrics for approximative source-to-source transformations (in german). Bachelor's Thesis, Dptm. of Computer Science, TU Dortmund (2016)

279. Khorramabadi, H.: ADC converters - pipelined ADCs. UC Berkeley, EECS 247, Lecture 22, http://www-inst.eecs.berkeley.edu/~ee247/fa05/lectures/L22_f05.pdf (2005)
280. Khorramabadi, H.: Oversampled ADCs. UC Berkeley, EECS 247, Lecture 23. http://www-inst.eecs.berkeley.edu/~ee247/fa05/lectures/L22_f05.pdf (2009)
281. Kienhuis, B., Rijjpkema, E., Deprettere, E.: Compaan: deriving process networks from Matlab for embedded signal processing architectures. In: Proceedings of the International Symposium on Hardware-Software Codesign (CODES), pp. 29–40 (2000)
282. Kim J, Lee S, Shin H (2013) Effective task scheduling for embedded systems using iterative cluster slack optimization. Circuits Syst 4:479–488
283. Klaiber, A.: The technology behind Crusoe™ processors. http://web.archive.org/web/20010602205826/www.transmeta.com/crusoe/download/pdf/crusoetechwp.pdf (2000)
284. Kleidermacher, D., Kleidermacher, M.: Embedded Systems Security - Practical Methods for Safe and Secure Software and Systems Development. Newnes (2012)
285. Klumpp, M., Clausen, U., ten Hompel, M.: Logistics Research and the Logistics World of 2050, pp. 1–6. Springer, Berlin (2013). doi:10.1007/978-3-642-32838-1_1
286. Könemann, B., Mucha, J., Zwiehoff, G.: Built-in logic block observer. In: IEEE International Test Conference, pp. 261–266 (1979)
287. Ko, M., Koo, I.: An overview of interactive video on demand system. www.ece.ubc.ca/~irenek/techpaps/vod/vod.html (1996)
288. Kobryn, C.: UML 2001: A standardization Odyssey. Communications of the ACM (CACM). http://www.omg.org/attachments/pdf/UML_2001_CACM_Oct99_p29-Kobryn.pdf, pp. 29–36 (2001)
289. Kocher, P., Lee, R., McGraw, G., Raghunathan, A.: Security as a new dimension in embedded system design. In: Proceedings of the 41st Annual Design Automation Conference, DAC '04, pp. 753–760. ACM, New York, NY, USA (2004). doi:10.1145/996566.996771. Moderator-Ravi, Srivaths
290. Kohavi Z, Jha NK (2010) Switching and Finite Automata Theory, 3rd edn. Cambridge University Press, Cambridge
291. Koopman, P.J., Upender, B.P.: Time division multiple access without a bus master. United Technologies Research Center, UTRC Technical report RR-9500470. http://www.ece.cmu.edu/~koopman/jtdma/jtdma.html (1995)
292. Kopetz H (2011) Real-Time Systems - Design Principles for Distributed Embedded Applications. Springer, Berlin
293. Kopetz H, Grunsteidl G (1994) TTP - a protocol for fault-tolerant real-time systems. IEEE Comput. 27:14–23
294. Korf, R.E.: A new algorithm for optimal bin packing. In: American Association for Artificial Intelligence Proceedings, pp. 731–736 (2002). www.aaai.org
295. Kranitis N, Paschalis A, Gizopoulos D, Zorian Y (2003) Instruction-based self-testing of processor cores. J. Electron. Testing 19:103–112
296. Krhovjak, J., Matyas, V.: Secure hardware - pv018. http://www.fi.muni.cz/~xkrhovj/lectures/2006_PV018_Secure_Hardware_slides.pdf (2006)
297. Krishna C, Shin KG (1997) Real-Time Systems. Computer Science Series. McGraw-Hill, New York
298. Krstić A, Cheng K (1998) Delay fault testing of VLSI circuits. Kluwer Academic Publishers, Norwell
299. Krstic, A., Dey, S.: Embedded software-based self-test for programmable core-based designs. In: IEEE Design and Test, pp. 18–27 (2002)
300. Krüger, G.: Automatic generation of self-test programs: a new feature of the MIMOLA design system. In: Proceedings of the Design Automation Conference (DAC), pp. 378–384 (1986)
301. Kuang, S.R., Chen, C.Y., Liao, R.Z.: Partitioning and pipelined scheduling of embedded system using integer linear programming. In: 11th International Conference on Parallel and Distributed Systems (ICPADS'05), vol. 2, pp. 37–41 (2005). doi:10.1109/ICPADS.2005.219
302. Kumar, R., Farkas, K.I., Jouppi, N.P., Ranganathan, P., Tullsen, D.M.: Single-isa heterogeneous multi-core architectures: The potential for processor power reduction. In: Proceedings

of the 36th Annual IEEE/ACM International Symposium on Microarchitecture, MICRO 36, pp. 81–. IEEE Computer Society, Washington, DC, USA (2003). http://dl.acm.org/citation.cfm?id=956417.956569

303. Labrosse J (2000) Embedded Systems Building Blocks - Complete and Ready-to-use Modules in C. Elsevier, Amsterdam

304. Lala P (1985) Fault tolerant and Fault Testable Hardware Design. Prentice Hall, Englewood Cliffs

305. Lam KY, Kuo TW (eds) (2001) Real-Time Database Systems: Architecture and Techniques. Kluwer Academic Publishers, Norwell, MA, USA

306. Lam, M.S., Rothberg, E.E., Wolf, M.E.: The cache performance and optimizations of blocked algorithms. In: Architectural Support for Programming Languages and Operating Systems (ASPLOS), pp. 63–74 (1991)

307. Lapinskii, V., Jacome, M.F., de Veciana, G.: Application-specific clustered VLIW datapaths: Early exploration on a parameterized design space. Technical report. UT-CERC-TR-MFJ/GDV-01-1, Computer Engineering Research Center, University of Texas at Austin (2001)

308. Laplante P (1997) Real-Time Systems: Design and Analysis - An Engineer's Handbook. IEEE Press, New York

309. Laprie, J.C. (ed.): Dependability: basic concepts and terminology in English, French, German, Italian and Japanese. IFIP WG 10.4, dependable computing and fault tolerance. In: Dependable Computing and Fault Tolerant Systems, vol. 5. Springer, Berlin (1992)

310. Latendresse, M.: The code compression bibliography. http://www.iro.umontreal.ca/~latendre/compactBib (2004)

311. Law AM (2006) Simulation Modeling and Analysis. McGraw-Hill, New York

312. Lawler EL (1973) Optimal sequencing of a single machine subject to precedence constraints. Manag. Sci. 19:544–546

313. Le Boudec, J., Thiran, P.: Network Calculus. LNCS, vol. 2050. Springer, Berlin (2001)

314. Lee, E.: Leveraging synchronized clocks in cyber-physical systems. In: Workshop on Synchronization in Telecommunications Systems. http://www.atis.org/wsts/papers/3-3-1_UCBerkeley_Lee_LeveragingClocks.pdf (2014)

315. Lee, E.A.: Embedded software – an agenda for research. Technical report, UCB ERL Memorandum M99/63 (1999)

316. Lee, E.A.: The future of embedded software. ARTEMIS Conference, Graz. http://ptolemy.eecs.berkeley.edu/presentations/06/FutureOfEmbeddedSoftware_Lee_Graz.ppt (2006)

317. Lee, E.A.: Computing foundations and practice for cyber-physical systems: a preliminary report. Tech. Rep. UCB/EECS-2007-72, EECS Department, University of California, Berkeley (2007). http://www.eecs.berkeley.edu/Pubs/TechRpts/2007/EECS-2007-72.html

318. Lee, E.A.: Absolutely positively on time. IEEE Comput. (2005)

319. Lee EA, Messerschmitt D (1987) Synchronous data flow. Proc. IEEE 75:1235–1245

320. Lee, E.A., Seshia, S.A.: Introduction to Embedded Systems: A Cyber-physical Systems Approach, 2 edn. (2015). http://LeeSeshia.org

321. Lee, S., Gerstlauer, A.: Fine grain word length optimization for dynamic precision scaling in DSP systems. In: VLSI-SoC, pp. 266–271 (2013)

322. Lelli, J., Scordino, C., Abeni, L., Faggioli, D.: Deadline scheduling in the Linux kernel. Software: Pract. Exper. **46**(6), 821–839 (2016). doi:10.1002/spe.2335. Spe.2335

323. Leupers R (1997) Retargetable Code Generation for Digital Signal Processors. Kluwer Academic Publishers, Norwell

324. Leupers R (2000) Code Optimization Techniques for Embedded Processors - Methods, Algorithms, and Tools. Kluwer Academic Publishers, Norwell

325. Leveson, N.: Safeware, System Safety and Computers. Addison Wesley, Reading (1995)

326. Lewis, J., Rashba, E., Brophy, D.: VHDL-2006-D3.0 Tutorial. Tutorial at Design, Automation, and Test in Europe (DATE). http://www.accellera.org/apps/group_public/download.php/934/date_vhdl_tutorial.pdf (2007)

327. Li, L., Wu, H., Feng, H., Xue, J.: Towards data tiling for whole programs in scratchpad memory allocation. In: Proceedings of the Asia-Pacific Conference on Advances in Computer Systems Architecture (ACSAC), pp. 63–74. Springer, Berlin (2007). doi:10.1007/978-3-540-74309-5_8

328. Li ZR (ed) (2015) Organic Light-Emitting Materials and Devices. CRC Press, Boca Raton

329. Liebisch, D.C., Jain, A.: Jessi common framework design management: the means to configuration and execution of the design process. In: Conference on European Design Automation (EURO-DAC), pp. 552–557. IEEE Computer Society Press, New York (1992)

330. LIN Consortium: LIN specification package - revision 2.2A. http://www.cs-group.de/fileadmin/media/Documents/LIN_Specification_Package_2.2A.pdf (2010)

331. Liu, C.L., Layland, J.W.: Scheduling algorithms for multi-programming in a hard real-time environment. J. Assoc. Comput. Mach. (JACM) 40–61 (1973)

332. Liu JW (2000) Real-Time Systems. Prentice Hall, Englewood Cliffs

333. Liu Y, Zhang W (2015) Scratchpad memory architectures and allocation algorithms for hard real-time multicore processors. J. Comput. Sci, Eng

334. Lohmann, D., Hofer, W., Schröder-Preikschat, W., Spinczyk, O.: CiAO: an aspect-oriented operating-system family for resource-constrained embedded systems. In: USENIX Annual Technical Conference (2009)

335. Lohmann, D., Scheler, F., Schröder-Preikschat, W., Spinczyk, O.: PURE embedded operating systems - CiAO. In: Proceedings of the International Workshop on Operating System Platforms for Embedded Real-Time Applications, OSPERT (2006)

336. Lohmann, D., Streicher, J., Hofer, W., Spinczyk, O., Schröder-Preikschat, W.: Configurable memory protection by aspects. In: Proceedings of the 4th Workshop on Programming Languages and Operating Systems (PLOS '07). ACM Press, New York, NY, USA (2007)

337. Lokuciejewski P, Marwedel P (2010) WCET-aware Source Code and Assembly Level Optimization Techniques for Real-Time Systems. Springer, Berlin

338. López JM, Díaz JL, García DF (2004) Utilization bounds for EDF scheduling on real-time multiprocessor systems. Real-Time Syst. 28(1):39–68. doi:10.1023/B:TIME.0000033378.56741.14

339. Lu, Y.H., Chung, E.Y., Šimunic, T., Benini, L., De Micheli, G.: Quantitative comparison of power management algorithms. In: Proceedings of Design, Automation and Test in Europe (DATE), pp. 20–26 (2000)

340. Luican, I.I., Zhu, H., Balasa, F.: Formal Model of Data Reuse Analysis for Hierarchical Memory Organizations. In: Proceedings of the International Conference on Computer-Aided Design (ICCAD), pp. 595–600. ACM, New York, NY, USA (2006). doi:10.1145/1233501.1233623

341. Machanik P (2002) Approaches to addressing the memory wall. Univ. Brisbane, November, Technical Report

342. Macii A, Benini L, Poncino M (2002) Memory Design Techniques for Low Energy Embedded Systems. Kluwer Academic Publishers, Norwell

343. Macii E (ed) (2004) Ultra Low-Power Electronics and Design. Springer, Berlin

344. Maculan, N., Porto, S.C.S., Ribeiro, C., de Souza, C.C.: A new formulation for scheduling unrelated processors under precedence constraints. Revue Francaise d'Automatique, d'informatique et de recherche Operationelle (RAIRO), pp. 87–92 (1999)

345. Magnusson PS, Christensson M, Eskilson J, Forsgren D, Hallberg G, Hogberg J, Larsson F, Moestedt A, Werner B (2002) Simics: a full system simulation platform. Computer 35(2):50–58. doi:10.1109/2.982916

346. Mallik, A., Mamagkakis, S., Baloukas, C., Papadopoulos, L., Soudris, D., Stuijk, S., Jovanovic, O., Schmoll, F., Cordes, D., Pyka, R., Marwedel, P., Capman, F., Collet, S.: An automated toolflow for parallelization and memory management in MPSoC platforms. In: Proceedings of the Design Automation Conference (DAC) (2011)

347. Man, H.D.: From the heaven of software to the hell of nanoscale physics: an industry in transition... Keynote, HiPEAC ACACES Summer School, L'Aquila (2007)

348. Manwell JF, McGowan JG (1993) Lead acid battery storage model for hybrid energy systems. Sol. Energy 50:399–405

349. Marian, N., Ma, Y.: Translation of Simulink models to component-based software models. In: 8th International Workshop on Research and Education in Mechatronics REM. http://seg.mci.sdu.dk/publications/Translation, pp. 262–267 (2007)

350. Marongiu, A., Benini, L.: Efficient OpenMP support and extensions for MPSoCs with explicitly managed memory hierarchy. In: Proceedings of Design, Automation and Test in Europe (DATE), pp. 809–814 (2009)

351. Martello, S., Toth, P.: Knapsack Problems – Algorithms and Computer Implementations, Chap. 6. Wiley, New York (1990). http://www.or.deis.unibo.it/kp/Chapter6.pdf

352. Martin G, Müller W (eds) (2010) UML™ for SOC Design. Springer, Berlin

353. Martin, S.M., Flautner, K., Mudge, T., Blaauw, D.: Combined dynamic voltage scaling and adaptive body biasing for lower power microprocessors under dynamic workloads. In: Proceedings of the 2002 IEEE/ACM International Conference on Computer-Aided Design ICCAD '02, pp. 721–725 (2002). doi:10.1145/774572.774678

354. Marwedel, P.: A software system for the synthesis of computer structures and the generation of microcode (in German). habilitation thesis, Universität Kiel, 1985, Reprint: Report Nr.356, CS Dept., TU Dortmund (1990)

355. Marwedel P (2003) Embedded System Design. Kluwer Academic Publishers, Norwell

356. Marwedel, P.: Towards laying common grounds for embedded system design education. In: ACM SIGBED Review, pp. 25–28 (2005)

357. Marwedel P (2008) MIMOLA - a fully synthesizable language. In: Mishra P, Dutt N (eds) Processor Description Languages - Applications and Methodologies. Morgan Kaufmann, San Francisco, pp 35–63

358. Marwedel, P., Engel, M.: Flipped classroom teaching for a cyber-physical system course - an adequate presence-based learning approach in the internet age. In: Proceedings of the Tenth European Workshop on Microelectronics Education (EWME) (2014)

359. Marwedel, P., Falk, H.: Memory-architecture aware compilation. The ARTIST2 Summer School 2008 in Europe. http://ls12-www.cs.tu-dortmund.de/daes/media/documents/publications/downloads/2008-artist2summerschool.pdf (2008)

360. Marwedel, P., Falk, H., Neugebauer, O.: Memory-aware optimization of embedded software for multiple objectives. In: Ha, S., Teich, J. (eds.) Handbook of Hardware/Software CoDesign (2017)

361. Marwedel P, Goossens G (eds) (1995) Code Generation for Embedded Processors. Kluwer Academic Publishers, Norwell

362. Marwedel, P., Schenk, W.: Cooperation of synthesis, retargetable code generation and test generation in the MSS. In: Proceedings of the European Design Automation Conference (Euro-DAC), pp. 63–69 (1993)

363. Marzario, L., Lipari, G., Balbastre, P., Crespo, A.: IRIS: a new reclaiming algorithm for server-based real-time systems. In: Proceedings of the Real Time and Embedded Technology and Applications Symposium (RTAS) (2004)

364. Massa AJ (2002) Embedded Software Development with eCos. Prentice Hall, Englewood Cliffs

365. MathWorks, T.: Stateflow 7.3. http://www.mathworks.com/products/stateflow (2016)

366. McIlroy, R., Dickman, P., Sventek, J.: Efficient dynamic heap allocation of scratch-pad memory. In: Proceedings of the International Symposium on Memory Management, pp. 31–40 (2008)

367. Mckusick, M.K., Karels, M.J.: A new virtual memory implementation for berkeley unix. In: EUUG Conference Proceedings (Autumn), pp. 451–458 (1986)

368. McLaughlin, M., Moore, A.: Real-time extensions to UML. http://www.ddj.com/184410749 (1998)

369. McNamee, D., Walpole, J., Pu, C., Cowan, C., Krasic, C., Goel, A., Wagle, P., Consel, C., Muller, G., Marlet, R.: Specialization tools and techniques for systematic optimization of system software. ACM Trans. Comput. Syst. **19**(2), 217–251 (2001). doi:10.1145/377769.377778

370. Meena JS, Sze SM, Chand U, Tseng TY (2014) Overview of emerging nonvolatile memory technologies. Nanoscale Res. Lett. 9(1):526. doi:10.1186/1556-276X-9-526

371. Meijer, S., Nikolov, H., Stefanov, T.: Throughput modeling to evaluate process merging transformations in polyhedral process networks. In: Proceedings of Design, Automation and Test in Europe (DATE) (2010)

372. Menard, D., Sentieys, O.: Automatic evaluation of the accuracy of fixed-point algorithms. In: Proceedings of Design, Automation and Test in Europe (DATE), pp. 529–535 (2002)

373. Merkel, A., Bellosa, F.: Event-driven thermal management in SMP systems. In: Proceedings of the Second Workshop on Temperature-Aware Computer Systems (TACS'05) (2005)

374. Mermet J, Marwedel P, Ramming FJ, Newton C, Borrione D, Lefaou C (1998) Three decades of hardware description languages in Europe. J. Electr. Eng. Inf. Sci. 3:106

375. Mesa-Martinez, F.J., Ardestani, E.K., Renau, J.: Characterizing processor thermal behavior. In: ASPLOS '10: Proceedings of the Fifteenth Edition of ASPLOS on Architectural Support for Programming Languages and Operating Systems, pp. 193–204 (2010). doi:10.1145/1736020.1736043

376. Message Passing Interface Forum: MPI: a message-passing interface standard - version 3.1. http://www.mpi-forum.org/docs/mpi-3.1/mpi31-report.pdf (2015)

377. Microsoft Inc.: Windows® embedded home. http://www.microsoft.com/windowsembedded (2016)

378. Miles, R., Chalmers, A. (eds.): Progress in Transputer and OCCAM Research. IOS Press (1994)

379. MISRA C Working Group: MISRA compliance:2016 – achieving compliance with MISRA coding guidelines. http://www.misra.org.uk/LinkClick.aspx?fileticket=w_Syhpkf7xA (2016)

380. Mittal, S.: A survey of techniques for approximate computing. ACM Comput. Surv. **48**(4), 62:1–62:33 (2016). doi:10.1145/2893356

381. Mittal, S., Vetter, J.S.: A survey of methods for analyzing and improving gpu energy efficiency. ACM Comput. Surv. **47**(2), 19:1–19:23 (2014). doi:10.1145/2636342

382. Modelica Association: Modelica® - A unified object-oriented language for systems modeling - language specification - version 3.3. https://www.modelica.org/documents/ModelicaSpec33.pdf (2012)

383. Möller S, Raake A (2014) Quality of Experience - Advanced Concepts. Applications and Methods. Springer, Berlin

384. Montoreano, M.: Transaction level modeling using OSCI TLM 2.0. http://accellera.org/images/downloads/standards/systemc/systemc-2.3.1.tgz (2007)

385. MOST Cooperation: MOST Worldwide. http://www.mostcooperation.com/ (2010)

386. Mosterman PJ (2007) Hybrid dynamic systems: modeling and execution. In: Fishwick PA (ed) Handbook of Dynamic System Modeling. CRC Press, Boca Raton

387. Moynihan, T.: CMOS is winning the camera sensor battle, and here's why. http://www.techhive.com/article/246931/cmos_is_winning_the_camera_sensor_battle_and_heres_why.html?page=0 (2011)

388. Muchnick SS (1997) Advanced Compiler Design and Implementation. Morgan Kaufmann Publishers Inc, San Francisco

389. Mukherjee S (2008) Architecture Design for Soft Errors. Morgan Kaufmann Publishers Inc., San Francisco, CA, USA

390. Müller W, Rosenstiel W, Ruf J (2003) SystemC - Methodologies and Applications. Kluwer Academic Publishers, Norwell

391. Muralimanohar, N., Balasubramonian, R., Jouppi, N.P.: CACTI 6.0: A tool to model large caches. In: International Symposium on Microarchitecture (2007). http://www.hpl.hp.com/techreports/2009/HPL-2009-85.pdf (2009)

392. National Research Council (2001) Embedded. Everywhere. National Academies Press, Washington

393. National Research Council (2015) Interim Report on the 21st Century Cyber-Physical Systems Education. The National Academies Press, Washington

394. National Science Foundation: Cyber-physical systems (CPS). http://www.nsf.gov/pubs/2013/nsf13502/nsf13502.htm (2013)

395. National Space-Based Positioning, Navigation, and Timing Coordination Office: Global positioning system. http://www.gps.gov (2016)

396. Navet N, Simonot-Lion F (2009) Automotive Embedded Systems Handbook. CRC Press, Boca Raton

397. Neugebauer, O., Libuschewski, P., Engel, M., Mueller, H., Marwedel, P.: Plasmon-based virus detection on heterogeneous embedded systems. In: Proceedings of the International Workshop on Software and Compilers for Embedded Systems (SCOPES) (2015)

398. Neumann, P.G.: Computer Related Risks. Addison Wesley, Reading (1995)

399. Nguyen, N., Dominguez, A., Barua, R.: Memory allocation for embedded systems with a compile-time-unknown scratch-pad size. In: Proceedings of the International Conference on Compilers, Architectures, and Synthesis for Embedded Systems (CASES), pp. 115–125 (2005)

400. Nicolescu G, Mosterman PJ (2010) Model-Based Design for Embedded System. CRC Press, Boca Raton

401. Nikolov, H., Thompson, M., Stefanov, T., Pimentel, A., Polstra, S., Bose, R., Zissulescu, C., Deprettere, E.: Daedalus: toward composable multimedia MP-SoC design. In: Proceedings of the Design Automation Conference (DAC), pp. 574–579 (2008)

402. Nilsen, K.: Adding real-time capabilities to Java. Commun. ACM **41**(6), 49–56 (1998). doi:10.1145/276609.276619

403. Noergard, T.: Embedded Systems Architecture: A Comprehensive Guide for Engineers and Programmers. Newnes (2012)

404. Northeast Sustainable Energy Association: Buildingenergy. http://www.nesea.org/ (2010)

405. Novosel, D.: Timing the power grid. http://www.pserc.wisc.edu/documents/general_information/presentations/smartr_grid_executive_forum/ (2009)

406. NXP Semiconductors N.V.: I^2C-bus specification and user manual. http://www.nxp.com/documents/user_manual/UM10204.pdf (2014)

407. Object Management Group (OMG): CORBA® basics. http://www.omg.org/gettingstarted/corbafaq.htm (2003)

408. Object Management Group (OMG): Real-time CORBA specification, version 1.2, jan. 2005. http://www.omg.org/cgi-bin/doc?formal/05-01-04.ps (2005)

409. Object Management Group (OMG): UML™ profile for schedulability, performance, and time specification, version 1.1. http://www.omg.org/cgi-bin/doc?formal/05-01-02.pdf (2005)

410. Object Management Group (OMG): OMG systems modeling language (OMG SysML™). http://www.omg.org/spec/SysML/1.1/changebar/PDF (2008)

411. Object Management Group (OMG): A UML™ profile for MARTE: Modeling and analysis of real-time embedded systems - 1.0. http://www.omg.org/spec/MARTE/1.0/PDF (2009)

412. Object Management Group (OMG): OMG® specifications. http://www.omg.org/spec/ (2016)

413. Object Management Group (OMG): Unified modeling language™ resource page. http://www.uml.org (2016)

414. O'Neill, A.: Analog to digital types. IEEE tv (for members only). https://ieeetv.ieee.org/conference-highlights/analog-to-digital-types? (2006)

415. Open Connectivity Foundation: about UPnP. https://openconnectivity.org/resources/specifications/upnp/upnp-resources (2016)

416. Open SystemC Initiative: IEEE 1666 LRM. http://www.systemc.org/downloads/lrm (2005)

417. OpenMP Architecture Review Board: OpenMP application program interface. http://www.openmp.org/mp-documents/spec30.pdf (2008)

418. Oppenheim, A.V., Schafer, R., Buck, J.R.: Digital Signal Processing. Pearson Higher Education (2009)

419. OSEK Group: OSEK/VDX - communication (version 3.0.3). http://portal.osek-vdx.org/files/pdf/specs/osekcom303.pdf (2004)

420. Otter, M., Winkler, D.: Modelica overview. https://www.modelica.org/education/educational-material/lecture-material/english/ModelicaOverview.ppt (2013)

421. Pallister, J., Kerrison, S., Morse, J., Eder, K.: Data Dependent Energy Modelling for Worst Case Energy Consumption Analysis (2015). arXiv preprint arXiv:1505.03374
422. Palumbo, M.: The ERTMS/ETCS signalling system – an overview on the Standard Interoperable signalling and train control system. http://www.railwaysignalling.eu/wp-content/uploads/2014/08/ERTMS_ETCS_signalling_system_MaurizioPalumbo1.pdf (2014)
423. Pan, S., Hu, Y., Li, X.: IVF: characterizing the vulnerability of microprocessor structures to intermittent faults. In: Proceedings of Design, Automation and Test in Europe (DATE) (2010)
424. Pappas, G.J.: Science of cyber-physical systems-bridging CS and control. http://cps-vo.org/node/5876 (2014)
425. Parker KP (1992) The Boundary Scan Handbook. Kluwer Academic Publishers, Norwell
426. Paulin, P., Knight, J.: Force-directed scheduling in automatic data path synthesis. In: Proceedings of the Design Automation Conference (DAC) (1987)
427. Petri, C.A.: Kommunikation mit Automaten. Schriften des Rheinisch-Westfälischen Institutes für instrumentelle Mathematik an der Universität Bonn (1962)
428. Peukert, W.: Über die Abhängigkeit der Kapacität von der Entladestromstärcke bei Bleiakkumulatoren. Elektrotechnische Zeitschrift 20 (1897)
429. Piao, X., Han, S., Kim, H., Park, M., Cho, Y., Cho, S.: Predictability of earliest deadline zero laxity algorithm for multiprocessor real-time systems. In: Ninth IEEE International Symposium on Object and Component-Oriented Real-Time Distributed Computing (ISORC'06), p. 6 (2006). doi:10.1109/ISORC.2006.64
430. Pinedo ML (2016) Scheduling - Theory, Algorithms, and Systems, 5th edn. Springer, Berlin
431. Pohl K, Böckle G, van der Linden F (2005) Software Product Line Engineering. Springer, Berlin. ISBN-10 3-540-28901-1
432. Popovici K, Rousseau F, Jerraya AA, Wolf M (2010) Embedded Software Design and Programming of Multiprocessor System-on-Chip. Springer, Berlin
433. Potop-Butucaru D, de Simone R, Talpin JP (2006) The synchronous hypothesis and synchronous languages. In: Zurawski R (ed) Embedded Systems Handbook. CRC Press, Boca Raton
434. Press D (2003) Guidelines for Failure Mode and Effects Analysis for Automotive. Aerospace and General Manufacturing Industries. CRC Press, Boca Raton
435. Ptolemaeus, C. (ed.): System Design, Modeling, and Simulation using Ptolemy II. Ptolemy.org (2014). http://ptolemy.org/books/Systems
436. Puaut, I.: Multi-core real-time scheduling. http://www.irisa.fr/alf/downloads/puaut/STR/STRmulticore.pdf (2012–2013)
437. Puaut, I.: Real-time systems – slides of a master course. https://team.inria.fr/pacap/members/isabelle-puaut/#teaching (2015)
438. Pyka, R., Faßbach, C., Verma, M., Falk, H., Marwedel, P.: Operating system integrated energy aware scratchpad allocation strategies for multi-process applications. In: Proceedings of the International Workshop on Software and Compilers for Embedded Systems (SCOPES), pp. 41–50 (2007)
439. Radetzki M (ed) (2009) Languages for Embedded Systems and their Applications. Springer, Berlin
440. Rajkumar, R.: Real-time synchronization protocols for shared memory multiprocessors. In: Proceedings of the 10th International Conference on Distributed Computing Systems (ICDCS), pp. 116–123 (1990)
441. Rajkumar, R., Sha, L., Lehoczky, J.: Real-time synchronization protocols for multiprocessors. In: Proceedings of the Real-Time Systems Symposium (RTSS), pp. 259–269 (1988)
442. Rajkumar R, Sha L, Lehoczky JP (1988) Real-time synchronization protocols for multiprocessors. Proc. Real-Time Syst. Symp. (RTSS) 88:259–269
443. Rao, R., Vrudhula, S., Rakhmatov, D.: Battery modeling for energy-aware system design. IEEE Comput. 77–87 (2003)
444. Reed, R.: SDL-2010: Background, rationale, and survey. http://sdl-forum.org/ftp/pub/SDL-2010/SDL2010Background.pdf (2010)

445. Reisig W (2013) Understanding Petri Nets - Modeling Techniques. Analysis Methods. Case Studies. Springer, Berlin

446. Riccobene, E., Scandurra, P., Rosti, A., Bocchio, S.: A UML™ 2.0 profile for SystemC: toward high-level SoC design. In: Proceedings of the International Conference on Embedded Software (EMSOFT), pp. 138–141 (2005). doi:10.1145/1086228.1086254

447. Ritchie DM, Thompson K (1974) The unix time-sharing system. Commun. ACM 17(7):365–375. doi:10.1145/361011.361061

448. Rixner, S., Dally, W.J., Khailany, B.J., Mattson, P.J., Kapasi, U.J.: Register organization for media processing. In: 6th High-Performance Computer Architecture (HPCA-6), pp. 375–386 (2000)

449. Rosenblum M, Ousterhout JK (1992) The design and implementation of a log-structured file system. ACM Trans. Comput. Syst. 10(1):26–52. doi:10.1145/146941.146943

450. RTCA: DO-178B, Software Considerations in Airborne Systems and Equipment Certification. RTCA Inc. (1992)

451. Ruggiero, M., Benini, L.: Mapping Task Graphs to the CELL BE Processor (2008). http://www.artist-embedded.org/docs/Events/2008/Map2MPSoC/Map2mpsoc-08-ruggiero.pdf

452. Sangiovanni-Vincentelli, A.: The context for platform-based design. IEEE Des. Test Comput. 120 (2002)

453. Sangiovanni-Vincentelli A, Zeng H, Natale MD, Marwedel P (eds) (2013) Embedded Systems Development - From Functional Methods to Implementations. Springer, Berlin. ISBN 978-1-4616-3878-6

454. Schmitz, M., Al-Hashimi, B., Eles, P.: Energy-efficient mapping and scheduling for DVS enabled distributed embedded systems. In: Proceedings of Design, Automation and Test in Europe (DATE), pp. 514–521 (2002)

455. Schneider, K.: The synchronous programming language Quartz. Technical report, Internal Report 375, Department of Computer Science, University of Kaiserslautern, Kaiserslautern, Germany (2009)

456. S. Cho, S.-K. Lee, A. Han, K.-J. Lin: Efficient real-time scheduling algorithms for multiprocessor systems. IEICE Trans. Commun. 2859–2867 (2002)

457. SDL Forum Society: Home page. http://www.sdl-forum.org (2016)

458. SDL Forum Society: list of commercial tools. http://www.sdl-forum.org/Tools/Commercial.htm (2016)

459. SGI: OpenGL Software Development Kit. https://www.opengl.org/sdk/libs/ (2016)

460. Sha, L., Rajkumar, R., Lehoczky, J.: Priority inheritance protocols: an approach to real-time synchronisation. IEEE Trans. Comput. 1175–1185 (1990)

461. Shi, C., Brodersen, R.: An automated floating-point to fixed-point conversion methodology. In: International Conference on Audio Speed and Signal Processing (ICASSP), pp. 529–532 (2003)

462. Siciliano B, Oussama O (2016) Handbook of Robotics. Springer, Berlin

463. Siemens: SIMATIC STEP 7 programming software. http://www.automation.siemens.com/simatic/industriesoftware/html_76/products/step7.htm (2016)

464. Sifakis, J.: A notion of expressiveness for component-based design. In: Workshop on Foundations and Applications of Component-based Design, ES-Week (2008). http://www.artist-embedded.org/docs/Events/2008/Components/SLIDES/12-JosephSifakis-WFCD-ArtistDesign-Oct192008.pdf

465. Simunic, T., Benini, L., Acquaviva, A., Glynn, P., De Micheli, G.: Energy efficient design of portable wireless devices. In: International Symposium on Low Power Electronics and Design (ISLPED), pp. 49–54 (2000)

466. Simunic, T., Benini, L., Acquaviva, A., Glynn, P., De Micheli, G.: Dynamic voltage scaling and power management for portable systems. In: Proceedings of the Design Automation Conference (DAC), pp. 524–529 (2001)

467. Simunic-Rosing, T., Coskun, A.K., Whisnant, K.: Temperature aware task scheduling in MPSoCs. In: Proceedings of Design, Automation and Test in Europe (DATE), pp. 1659–1664 (2007)

468. Singh, A.K., Shafique, M., Kumar, A., Henkel, J.: Mapping on multi/many-core systems: survey of current and emerging trends. In: Proceedings of the Design Automation Conference (DAC), pp. 1:1–1:10. ACM, New York, NY, USA (2013). doi:10.1145/2463209.2488734

469. Sipser M (2006) Introduction to the Theory of Computation. Thomson Course Technology, Parts One and Two

470. Sirocic, B., Marwedel, P.: Levi Flexray® Simulation Software (2007). https://ls12-www.cs.tu-dortmund.de/daes/media/documents/teaching/downloads/levi/download/leviFRP.zip

471. Sirocic, B., Marwedel, P.: Levi KPN Simulation Software (2007). https://ls12-www.cs.tu-dortmund.de/daes/media/documents/teaching/downloads/levi/download/leviKPN.zip

472. Sirocic, B., Marwedel, P.: Levi RTS simulation software. https://ls12-www.cs.tu-dortmund.de/daes/media/documents/teaching/downloads/levi/download/leviRTS.zip (2007)

473. Sirocic, B., Marwedel, P.: Levi TDD simulation software. https://ls12-www.cs.tu-dortmund.de/daes/media/documents/teaching/downloads/levi/download/leviTDD.zip (2007)

474. Skadron, K., Stan, M.R., Ribando, R.J., Gurumurthi, S., Huang, W., Sankaranarayanan, K., Tarjan, D., Burr, J., Ghosh, S., Velusamy, S., Link, G.: Hotspot 5.0. http://lava.cs.virginia.edu/HotSpot/index.htm (2009)

475. Skadron K, Stan MR, Sankaranarayanan K, Huang W, Velusamy S, Tarjan D (2004) Temperature-aware microarchitecture: modeling and implementation. ACM Trans. Archit. Code Optim. 1(1):94–125. doi:10.1145/980152.980157

476. Skjellum, A., Kanevsky, A., Dandass, Y.S., Watts, J., Paavola, S., Cottel, D., Henley, G., Hebert, L.S., Cui, Z., Rounbehler, A.: The real-time message passing interface standard (mpi/rt-1.1). Concurr. Comput. Pract. Exper. **16**(S1), Si–S322 (2004). doi:10.1002/cpe.744

477. Smith JJ, Nair R (2005) Virtual Machines: Versatile Platforms For Systems And Processes. Morgan Kaufmann Publishers, San Francisco

478. Society for Display Technology: Home page. http://www.sid.org (2003)

479. Souza, A.: Combinatorial algorithms. Lecture Notes, Winter Term 10/11, Humboldt University Berlin. https://www2.informatik.hu-berlin.de/alcox/lehre/lvws1011/coalg/combinatorial_algorithms.pdf (2010)

480. Spivey, M.: The Z Notation: A Reference Manual. International Series in Computer Science, 2nd edn. Prentice Hall, Englewood Cliffs (1992)

481. Sprint Consortium: Open SoC design platform for reuse and integration of IPs. http://ecsi.org/sprint (2008)

482. Sridhar, A., Vincenzi, A., Atienza, D., Brunschwiler, T.: 3D-ICE: a compact thermal model for early-stage design of liquid-cooled ICs. IEEE Trans. Comput. **63** (2014)

483. Stallings W (2015) Operating Systems: Internals and Design Principles, 8th edn. Prentice Hall, Englewood Cliffs

484. Stankovic J, Ramamritham K (1991) The Spring kernel: a new paradigm for real-time systems. IEEE Softw. 8:62–72

485. Steinke, S.: Analysis of the potential for saving energy in embedded systems through energy-aware compilation (in German). Ph.D. thesis, TU Dortmund. http://hdl.handle.net/2003/2769 (2003)

486. Steinke, S., Grunwald, N., Wehmeyer, L., Banakar, R., Balakrishnan, M., Marwedel, P.: Reducing energy consumption by dynamic copying of instructions onto onchip memory. In: Proceedings of the International Symposium on System Synthesis (ISSS), pp. 213–218 (2002)

487. Steinke, S., Knauer, M., Wehmeyer, L., Marwedel, P.: An accurate and fine grain instruction-level energy model supporting software optimizations. In: International Workshop on Power and Timing Modeling, Optimization and Simulation (PATMOS) (2001)

488. Steinke, S., L.Wehmeyer, Lee, B.S., Marwedel, P.: Assigning program and data objects to scratchpad for energy reduction. In: Proceedings of Design, Automation and Test in Europe (DATE), pp. 409–417 (2002)

489. Storey, N.: Safety-critical Computer Systems. Addison Wesley, Reading (1996)

490. Stritter E, Gunter T (1979) Microprocessor architecture for a changing world: the Motorola 68000. IEEE Comput. 12:43–52

491. Stuijk, S.: Predictable mapping of streaming applications on multiprocessors. Ph.D. thesis, TU Eindhoven (2007)

492. Sun: Oracle technology network > Java. http://www.oracle.com/technetwork/java (2016)

493. Sundmaeker H, Guillemin P, Friess P, Woelfflé S (2010) Vision and challenges for realising the internet of things. Cluster of European Research Projects on the Internet of Things, European Commision

494. Sutherland, S.: An overview of SystemVerilog 3.1. EEdesign, May. http://www.eetimes.com/news/design/features/showArticle.jhtml?articleID=16501063 (2003)

495. Synopsys: System studio. http://www.synopsys.com/Prototyping/VirtualPrototyping/DigitalSignalProcessing/Pages/system-studio.aspx (2016)

496. Synopsys® Inc.: HSPICE®: The gold standard for accurate circuit simulation. https://www.synopsys.com/Tools/Verification/AMSVerification/CircuitSimulation/HSPICE/Documents/hspice_ds.pdf (2010)

497. SYSGO AG: PikeOS RTOS and virtualization concept. http://www.sysgo.com (2016)

498. SystemC: Home page. http://www.SystemC.org (2016)

499. Taha W, Cartwright R (2013) Some challenges for model-based simulation. The 4th Analytic Virtual Integration of Cyber-Physical Systems Workshop. Vancouver, Canada, December 3, 2013. Linköping University, Electronic Press, pp 1–4

500. Takada, H.: Real-time operating system for embedded systems. In: Imai, M., Yoshida, N. (eds.) Tutorial 2 – Software Development Methods for Embedded Systems, Asia South-Pacific Design Automation Conference (ASP-DAC) (2001)

501. Tan, T.K., Raghunathan, A., Jha, N.K.: Software architectural transformations: a new approach to low energy embedded software. In: Proceedings of Design, Automation and Test in Europe (DATE), pp. 11046–11051 (2003)

502. Tanenbaum A (2014) Modern Operating Systems. Prentice Hall, Englewood Cliffs

503. Tartler, R., Lohmann, D., Sincero, J., Schröder-Preikschat, W.: Feature consistency in compile-time-configurable system software: facing the linix 10,000 feature problem. In: Sixth Conference on Computer Systems (EuroSys) (2011)

504. Tehrani, S.N.: Functional safety and IEC 61508 – a basic guide. http://www.ida.liu.se/~simna73/teaching/SCRTS/IEC61508_Guide.pdf (2004)

505. Teich J (1997) Digitale Hardware/Software-Systeme. Springer, Berlin

506. Tewari A (2001) Modern Control Design with MATLAB and SIMULINK. Wiley Ltd, New York

507. Texas Instruments Inc.: 66AK2x Multicore DSP + ARM Processors. http://www.ti.com/lsds/ti/processors/dsp/c6000_dsp-arm/66ak2x/overview.page (2016)

508. Thayer, D., Miller, K.: Four UNIX programs in four UNIX collections: seeking consistency in an Open Source icon. In: Proceedings of the 37th Midwest Instruction and Computing Symposium (2004)

509. The Dobelle Institute: Home page. http://www.dobelle.com (no longer accessible) (2003)

510. The MathWorks Inc.: Simulink - simulation and model-based design. http://www.mathworks.com/products/simulink (2016)

511. Thesing, S.: Safe and Precise WCET Determination by Abstract Interpretation of Pipeline Models. Pirrot Verlag (2004)

512. Thiele, L.: Design space exploration of embedded systems. Artist Spring School on Embedded Systems, Xi'an. http://www.artist-embedded.org/docs/Events/2006/ChinaSchool/4_DesignSpaceExploration.pdf (2006)

513. Thiele, L.: Performance analysis of distributed embedded systems. In: R. Zurawski (Hrg.): Embedded Systems Handbook. CRC Press, Boca Raton (2006)

514. Thiele, L. et al.: SHAPES TIK. http://www.tik.ee.ethz.ch/~shapes/dol.html (2009)

515. Thoen, F., Catthoor, F.: Modelling, Verification and Exploration of Task-Level Concurrency in Real-Time Embedded Systems. Kluwer Academic Publishers, Norwell (2000)

516. Tian, T., Shih, C.P.: Software techniques for shared-cache multi-core systems. https://software.intel.com/en-us/articles/software-techniques-for-shared-cache-multi-core-systems (2012)

517. Tiller, M.M.: Modelica by example. http://book.xogeny.com/ (2016)

518. Tiwari, A., Ballal, P., Lewis, F.L.: Energy-efficient wireless sensor network design and implementation for condition-based maintenance. ACM Trans. Sen. Netw. **3**(1) (2007). doi:10.1145/1210669.1210670

519. Tiwari, V., Malik, S., Wolfe, A.: Power analysis of embedded software: a first step towards software power minimization. IEEE Trans. VLSI Syst. 437–445 (1994)

520. Tomasulo RM (1967) An efficient algorithm for exploiting multiple arithmetic units. IBM J. Res. Dev. 11(1):25–33. doi:10.1147/rd.111.0025

521. Topcuoglu H, Hariri S, Wu MY (2002) Performance-effective and low-complexity task scheduling for heterogeneous computing. IEEE Trans. Parallel Distrib. Syst. 13(3):260–274. doi:10.1109/71.993206

522. TriQuint Semiconductor Inc.: FAQ 11: What is the MTBF for gallium arsenide devices? http://www.triquint.com/about-us/quality/reliability-faq (2016)

523. Tsai J, Yang SJH (1995) Monitoring and Debugging of Distributed Real-Time Systems. IEEE Computer Society Press, New York

524. Udayakumararan, S., Dominguez, A., Barua, R.: Dynamic allocation for scratch-pad memory using compile-time decisions. ACM Trans. Embed. Comput. Syst. (TECS) 472–511 (2006)

525. University of Cambridge: HOL -interactive theorem prover. http://hol.sourceforge.net (2016)

526. US Department of Transportation - Federal Aviation Administration: Advisory circular - system design and analysis. http://www.faa.gov/documentLibrary/media/Advisory_Circular/AC25.1309-1A.pdf (1988)

527. V-Modell XT Authors: V-Modell® XT. ftp://ftp.heise.de/pub/ix/projektmanagement/vmodell/V-Modell-XT-Gesamt-Englisch-V1.3.pdf (2006)

528. Vahid, F.: Procedure exlining. In: Proceedings of the International Symposium on System Synthesis (ISSS), pp. 84–89 (1995)

529. Vahid, F., Givargis, T.: Programming embedded systems - an introduction to time-oriented programming. zyBooks. https://zybooks.zyante.com/#/catalog/zybook/ProgrammingEmbeddedSystemsR25 (2011)

530. Verachtert, W.: Introduction to parallelism. Tutorial at Design, Automation, and Test in Europe (DATE) (2008)

531. Verma, M., Marwedel, P.: Overlay techniques for scratchpad memories in low power embedded processors. IEEE Trans. Very Large Scale Integr. Syst. **14**(8) (2006)

532. Verma, M., Petzold, K., Wehmeyer, L., Falk, H., Marwedel, P.: Scratchpad sharing strategies for multiprocess embedded systems: a first approach. In: IEEE 3rd Workshop on Embedded System for Real-Time Multimedia (ESTIMedia), pp. 115–120 (2005)

533. Vladimirescu A (1987) SPICE user's guide. Northwest Laboratory for Integrated Systems, Seattle

534. Vogels, M., Gielen, G.: Architectural selection of a/d converters. In: Proceedings of the Design Automation Conference (DAC), pp. 974–977 (2003). doi:10.1145/775832.776076

535. Wägemann, P., Distler, T., Hönig, T., Janker, H., Kapitza, R., Schröder-Preikschat, W.: Worst-case energy consumption analysis for energy-constrained embedded systems. In: 2015 27th Euromicro Conference on Real-Time Systems, pp. 105–114. IEEE (2015)

536. Wandeler, E., Thiele, L.: Real-Time Calculus (RTC) Toolbox. http://www.mpa.ethz.ch/Rtctoolbox (2006)

537. Wang, C., Li, X., Zhang, J., Zhou, X., Nie, X.: MP-Tomasulo: a dependency-aware automatic parallel execution engine for sequential programs. ACM Trans. Archit. Code Optim. **10**(2), 9:1–9:26 (2013). doi:10.1145/2459316.2459320

538. Wang, P., Sun, G., Wang, T., Xie, Y., Cong, J.: Designing scratchpad memory architecture with emerging STT-RAM memory technologies. In: Proceedings of the International Symposium on Circuits and Systems (ISCAS), pp. 1244–1247 (2013). doi:10.1109/ISCAS.2013.6572078

539. Wang Z, Bovik AC (2002) A universal image quality index. IEEE Sig. Process. Lett. 3:81–84

540. Wang Z, Bovik AC, Sheikh HR, Simoncelli EP (2004) Image quality assessment: from error visibility to structural similarity. IEEE Trans. Image Process. 13:600–612

541. Wang, Z., Gu, Z., Yao, M., Shao, Z.: Endurance-Aware Allocation of Data Variables on NVM-Based Scratchpad Memory in Real-Time Embedded Systems. IEEE Trans. Comput. Aided Des. Integr. Circuits Syst. **34**(10), 1600–1612 (2015). doi:10.1109/TCAD.2015.2422846

542. Waxman R, Bergé JM, Levia O, Rouillard J (1996) High-Level System Modeling. Springer, Berlin

543. Weast R (ed) (1977) CRC Handbook of Chemistry and Physics. CRC Press, Boca Raton

544. Wedde, H., Lind, J.: Integration of task scheduling and file services in the safety-critical system MELODY. In: EUROMICRO '98 Workshop on Real-Time Systems, pp. 18–25. IEEE Computer Society Press, New York (1998)

545. Weddell, A.S., Magno, M., Merrett, G.V., Brunelli, D., Al-Hashimi, B.M., Benini, L.: A survey of multi-source energy harvesting systems. In: Proceedings of the Conference on Design, Automation and Test in Europe, DATE '13, pp. 905–908. EDA Consortium, San Jose, CA, USA (2013). http://dl.acm.org/citation.cfm?id=2485288.2485505

546. Wegener, I.: Branching programs and binary decision diagrams – theory and applications. In: SIAM Monographs on Discrete Mathematics and Applications (2000)

547. Wehmeyer L, Marwedel P (2006) Fast, Efficient and Predictable Memory Accesses. Springer, Berlin

548. Weiser M (1999) The computer for the 21st century. Mob. Comput. Commun. Rev. 3(3):3–11

549. Wellings A (2004) Concurrent and Real-Time Programming in Java. Wiley, New York

550. Weste, N.H.H., Eshraghian, K., Michael, S., Michael, J.S., Smith, J.S.: Principles of CMOS VLSI Design: A Systems Perspective. Adision-Wesley, Reading (2000)

551. Wikipedia: Energy harvesting. https://en.wikipedia.org/wiki/Energy_harvesting (2016)

552. Wikipedia: Heat capacity. https://en.wikipedia.org/wiki/Heat_capacity (2016)

553. Wikipedia: Safety integrity level. https://en.wikipedia.org/wiki/Safety_integrity_level (2016)

554. Wikipedia: Structured systems analysis and design method. http://en.wikipedia.org/wiki/Structured_Systems_Analysis_and_Design_Methodology (2016)

555. Wikipedia: Thermal conductivity. https://en.wikipedia.org/wiki/Thermal_conductivity (2016)

556. Wikipedia: Thermal design power. https://en.wikipedia.org/wiki/Thermal_design_power (2016)

557. Wikipedia: Safety. https://en.wikipedia.org/wiki/Safety (2017)

558. Wikipedia: Zeno's paradoxes. https://en.wikipedia.org/wiki/Zeno's_paradoxes#Achilles_and_the_tortoise (2017)

559. Wilhelm R (2006) Determining bounds on execution times. In: Zurawski R (ed) Embedded Systems Handbook. CRC Press, Boca Raton

560. Willems, M., Bürsgens, V., Keding, H., Grötker, T., Meyr, H.: System level fixed-point design based on an interpolative approach. In: Proceedings of the Design Automation Conference (DAC), pp. 293–298 (1997)

561. Wilton S, Jouppi N (1996) CACTI: an enhanced access and cycle time model. Int. J. Solid State Circuits 31(5):677–688

562. Wind River: VxWorks. http://www.windriver.com/products/vxworks (2016)

563. Wind River: Web pages. http://www.windriver.com (2016)

564. Winkler, J.: The CHILL homepage. http://psc.informatik.uni-jena.de/languages/chill/chill.htm (2002)

565. Wolf W (2001) Computers as Components. Morgan Kaufmann Publishers, San Francisco

566. Wolsey L (1998) Integer Programming. Wiley, New York

567. Wong, C., Marchal, P., Yang, P., Prayati, A., Catthoor, F., Lauwereins, R., Verkest, D., Man, H.D.: Task concurrency management methodology to schedule the MPEG4 IM1 player on a highly parallel processor platform. In: Proceedings of the International Symposium on Hardware-Software Codesign (CODES), pp. 170–177 (2001)

568. Woodhouse D (2001) Jffs: The journalling flash file system. Red Hat Inc, White Paper

569. ws4d: Web services for devices. http://www.ws4d.org (2016)

570. Xilinx: MicroBlaze processor reference guide. http://www.xilinx.com/support/documentation/sw_manuals/mb_ref_guide.pdf (2008)

571. Xilinx: Device reliability report - second half 2015. http://www.xilinx.com/support/documentation/user_guides/ug116.pdf (2016)

572. Xilinx Inc.: UltraScale+ FPGAs - product tables and product selection guide. http://www.xilinx.com/support/documentation/selection-guides/ultrascale-plus-fpga-product-selection-guide.pdf (2016)
573. Xilinx Inc.: UltraScale architecture and product overview. http://www.xilinx.com/support/documentation/data_sheets/ds890-ultrascale-overview.pdf (2016)
574. Xilinx Inc.: UltraScale architecture configurable logic block. http://www.xilinx.com/support/documentation/user_guides/ug574-ultrascale-clb.pdf (2015)
575. XMOS Ltd.: Home page. http://www.xmos.com/ (2016)
576. Xu J, Parnas DL (1993) On satisfying timing constraints in hard real-time systems. IEEE Trans. Softw. Eng. 19:70–84
577. Xu, Q., Huang, L., Yuan, F.: Lifetime reliability-aware task allocation and scheduling for MPSoC platforms. In: Proceedings of Design, Automation and Test in Europe (DATE), pp. 51–56 (2009)
578. Xue J (2000) Loop tiling for parallelism. Kluwer Academic Publishers, Norwell
579. Yodaiken, V.: Adding real-time support to general purpose operating systems. US Patent 5,995,745 (1997)
580. Yodaiken, V.: Real-time applications in real-time linux. In: USENIX'99 Tutorial 3: Linux on the Edge (1999)
581. Young, S.: Real Time Languages – Design and Development –. Ellis Horwood (1982)
582. Zanini, F., Atienza, D., Jones, C.N., Benini, L., De Micheli, G.: Online thermal control methods for multiprocessor systems. ACM Trans. Des. Autom. Electron. Syst. 18(1), 6:1–6:26 (2013). doi:10.1145/2390191.2390197
583. Zhang, L., Tiwana, B., Qian, Z., Wang, Z., Dick, R.P., Mao, Z.M., Yang, L.: Accurate online power estimation and automatic battery behavior based power model generation for smartphones. In: Proceedings of the Eighth IEEE/ACM/IFIP International Conference on Hardware/Software Codesign and System Synthesis, pp. 105–114. ACM, New York (2010)
584. Zhang, W., Ding, Y.: Hybrid SPM-cache architectures to achieve high time predictability and performance. In: Proceedings of the Conference on Application-Specific Systems, Architectures and Processors (ASAP), pp. 297–304 (2013). doi:10.1109/ASAP.2013.6567593
585. Zhuo, C., Sylvester, D., Blaauw, D.: Process variation and temperature-aware reliability management. In: Proceedings of Design, Automation and Test in Europe (DATE) (2010)
586. Zurawski R (ed) (2006) Embedded Systems Handbook. CRC Press, Boca Raton

Index

A

Abstraction, 28, 112
AC-AMC 25.1309, 268
Acatech report, 7
Accuracy, 248
ACID-property, 228
Actor, 65, 72
Actuator, 15, 173, 174, 184
Ada, 37, 38, 55, 67, 95, 104, 107, 108, 116, 208
Adaptive sampling, 11
ADC, 138, 139
   delta-sigma ~, 137
   folding ~, 137
   integrating ~, 137
   pipelined ~, 136
Address generation unit, 149
Addressing mode, 148
Algorithm
   critical-path-on-a-processor (CPOP) ~, 327
   evolutionary ~, 331
   heterogeneous-earliest-finish-time (HEFT) ~, 327
   latest deadline first ~(LDF), 299
   reasonable allocation ~, 309
Aliasing, 131, 132
AMBA-bus, 173
Ambient intelligence, 1
Analog-to-Digital Converter (ADC), 126, 133
Analyzability, 114
Answering machine, 39, 45
Anti-aliasing, 132
Application area, 19, 30
Application mapping, 283

Application Programming Interface (API), 110, 197, 200, 203, 225, 227, 228, 284
Application-Specific Integrated Circuit (ASIC), 139, 142, 161, 185, 186
Approximate computing, 245
Arithmetic
   fixed-point ~, 150, 345
   floating-point ~, 345
   saturating ~, 149, 150
ARM®, 145, 156, 164, 173
Arrival curves, 241
Artificial eye, 128
ARTIST guidelines, viii, 125
Audience, ix
Automata, 29, 47, 57, 60
   timed ~, 45
Automotive electronics, 4
AUTOSAR, 15, 285
Availability, 9, 273
Avionics, 4
Awareness
   code size ~, 13
   energy ~, 12
   run-time ~, 13

B

Babbling idiot, 174
Basic block, 36, 239, 353, 354
Bath tub curve, 271
Battery, 189
Behavior
   determinate ~, 53, 57, 93, 94
   deterministic ~, 45, 54
   non-functional ~, 31
   real-time ~, 172

© Springer International Publishing AG 2018
P. Marwedel, *Embedded System Design*, Embedded Systems,
DOI 10.1007/978-3-319-56045-8